Juan R. González, David Alejandro Pelta, Carlos Cruz, Germán Terrazas, and Natalio Krasnogor (Eds.)

Nature Inspired Cooperative Strategies for Optimization (NICSO 2010)

Studies in Computational Intelligence, Volume 284

Editor-in-Chief

Prof. Janusz Kacprzyk
Systems Research Institute
Polish Academy of Sciences
ul. Newelska 6
01-447 Warsaw
Poland
E-mail: kacprzyk@ibspan.waw.pl

Juan R. González, David Alejandro Pelta, Carlos Cruz,
Germán Terrazas, and Natalio Krasnogor (Eds.)

Nature Inspired Cooperative Strategies for Optimization (NICSO 2010)

 Springer

Mr. Juan R. González
Dept. of Computer Science and A.I.
E.T.S. Ingeniería Informática y de
Telecomunicación
C/ Periodista Daniel Saucedo Aranda s/n
University of Granada
18071 Granada, Spain

E-mail: jrgonzalez@decsai.ugr.es

Mr. David Alejandro Pelta
Dept. of Computer Science and A.I.
E.T.S. Ingeniería Informática y de
Telecomunicación, C/ Periodista Daniel
Saucedo Aranda s/n
University of Granada
18071 Granada, Spain

E-mail: dpelta@decsai.ugr.es

Mr. Carlos Cruz
Dept. of Computer Science and A.I.
E.T.S. Ingeniería Informática y de
Telecomunicación
C/ Periodista Daniel Saucedo Aranda s/n
University of Granada, 18071 Granada, Spain
E-mail: carloscruz@decsai.ugr.es

Mr. Germán Terrazas
School of Computer Science
University of Nottingham
Jubilee Campus
Wollaton Road
Nottingham, NG8 1BB, UK
E-mail: gzt@cs.nott.ac.uk

Mr. Natalio Krasnogor
School of Computer Science
University of Nottingham
Jubilee Campus
Wollaton Road
Nottingham, NG8 1BB, UK
E-mail: nxk@cs.nott.ac.uk

ISBN 978-3-642-12537-9   e-ISBN 978-3-642-12538-6

DOI 10.1007/978-3-642-12538-6

Studies in Computational Intelligence   ISSN 1860-949X

Library of Congress Control Number: 2010924760

© 2010 Springer-Verlag Berlin Heidelberg

Typeset & Cover Design: Scientific Publishing Services Pvt. Ltd., Chennai, India.

Printed on acid-free paper

9 8 7 6 5 4 3 2 1

springer.com

Preface

Nature has become a source of inspiration for many areas related with Computer Science. For instance, Neural Networks were inspired on the brain cells behaviour, natural evolution have served as inspiration of Genetic and Evolutive Algorithms, the collective behaviour of insects or animals has inspired Ant Colony algorithms, Swarm-based algorithms and so on. In general, many aspects of Nature, Biology or even from Society have become part of the techniques and algorithms used in computer science or they have been used to enhance or hybridize several techniques through the inclusion of advanced evolution, cooperation or biologically based additions.

The previous editions of the International Workshop on Nature Inspired Cooperative Strategies for Optimization (NICSO) were held in Granada, Spain, 2006, Acireale, Italy, 2007, and in Tenerife, Spain, 2008, respectively. As in these three previous editions, the aim of NICSO 2010, held in Granada, Spain, was to provide a forum were the latest ideas and state of the art research related to nature inspired cooperative strategies for problem solving were discussed. The contributions collected in this book were strictly peer reviewed by at least two members of the international programme committee, to whom we are indebted for their support and assistance. The topics covered by the contributions include nature-inspired techniques like Genetic Algorithms, Evolutionary Algorithms, Ant and Bee Colonies, Particle Swarm Optimization and other Swarm Intelligence approaches, Neural Networks, several Cooperation Models, Structures and Strategies, Agents Models, Social Interactions, as well as new algorithms based on the behaviour of fireflies or bats.

NICSO 2010 had three plenary lectures given by Prof. Pier Luca Lanzi, *Learning to Play, Learning to Program, Learning to Learn (Experiences with Computational Intelligence for Simulated Car Racing)*, Dr. Julian F. Miller, *Evolving the brain inside the brain,* and Prof. Alan F.T. Winfield, *Adaptive Swarm Foraging: a case study in self-organised cooperation.*

As Workshop Chairs we wish to thank the support given by several people and institutions. We want to thank the Spanish Ministry of Science and

Innovation (projects TIN2008-01948, TIN2008-06872-C04-04 and TIN2009-08341-E), the Andalusian Government (project P07-TIC-02970), and the EP-SRC (Grant EP/D061571/1 Next Generation Decision Support: Automating the Heuristic Design Process) for their financial support. We also wish to thank the members of the Models of Decision and Optimization Research Group for their help in the local organization tasks.

Our experience after four editions of NICSO demonstrates that there is an emerging and thriving community of scholars doing research on Nature Inspired Cooperative Strategies for Optimization. It is to these scholars, both authors and reviewers, to whom the organisers are indebted for the success of the NICSO series.

Spain,   Juan R. González
Spain,   David Alejandro Pelta
Spain,   Carlos Cruz
UK,   Germán Terrazas
UK,   Natalio Krasnogor
May 2010

Organization

Steering Committee

David A. Pelta   University of Granada
Natalio Krasnogor   University of Nottingham

Programme Chair

Germán Terrazas   University of Nottingham

Organizing Committee

Carlos Cruz   University of Granada
Juan R. González   University of Granada

Programme Committee

Belen Melian   University of La Laguna, Spain
Carlos Coello Coello   CINVESTAV-IPN, Mexico
Carlos Garcia Martinez   University of Cordoba, Spain
Cecilio Angulo   Technical University of Catalunya, Spain
Dario Landa-Silva   University of Nottingham, UK
Davide Anguita   University of Genova, Italy
Francisco Herrera   University of Granada, Spain
Gabriela Ochoa   University of Nottingham, UK
Gianluigi Folino   Istituto di Calcolo e Reti ad Alte
                                 Prestazioni, Italy
Giuseppe Scollo   University of Catania, Italy
Graham Kendall   University of Nottingham, UK
Ignacio G. del Amo   University of Granada, Spain

Plenary Lectures

Prof. Pier Luca Lanzi

Politecnico di Milano, Italy

Learning to Play, Learning to Program, Learning to Learn (Experiences with Computational Intelligence for Simulated Car Racing)

Modern computer games are fun to watch and to play. Students love them! They are also difficult to program which makes them an excellent way to challenge students with difficult yet highly-rewarding problems. They pose great challenges to most methods of computational intelligence which makes them very good testbeds for research.

Car racing games are particularly attractive in that any car driver is a domain expert! In this talk, I will provide an overview of the recent research on the applications of computational intelligence to simulated car racing, including, development of drivers and driving behaviors through evolution, imitation and hand-coded design, evolution of tracks, and lessons learned from the recent scientific competitions.

Dr. Julian F. Miller

Department of Electronics, University of York

Evolving the brain inside the brain

Most of evolutionary history is the history of single cells. One of these cells is very special, it is called a neuron. Like other cells neurons are far from simple. In fact, a neuron is a miniature brain in itself. Neurons are not only very complex on the inside they also come in a vast range of complex morphologies.

Of course, natural evolution does not evolve brains directly. Instead it evolves genes. These genes represent complex 'programs' that cause the development of the entire organism (including the brain). All learning in the brain occurs during the development process.

So why do conventional Artificial Neural Networks (ANNs) represent neurons as extremely simple computational units in static networks? Why do they represent memory as synaptic weights?

Great advances in neuroscience have been made in recent decades and We argue that the time has come to create new models of neural networks in which the neuron is much more complex and dynamic. In such models, neural structures will grow and change in response to internal dynamics and environmental interactions. Like real brains, they should be able to learn across multiple domains without unlearning.

We review previous models and discuss in detail a recent new model and show that complex neural programs can be evolved that allow a developing 'brain' to learn in a number of problem domains.

Prof. Alan F.T. Winfield

Faculty of Environment and Technology, University of the West of England, Bristol

Adaptive Swarm Foraging: a case study in self-organised cooperation

Inspired by the division of labour observed in ants, collective foraging has become a benchmark problem in swarm robotics. With careful design of the individual robot behaviours we can observe adaptive foraging for energy in which the swarm automatically changes the ratio of foraging to resting robots, in response to a change in the density of forage available in the environment, even though individual robots have no global knowledge of either the swarm or the environment. Swarm robotics thus provides us with both an interesting model of collective foraging, illuminating its processes and mechanisms, and a possible engineering solution to a broad range of real world applications, for example, in cleaning, harvesting, search and rescue, landmine clearance or planetary astrobiology. This talk will introduce the field of swarm robotics, using adaptive swarm foraging as a case study; the talk will address both the engineering challenges of design, mathematical modelling and optimisation, and the insights offered by this case study in self-organised cooperation.

Contents

A Metabolic Subsumption Architecture for Cooperative Control of the e-Puck

Verena Fischer and Simon Hickinbotham

Abstract. Subsumption architectures are a well-known model for behaviour-based robotic control. The overall behaviour is achieved by defining a hierarchy of increasingly sophisticated behaviours. We are interested in using evolutionary algorithms to develop appropriate control architectures. We observe that the layered arrangement of behaviours in subsumption architectures are a significant obstacle to automating the development of control systems. We propose an alternative subsumption architecture inspired by the bacterial metabolism, that is more amenable to evolutionary development, where communities of simple reactive agents combine in a stochastic process to confer appropriate behaviour on the robot. We evaluate this approach by developing a traditional and a metabolic solution to a simple control problem using the e-puck educational robot.

1 Introduction

The behaviour-based approach to robotics and artificial intelligence [4] has given a new spirit to a field that seemed lost in abstractions of the real world. While "traditional" robotics built complicated reasoning systems that created models of the real world and successfully produced reasonable behaviour in simple and static environments, it seemingly failed to extend these systems to deal with dynamic real world situations [11]. Behaviour-based robotics works on the assumption that internal representations of the real world are unnecessary to produce reasonable behaviour in dynamic environments and proves this to be true with many examples described in several of Brooks' papers [2, 3].

Verena Fischer
Department of Informatics, University of Sussex, Falmer, Brighton BN1 9QJ
e-mail: vf37@sussex.ac.uk

Simon Hickinbotham
YCCSA, University of York, Heslington, York YO1 5DD, UK
e-mail: sjh@cs.york.ac.uk

J.R. González et al. (Eds.): NICSO 2010, SCI 284, pp. 1–12, 2010.
springerlink.com   © Springer-Verlag Berlin Heidelberg 2010

Subsumption architectures are as highly engineered as their traditional counterparts. The approach identifies a hierarchy of autonomous behavioural layers with simpler behaviours placed at the lower layers. Every layer produces some behaviour in the robot and the higher layers can subsume the behaviour of lower layers, while lower layers are not aware of the higher ones. During the design of the controller, each layer is implemented as an autonomous system before ascending the hierarchy. The upper layers can override instructions from the lower layers should the situation demand it. Control modules are then assigned to appropriate layers, each of which can connect sensors to actuators in different ways.

The problem of designing any form of layered control remains challenging. For sophisticated environments, the number of layers can proliferate and it becomes unclear where a control module should be placed and what the interconnectedness should be. Attempts to automate the process have tended to simplify the problem by evolving the system a layer at a time [9, 13]. However, it is difficult to ensure that the entire system is optimised, since the overall control rarely depends on a single layer.

We note that as the behaviours get richer, more and more internal modules are connected only to other internal modules rather than being connected to sensors or actuators. There is potential to make such modules and the connections between them subject to adaptation via evolutionary algorithms since as long as the connections to the outside world are preserved, the internal processing can change. Evolving this sort of network is a difficult challenge however, particularly if the role of modules and their connections is predefined (e.g. connections relating to "feel-force", "heading" and "turn"). We propose a finer-grained solution, in which control is shared amongst a community of very simple processing agents that behave like molecular species in biological reaction networks [7], and whose connections are set by simple reaction rules that can be changed arbitrarily. This metabolic representation allows a high level of interconnectedness between control layers, which is more akin to biological reaction networks than control engineering. The metabolism can be thought of as a community of control agents, which through their interaction rates, network topology and concentrations give rise to emergent behaviour.

This paper compares an implementation of a subsumption architecture controller with a controller based on a model metabolism. We refer to the two systems as "subsumption control" and "metabolic control" respectively. We favour the latter approach because we believe it lends itself more readily to solutions which can be found through artificial evolution [8]. The work we present here shows how an engineered control system can be implemented in an evolvable community control system, and compares the performance of the two.

2   The Robot Model

The platform for our robot experiments is the e-puck, which is a readily available open-source research platform [10]. In addition to the physical hardware being available, a simulator model is available for the open-source player/stage platform [1, 6]. We developed our control software using the simulator, but we used the physical characteristics of the real robot to constrain the design. The e-puck is equipped with a variety of sensors and actuators. There are eight infrared sensors, as shown in figure 1(a). We combine these into four channels to produce a sufficiently fine-grained reaction to the environment: front (S_F); left-of-front (S_L), right-of-front (S_R), and back (S_B). The e-puck is driven by two wheels, which are controlled by actuator behaviours: Forward speed (A_F); Backward speed (A_B); Left turn speed (A_L); Right turn speed (A_R). In the two control architectures we investigate here, we define two functions called SensorHandler and a RobotUpdater for the sensors and actuators respectively, to carry out any signal transduction between the e-puck and the control system. The control challenge is thus to link the sensor data to the actuator instructions, as illustrated in figure 1(b).

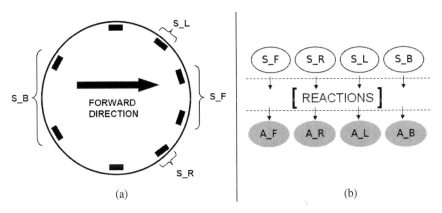

(a)   (b)

Fig. 1 (a) coupling of sensors on the e-puck simulation. (b) the control task: incoming sensor data must be coupled with actuator controllers to determine the speed and heading of the robot

E-pucks have a range of control settings. The proximity sensor range is 0.1 metres, which imposes constraints on the responsiveness of the control system if a collision is to be avoided. The maximum speed of the real e-puck is 0.12m/s, therefore the robot has 0.83s to respond appropriately to an obstacle detected by its sensors. We are of course free to change this speed in the simulation.

As an experimental framework we used the open source platform Player/Stage. Player is a network server that handles the actual low level robot control and provides a clean and simple interface to the robot's sensors and actuators. Stage simulates mobile robots, sensors and objects in a 2D environment and therefore provides the hardware basis for the robot control handled by Player.

3   Subsumption Architecture

A subsumption architecture is a layered control system. Our subsumption architecture control system contains the following layers:

Layer 1: AVOID responds to obstacles flagged by the SensorHandler and changes the speed of the robot accordingly. Each sensor produces a different hard coded behaviour, although the behaviour has a small noise component built into the design. The avoid behaviour sets a new heading as soon as a sensor flags an obstacle. Details are given in table 1

Layer 2: WANDER pseudo-randomly produces a new heading and speed for the robot after a set number of time steps. The goal is to induce a behaviour which allows the robot to explore its world. When wandering, headings are set every 15 time steps by selecting a turn value in the range $+/-15$ degrees, while the speed is set to a pseudo-random value between 0.05 - 0.095 m/s.

Table 1 Reactions specified by the AVOID layer in the subsumption architecture

Sensor direction	speed	turn
S_F	A_B: -0.15 m/s	A_R: 90-180 degrees
S_L	A_B: -0.1 m/s	A_R: -60 degrees
S_R	A_B: -0.1 m/s	A_L: 60 degrees
S_B	A_F: 0.15 m/s	

4   Artificial Metabolomes

Our endeavours to create a mobile robot control system using an artificial metabolism are built upon the particle metabolome developed within the Plazzmid project [12]. The metabolic model is composed of four components. Firstly, there exists a *container*, which specifies the volume and dimensionality of the space in which the agents exist. We specify a simple 2D container of area $v_c = 40$ units. Secondly, we have a set of metabolite *agents*, of area $v_a = 9$ units which are present in varying quantities in the container, analogous to the various quantities of different molecular species in a biological system. Thirdly we have a stochastic *mixer*, which governs the movement and changes in adjacency of the elements within the container. For a bimolecular reaction such as the bind B, our mixer utilises a simple propensity function $P(B)$, which estimates the probability of two agents being sufficiently close enough for the reaction to occur. For any one agent in a bimolecular reaction, the chance of the second agent in the reaction being close enough to react is:

$$P(B|v_c, v_a, n) = 1 - (1 - (v_a/v_c))^n \qquad (1)$$

where n is the number of instances of the second agent in the metabolism. Space in the system is represented abstractly via the ratio of container area to agent area. Apart from this consideration, the model is aspatial. Fourthly, agents react according to a set of *rules*, which specify the reactions in the system. There are four types of rules in the system as shown in table 2. Each rule has a rate, which governs how often the reaction occurs when it is selected via a stochastic process. *Influx* is the spontaneous generation of new agents in the system. In our case, objects detected by sensors cause production of corresponding sensor agents to be generated in the metabolism. *Binding* occurs when two reactants combine to create a single product. Binding is the only bimolecular reaction permitted. Bimolecular reaction rates are governed by the concentration of the two reactants in the system (via $P()$) and a further reaction rate specified by the reaction rule. Behavioural switching is caused by a sensor agent binding with a WANDER enzyme to produce an AVOID enzyme. *Dissociation* is the splitting of a single agent into two agents. A dissociation rule which has the same agent type on either side of the reaction (for example $A \rightarrow A + X$) can be thought of as representing the production of new agents using materials that are available at saturation in the metabolism, and whose concentrations are not modelled for computational expediency. *Decay* is the spontaneous removal of an agent from the system, and is important for sensor and actuator molecules, which must decay quickly in order for the system to be responsive. Note that uni-molecular changes from one molecular species to another are not permitted. The probability of a bimolecular reaction is the product of the propensity and the reaction rate. Unimolecular reactions are governed by their reaction rate alone, since adjacency does not need to be considered.

Table 2 The four types of reaction rule in the metabolic controller

Reaction	Rule format	network symbol
Influx:	$\rightarrow A$	□
Binding:	$A + B \rightarrow C$	●
Dissociation: A	$\rightarrow B + C$	○
Decay: A	$\rightarrow$	■

These ingredients allow us to specify a metabolic control model for our e-pucks. For a more detailed overview of this metabolic model see [7].

In our metabolic controller, there are three classes of agents which possess different qualities within this framework. *Sensor* agents are generated when a sensor detects an obstacle. These are shown in white on the network diagrams below. *Actuator* agents are used to govern the speed and turning rate of the robot. These are shown in grey. Both sensors and actuators decay quickly, in order to allow the robot to be responsive. *Enzyme* agents form the connectivity between sensors and actuators. They are shown in black. Enzymes do not decay, but can be changed into other enzymes by reacting with other agents in the system. Reactions must be designed such that the total number of enzymes in the system is conserved.

5   Metabolic Subsumption

The metabolic network that we have designed for the robot control is based on the control layers described in section 3. We describe here the reaction system that we use to build behaviours that emulate these layers.

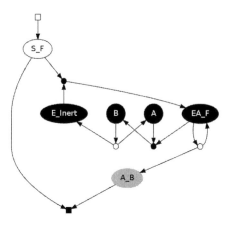

Fig. 2 Change in behaviour as a result of a sensor event. The metabolism switches from an inert behaviour to an avoidance behaviour on the influx of S_F agents by producing A_B agents

Figure 2 shows the network of an AVOID behaviour for a single sensor/actuator pair, which illustrates the basic reaction system we use for metabolic robot control. Symbols for reactions are described in table 2 and associated text. The system exists in an inert state until sensor information is received. In this state, the only agent types present in the system are the enzyme E_Inert and the deactivator enzyme A. When the sensor is activated, sensor agents of type S_F are generated in the metabolism. S_F binds with E_Inert to create EA_F. This enzyme uses a dissociation rule to create a copy of itself and the actuator agent A_B, which instructs the robot to move backwards. Once sensor agents have decayed out of the system, enzyme EA_F binds with A to produce an intermediate B. B then dissociates back to E_Inert and A.

The network in figure 2 shows an AVOID behaviour for a single sensor and a single actuator. We extend this model in figure 3 to show the metabolic network for one sensor type that produces AVOID behaviour subsumed by a WANDER behaviour appropriate to the input. Since wandering involves moving in a particular direction, actuators for turning are required. Information from the sensor is represented as quantities of S_F agents, which bind with the enzyme EW_F to create the EA_F. The avoid enzyme produces the signalling agents A_B that instruct the robot to reverse away from the obstacle. Note that EW_F and EA_F produce different actuator enzymes, whereas EW_L and EA_L produce A_L at different rates, appropriate to the dominant behaviour. (We have not represented these different rates on

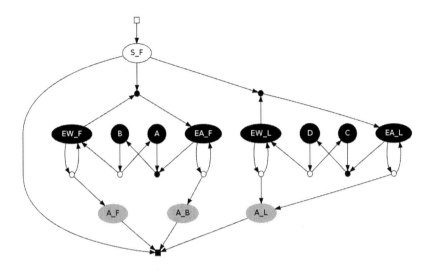

Fig. 3 A metabolic network for input coming from a front sensor

the diagram for clarity.) This metabolic approach to control emulates a subsumption architecture since the **AVOID** behaviour is autonomous and there is a switching mechanism from one behaviour to another as the situation demands it. As long as an avoid reaction is needed the wander behaviour is inhibited, because the wander enzymes are turned into avoid enzymes. When it is not needed anymore, i.e. no sensor agents are injected, the avoid enzymes are turned back into wander enzymes. For every type of sensor agent, the binding strengths are different to produce a different change of speed. For example, if there is an obstacle directly in front of the robot, there needs to be a wider turn than would be needed if the obstacle were slightly to the left or right.

The complete control network for the metabolic subsumption is shown in figure 4. Such a network becomes necessarily complex when information from four sensors is combined using a simple reaction rule set. Although the control architecture for each sensor follows the same basic pattern, there are subtle differences. The most striking of these is the sub-network for the rear sensor **S_B**. This is for two reasons. Firstly, the **WANDER** behaviour has no connection to the actuator enzyme **A_B** since when wandering the e-puck always moves in the forward direction as governed by **A_F**. Secondly, if **S_B** is present, the e-puck should move forward just as in the **WANDER** behaviour, but remaining **A_B** agents from previous reactions might have to be counteracted, so that more **A_F** have to be produced to ensure a forward movement.

We use the graphviz program [5] to visualise the network that our reaction rules represent. Although this is a useful tool, the high level of connectedness in the network prevents the automatic creation of network visualisations that makes the two control layers distinct. Although the concept of control layers is essential to the design of the subsumption architecture, the embodiment of the layers in the metabolic

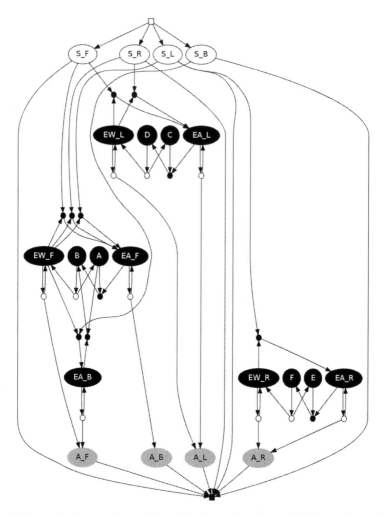

Fig. 4 The metabolic network for 4 input directions: front, front-left, front-right and back

network exhibits strong connectivity between the layers. This rich connectedness means that small changes at the level of nodes and reactions between them have the potential to cause larger changes at the emergent level.

When the metabolism is initialised, all that is present in the network are 10 of each of the **WANDER** enzymes, and 1 of the return enzymes **A**, **C**, and **E**. Actuator agents for the wander behaviour are created as the metabolism runs. When a sensor detects an obstacle, 10 of the corresponding sensor enzymes are placed in the metabolism. The reaction rates that were used in our experiments are shown in table 3.

Table 3 Reaction rates of a metabolic control network for the e-puck

Reaction	Rate	Reaction	Rate
Actuator signals for WANDER		*Actuator signals for AVOID*	
EW_F → EW_F + A_F	0.8	EA_F → EA_F + A_B	0.9
EW_L → EW_L + A_L	0.16	EA_B → EA_B + A_F	0.9
EW_R → EW_R + A_R	0.16	EA_L → EA_L + A_L	0.9
		EA_R → EA_R + A_R	0.9
Switch to AVOID behaviour		*Reversion to WANDER*	
EW_F + S_F → EA_F	0.9	EA_F + A → B	0.1
EW_L + S_F → EA_L	0.9	EA_B + A → B	0.1
		B → A + EW_F	0.1
EW_F + S_R → EA_F	0.7		
EW_R + S_R → EA_R	0.7	EA_L + C → D	0.1
		D → C + EW_L	0.1
EW_F + S_L → EA_F	0.7		
EW_L + S_L → EA_L	0.7	EA_R + E → F	0.1
		F → E + EW_R	0.1
EW_F + S_B → EA_B	0.1		
Decay of actuators		*Decay of sensors*	
A_F →	0.15	S_F →	1
A_B →	0.05	S_B →	1
A_L →	0.1 2	S_L →	1
A_R →	0.1 2	S_R →	1

6 Experimental Evaluation

An appropriate **WANDER** behaviour should allow the robot to explore the arena without giving it any particular strategy of exploration. The two control strategies were manually tuned such that their average speeds were approximately equivalent. We evaluated this via a visual inspection of the routes of the e-puck using the two different controllers. Sample traces for both controllers during a 5 minute run (simulated time) are shown in figure 5. It is clear that both controllers induce behaviour that can be interpreted as "wandering". However, it is difficult to obtain a quantitative evaluation of the pattern of exploration that the two control strategies confer on the e-puck.

Successful behaviours should prevent the e-puck from colliding with obstacles and walls. To compare the performance of the control systems we looked at 50 wall encounters for each set-up and counted the number of collisions. Since the metabolic controller was more difficult to tune, we compared a single metabolic controller with three subsumption architectures with different average and maximum speeds. Collision events for both controllers are shown in table 4. It is clear that the subsumption controller is more successful at responding to obstacles since the number of

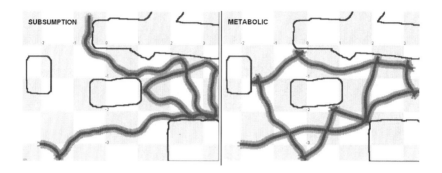

Fig. 5 A trace of the subsumption robot control running for a simulated time of 5 minutes

collisions is the same when it travels at nearly twice the speed of an e-puck that uses the metabolic controller.

Both control systems were designed to reverse away from an obstacle recorded on the sensors. While the subsumption architecture is able to change the speed and heading immediately upon receipt of a signal from a sensor, the metabolic model suffers from latency in its response. The wall encounters for the metabolic controller shown on the right of figure 5 are different from those on the right for the subsumption controller because of the latency in response between sensors and actuators. Latency is caused by the actuator agents that are present in the metabolism as the sensor data comes in. When an obstacle is encountered, actuator agents must be generated to counteract the actuator agents from the wander enzymes extant in the metabolism. This is illustrated in figure 6, which shows the changes in enzyme and actuator levels after an obstacle is encountered on sensor S_F. It is clear that this configuration of the metabolic controller cannot respond immediately to an obstacle since there are about 50 A_F agents in the system which instruct the robot to move forward. This situation could be changed by tuning the disassociation rate of A_F and making the enzyme EW_F produce A_F more quickly and so maintain a similar number of A_F whilst the WANDER behaviour is dominant.

Table 4 Area (in pixels) covered by the control systems and collisions for 50 wall encounters

	Subsumption			metabolic
Max speed (preset)	0.3	0.15	0.15	0.15
Average speed (recorded)	0.15	0.15	0.075	0.075
Median area covered for 5 runs of duration 5 minutes	6,264	6,491	3,460	3,574
collisions out of 50 wall encounters	11	4	0	4

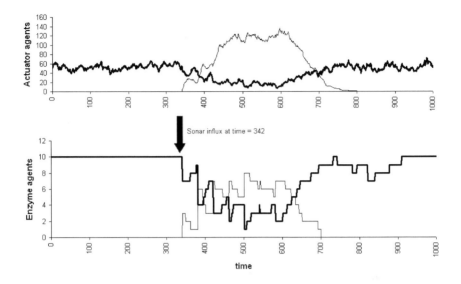

Fig. 6 Change in levels of actuator agents (top) and enzyme agents (bottom) as a result of a sensor event, indicated by the black arrow. The wander enzyme EW_F (thick line) is converted to the avoid enzyme EW_F (thin line) by binding with sensor agent S_F. This results in a change in the levels of two actuator agents in the metabolism: A_F (thick line) diminishes in quantity as A_B (thin line) accumulates, resulting in a change of direction from a forward to a backward motion

7 Conclusions

We have succeeded in our main objective to create a metabolic control system that emulated a simple subsumption architecture. This was motivated by the concept that a chemical reaction network would be more amenable to evolutionary adaptation.

A key difference between the two network types is that the metabolic architecture has *node multiplicity* - each node is represented by a quantity of autonomous agents. Each agent is capable of reacting with agents representing other nodes in the network. The metabolic approach lends itself to evolutionary adaptation [12], since agents for each node in the network can play a number of roles, allowing for duplication and divergence of function on evolutionary timescales. In this work, both the subsumption architecture and the metabolic architecture had to be "engineered" in the sense that the actual avoiding reaction and speeds needed to be optimised by hand. This approach allowed us to establish that an appropriate metabolic control could actually be produced within this framework.

Although the behaviour of both systems is qualitatively similar, the metabolic system suffers from latency in its reaction to obstacles. While the subsumption architecture basically reacts immediately to sensor inputs, the metabolic control needs some time to perform the necessary reactions and produce a sufficient metabolic response. This means that the metabolic control reacts inherently slower than the subsumption architecture. However, it should be noted that our metabolic controllers are more difficult to engineer by hand, since they have been designed to be trained *a posteriori* by an evolutionary system. Our goal was not to implement a system that performs "better" than a traditional subsumption architecture. Instead we focussed on creating a system that lends itself more readily to evolutionary adaptation than a traditional subsumption architecture. Our future work with the e-puck will therefore concentrate on implementing evolutionary adaptation in the metabolic controller.

Acknowledgements

The authors thank Susan Stepney, Peter Young, Tim Clarke, Edward Clark and Adam Nellis for comments and suggestions during the preparation of this manuscript. Verena Fischer is funded by the TRANSIT project, EPSRC grant EP/F032749/1. Simon Hickinbotham is funded by the Plazzmid project, EPSRC grant EP/F031033/1.

References

[1] Anon: Player-driver for e-puck robots (2009),
    http://code.google.com/p/epuck-player-driver/
[2] Brooks, R.A.: Elephants don't play chess. Robotics and Autonomous Systems 6(1&2), 3–15 (1990)
[3] Brooks, R.A.: Intelligence Without Reason. In: IJCAI 1991, pp. 569–595 (1991)
[4] Brooks, R.A.: Cambrian intelligence. MIT Press, Cambridge (1999)
[5] Ellson, J., Gansner, E.R., Koutsofios, E., North, S.C., Woodhull, G.: Graphviz - open source graph drawing tools. Graph Drawing, 483–484 (2001)
[6] Gerkey, B.P., Vaughan, R.T., Howard, A.: The player/stage project: Tools for multi-robot and distributed sensor systems. In: ICAR 2003, pp. 317–323 (2003)
[7] Hickinbotham, S., Clark, E., Stepney, S., Clarke, T., Young, P.: Gene regulation in a particle metabolome. In: CEC 2009, pp. 3024–3031. IEEE Press, Los Alamitos (2009)
[8] Hickinbotham, S., Clark, E., Stepney, S., Clarke, T., Nellis, A., Pay, M., Young, P.: Molecular microprograms. In: ECAL 2009 (2009)
[9] Liu, H., Iba, H.: Multi-agent learning of heterogeneous robots by evolutionary subsumption. In: Cantú-Paz, E., Foster, J.A., Deb, K., Davis, L., Roy, R., O'Reilly, U.-M., Beyer, H.-G., Kendall, G., Wilson, S.W., Harman, M., Wegener, J., Dasgupta, D., Potter, M.A., Schultz, A., Dowsland, K.A., Jonoska, N., Miller, J., Standish, R.K. (eds.) GECCO 2003. LNCS, vol. 2724, pp. 1715–1718. Springer, Heidelberg (2003)
[10] Mondada, F., Bonani, M., Raemy, X., Pugh, J., Cianci, C., Klaptocz, A., Magnenat, S., Zufferey, J., Floreano, D., Martinoli, A.: The e-puck, a robot designed for education in engineering. In: Robotica 2009, pp. 59–65 (2009)
[11] Pfeifer, R., Scheier, C.: Understanding Intelligence. MIT Press, Cambridge (1999)
[12] Stepney, S., Clarke, T., Young, P.: Plazzmid: An evolutionary agent-based architecture inspired by bacteria and bees. In: Almeida e Costa, F., Rocha, L.M., Costa, E., Harvey, I., Coutinho, A. (eds.) ECAL 2007. LNCS (LNAI), vol. 4648, pp. 1151–1160. Springer, Heidelberg (2007)
[13] Togelius, J.: Evolution of a subsumption architecture neurocontroller. Journal of Intelligent and Fuzzy Systems 15(1), 15–20 (2004)

Social Target Localization in a Population of Foragers

Héctor F. Satizábal M., Andres Upegui, and Andres Perez-Uribe

Abstract. Foraging has been identified as a benchmark for collective robotics. It consists on exploring an area and gathering prespecified objects from the environment. In addition to efficiently exploring an area, foragers have to be able to find special targets which are common to the whole population. This work proposes a method to cooperatively perform this particular task. Instead of using local or global localization strategies which can rely on the infrastructure installed in the environment, the proposed approach takes advantage of the knowledge gathered by the population about the localization of the targets. Robots communicate in an instrinsic way the estimation about how near they are from a target, and these estimations guide the navigation of the whole population when looking for these specific areas. The results comprehend some tests assessing the performance, robustness, and scalability of the approach. The proposed approach efficiently guides the robots towards the prespecified targets while allowing the modulation of their speed.

1 Introduction

It has been estimated that one-third of the animal biomass of the Amazon rain forest consists of ants and termites [17]. Their success might come from the fact that social interactions can compensate for individual limitations, both in terms of physical and cognitive capabilities. Indeed, herds and packs allow animals to attack larger prey and increase their chances for survival and mating [10], while organizations and teams facilitate information sharing and problem solving. The complexity of any society results from the local interactions among its members. Synthesizing and

Héctor F. Satizábal M.
ISI, Université de Lausanne
e-mail: Hector.SatizabalMejia@unil.ch

Andres Upegui · Andres Perez-Uribe
REDS, University of Applied Sciences Western Switzerland
e-mail: andres.upegui@heig-vd.ch, andres.perez-uribe@heig-vd.ch

J.R. González et al. (Eds.): NICSO 2010, SCI 284, pp. 13–24, 2010.
springerlink.com © Springer-Verlag Berlin Heidelberg 2010

analysing coherent collective behaviour from individual interactions is one of great challenges in both ethology and artificial intelligence.

Accomplishing tasks with a system of multiple robots is appealing because of its analogous relationship with populations of social insects. The hypothesis behind this approach is that such a synergistic robot system -one whose capabilities exceed the sum of its parts- can be created. Researchers argue that by organizing simple robots into cooperating teams, useful tasks may be accomplished otherwise impossible using a single robot[11].

Collective robotics has been used for a diversity of tasks like object manipulation, obstacle overpassing, and stair climbing. In all cases, collective robotics targets the execution of tasks that, when performed by a single robot, are impossible or inefficient. The goal is thus to find a strategy that allows a set of robots to, somehow, interact among them in order to find the solution in a more efficient manner than the same set of robots performing the task simultaneously but independently. Including such interaction implies additional costs in terms of robot set-up and computation, like addition of communicating capabilities or attachment mechanisms. In spite of this additional cost, the collective solution must still be more efficient than the individual one.

There is a large amount of real world tasks where a group of robots performs better or more efficiently than a single robot [2]. Collective robotics has gained the interest of a large number of researchers in the last decades. This situation demands the creation of new control strategies capable of taking advantage of the fact of having more than one individual, i.e. the presence of neighbours which can cooperate to ease the execution of the task.

Different approaches have been used in designing control strategies for groups of robots. There is a classical approach where a central planning unit coordinates the actions of robots. This unit sends commands according to the state of each unit in order to make them cooperate. The distribution of labours can be hierarchical, and each individual must be capable of replacing the planner unit if it fails due to malfunction [4]. This approach, while being the more intuitive and understandable, is often not scalable and difficult to implement due to the communication requirements of a central coordination, which in addition makes the system less robust. An alternative to this approach consists on endowing the system with self-organization properties, allowing individual units to cooperate without a central planner. Self-organization is frequently achieved by taking inspiration from biology [9], and in particular from the behaviour of social species. Stigmergy in ants [8] and trophallaxis in bees [7] are examples of strategies found in biology that have served as inspiration to develop controllers for collective robotics. There is a large number of different implementations of these concepts in robotics, each one requiring different levels of complexity of the robots, and different types of communication between the units.

This paper describes a novel approach for the localization of targets in a population of foragers. The control of the population of robots is performed in a distributed way. Our robots have two possible states which are "work" and "search". In the "work" state robots perform a certain foraging task and are distributed on the arena.

In the case of the work presented in this paper, we have a dummy foraging task consisting on navigating on the arena avoiding obstacles. The main interest is in the "search" state, where a robot will try to arrive to a specific target region on the arena. This target region can be a battery charging station, an area for garbage disposal, or the output of a maze. Whatever the robot may search, the goal will be to exploit the collective knowledge, given that there may be other robots that can estimate how far they are from the target region, and will somehow help the searching robot to achieve its goal. The proposed target localization avoids the use of global positioning systems, that might be difficult to deploy in unknown or hostile environments, and avoids also the use of odometry, which is sensitive to cumulated errors after large running periods. Our approach uses colour LEDs and omnidirectional cameras in order to indicate to other robots the shortest path to a desired target, based on a principle of disseminating the information gathered by the robots through the population.

This paper is structured as follows. Section 2 introduces the use of topological navigation and the use of landmarks, and describes the use of state comunication in coordinating a population of robots. Section 3 describes the simulation framework that was used in order to test the target localization strategy, and the robots and sensors implemented. Section 4 shows the results of the performed tests, and section 5 gives some conclusions.

2   Localizing a Target

Foraging is a common collective robotics task. In such a task, robots have to navigate in their environment while collecting items and depositing them at specific locations [4, 19]. In order to perform this type of task, robots need to be able to explore their environment in an efficient way, and at any moment, find target locations which are common to the whole population, such as storage places where the collected objects have to be stacked, battery charging stations, or specific sites if a fixed path has been stablished. Finding a target zone is thus a crucial behaviour for a robot being part of a swarm of foragers. Several approaches have been used with this purpose e.g. omniscient planners, sensing absolute position/orientation, following global beacons, using landmarks, pheromones, beacon chains or contact chains, etc. [18]. Global strategies like the use of centralized planners or GPS-like systems are expensive and difficult to implement or unreliable when the number of robots increases or when robots are placed in changing or hostile environments. Conversely, local strategies like the use of local beacons or landmarks, or bio-inspired methods like pheromones, are more easily scalable and allow the implementation of self-organized systems which can adapt to unknown environments.

2.1   *Landmarks and Beacons*

The use of landmarks in robot navigation is a widely used approach which has been called *topological navigation* [4]. Robots using this strategy do not use precise measurements of position but have to infer their own location from the perception of known marks in the environment such as doors or intersections in the case of indoor navigation. Topological navigation is common for us since most of the information we use to locate ourselves and target directions are relative to objects in our landscape. Nevertheless, using landmarks is not exclusive to superior animals. Some researchers have taken inspiration from small social animals like insects which employ similar strategies to find their way back to home after exploration journeys. Some species of desert ants, for instance, use visual landmarks in order to return to important places in their environment [14] when other methods like the use of pheromones is not possible. Bees, can use the physical contact with other individuals of the hive in order to regulate the behaviour of foragers [7]. This form of communication, where individuals employ other members of the population as landmarks or beacons for locating a target has also been a source of inspiration for robotics navigation [16, 18].

2.2 Social Localization

In this paper, we present a novel approach for finding a common target location based on the knowledge gathered by a population of robots. It supposes the existence of a population of robots performing a foraging task. A group of robots is thus distributed in the environment while searching for some kind of resource, and at any moment, any individual has to find a specific place which is common for the whole population e.g. a charging station, or a depot where gathered object have to be stacked. Robots are not provided with their positions when looking for the targets, instead, each individual has to use imprecise nearness estimations of neighbours which are transfered through *state communication* [3].

In *state communication* robots communicate through their behaviour. Hence, robots have to be able to interpret the behaviour of other robots by using their sensory capabilities. The communication can be explicit or implicit whether the sender is aware of the receiver or not. State communication has been succesfully used for coordinating tasks in collective robotics [12, 13, 16] and it has proven to be robust and scalable.

State comunication can be used to transfer information about the location of a specific place. Imagine you enter in a shopping mall where there are lots of stores and lots of people buying in such stores. You want to buy something in a store called *RoboShop*, but you do not know where the store is located. There are several possibilities to find *RoboShop*; you can explore the place without having any information of the location of the target, and perform a random search covering all stores in the building. Or, you can ask at the information desk, and analogously to having a central planner, you can ask for the location of *RoboShop* and go directly to the target. The store *RoboShop* could also be located by using the indications shown on the walls of the building as landmarks, and following them until arriving to the target. All these strategies imply the presence of a certain infrastructure on the building i.e. an information desk as a central source of information, or indications about the location of the shops as landmarks to guide the costumers. Alternative strategies must be adopted if such information is not available. One simple method could be, if *RoboShop* delivers its products within green plastic bags, all you have to do in order to find the target, is to follow the opposite direction of people having green plastic bags, and you will eventually find the *RoboShop* store. Here, a costumer must only know that people having green bags are likely to come from *RoboShop*, so that they can be used as dynamic beacons to guide the searching process. No infrastucture on the building is needed. Instead, state communication is used, and it is performed in an intrinsic manner because people carrying the green bags are not aware of the information they are sending to others.

The solution we present here has been inspired from the aforementioned strategy. Robots are distributed in an arena like people were distributed in the building, and robots have to find target places analogously to people looking for the *RoboShop* store in the example. Vision was chosen to perform state communication in the case of robots. Every robot can display a colour, and that colour reflects an internal state of the robot which is directly related to the certitude of being near the target. Each robot has limited vision which allows it to detect other robots as well as obstacles. Thus, as in the previous example, if a robot needs to go to a specific place, it has to follow robots showing the colour that was assigned to this place. These coloured robots act as moving beacons to guide other members of the population to the specific goals. Once arrived to the target, the robot must update its colour in order to cooperate with the rest of the population serving as beacon for other robots while linearly decreasing its colour.

However, one main modification was done to the initial setup inspiring the algorithm. Robots were programmed to copy a proportion of the colour of other robots, and as a consequence, an emerging colour gradient is formed in the population. The addition of this behaviour improves the dissemination of information through the robots, facilitating the task of looking for a target. Any robot in the population behaves as a mobile beacon, and cooperates with the execution of the task by guiding other robots to the target, even if the exact position of the target is unknown. The details of the implemmentation of the strategy are shown in section 3.2.

3   Experimental Setup

We used a simulator called Enki [15] to evaluate the performance of the robots in the task of finding the targets distributed in the arena. Enki is a 2D robot simulator written in C++. It is open source and provides collision and limited physics support for robots evolving on a flat surface.

3.1   The Arena

The flat space where robots evolve is a square arena with 300 cm of side length, limited by dark gray[1] walls of 15 cm height. There are two RFID tags located within

---

[1] R=30%, G=30%, B=30%.

this area at positions (40, 260)cm and (260, 40)cm. Each tag can be detected at a maximal distance of 21.5 cm. These areas are shown by the two light gray circles located at the top-left corner for tag number 1, and bottom-right corner for tag number 2. This setup is shown in figure 1.

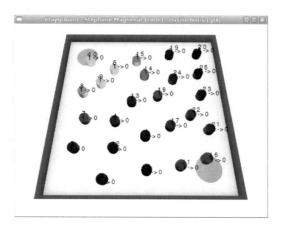

Fig. 1 Arena where the experiments evolve. The numbered cylinders are the mobile robots and the two gray circles represent the zones from where RFID tags can be detected. Tag 1 is placed in the top-left corner, and tag 2 is placed in the bottom-right corner

3.2   The Robots

The robots implemented on Enki simulate a real robot called marXbot [5] which is endowed with two wheels for locomotion, RGB LED for displaying colours, omnidirectional camera, infrared bumpers, rotating distance sensor scanner, and a RFID tag detector. Additionally, the behaviour of the omnidirectional camera was modified in a way that the detected colours are not only function of the colour of the object, but also a function of the distance to the object. For doing so, we used the information provided by the rotating distance sensor scanner modulating each one of the components (R, G, B) of the colours as shown in figure 2.

Navigation was performed in a pure reactive manner as in Braintenberg vehicles [6], and the integration of sensor information was based on a strategy called *motor schema-based navigation* [1]. Hence, 24 infrared sensors were used as bumpers, and a 180 pixels omnidirectional linear camera was used in order to detect mid-range and distant obstacles and colours. The steer direction **S** was calculated by adding 4 components:

- Bumpers (**b**): The vector pointing in the direction where there are no obstacles detected by the bumpers.
- Free Area (**f**): The vector pointing in the direction where there are no obstacles detected by the camera.

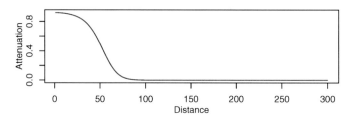

Fig. 2 Attenuation of the colours detected by the omnidirectional camera with respect to the distance to the object

- Attraction to Landmark (**t**): The vector pointing in the direction where there are objects having the colour associated to the target.
- Repulsion (**nt**): The vector pointing in the direction where there are no objects having a colour associated to a different target

$$S_x = b_x + f_x + t_x + nt_x \tag{1}$$
$$S_y = b_y + f_y + t_y + nt_y \tag{2}$$

Each one of the aforementioned components was calculated as the dot product of the vector having the response of the sensor group, and the vector of positions of each individual sensor. Taking the bumpers as an example, we have a vector **bumper** with the signals of the 24 infrared sensors, and a vector **A** compiling the angle α of each sensor. The vector **bo** summarizing the activation of the 24 bumpers, and pointing in the direction where there is an obstacle[2] is calculated as follows:

$$bo_x = \textbf{bumper} \cdot \cos \textbf{A} \tag{3}$$
$$bo_y = \textbf{bumper} \cdot \sin \textbf{A} \tag{4}$$

In the case of the camera, some masks were applied to the image in order to eliminate the influence of walls when calculating the **t** and **nt** components.

4 Testing Dynamic Landmarks and the Use of Population Knowledge

Several experiments were conducted in order to evaluate the changes in performance when using the robots as landmarks to perform target localization. In each case, we measure the time a robot spent in going from one RFID tag to the other one.

---

[2] If there are several obstacles this method returns the direction where there are more detections.

4.1 From Random Search to Social Search

This section shows the results of a series of five experiments with increasing information about the target position. A histogram of the distribution of the time a robot spent in going from one tag to the other was generated for each case, (see figure 3). The whole set of robots performed the task of sequentially finding the targets.

- **Random search:** The sequence of experiments starts with a searching strategy where robots freely navigate, and find the targets by chance i.e. components **t** and **nt** were set to 0. This strategy without any information about the location of the targets give us a reference to compare with other strategies where more information is provided. Figure 3a) shows an histogram of the time spent by the robots in finding one tag after the other.

- **Static landmarks:** Coloured landmarks placed in the corners of the arena were used as the next step to improve performance in target localization. Two green panels were placed behind target 1, and two red panels behind target 2. In this experiment the robots do not display any colour. Their only guide for finding the tags are the static landmarks, that is, the panels placed in the corners of the arena. Figure 3b) shows an histogram with the results of this experiment. It can be seen that the average time spent by a robot in going from one tag to the other was reduced to near 50% with respect to the random strategy.

- **Static landmarks and robots as landmarks:** In addition of having static landmarks, robots were enabled to display a colour when reaching a target. A robot reaching tag 1 sets its green component to 100%, and a robot reaching tag 2 sets its red component to 100%. Hereafter, the robot starts to decrease its colour components linearly. As a result, a robot reaching a tag serves as landmark for other robots in its proximity for some time. Figure 3c) shows an histogram of the time the robots spent in going from one tag to the other when using this strategy. In this case, the average time for performing the task was reduced to near 70% of the previous case.

- **Robots as landmarks:** In order to pursue the exploration on using population knowledge in guiding target localization, the static targets were removed from the arena, and the population performance was assesed when using only robots as targets. As in the preceding test, robots reaching a tag become a landmark of this tag for a certain time. Figure 3d) shows an histogram of the performance of the robots when finding one tag after another by using this strategy. The performance of the algorithm is not reduced even if there are no static landmarks in the corners.

- **Using a gradient of colours - social localization:** Finally, the feature of copying colours was implemmented. Each robot behave as a landmark in the proximities of a tag (as in the last two cases), and in addition, robots always compare each one of their [R, G, B] colour components against the colours detected by the omnidirectional camera. Then, if the component detected is greater than the own one, the component is copied and displayed by the robot. This behaviour allows the propagation of the information about the position of a tag through the population. The resulting gradient of colours is shown in figure 1. A histogram with the time spent by the robots in going from one tag to the other when using this strategy is shown in figure 3e). Again, a reduction of near 50% in the average time is achieved with respect to the previous case.

Figure 3 summarizes the results of the tests performed so far. It can be seen that using landmarks yields smaller times than navigating randomly, and that this improvement can be even larger if landmarks can move in the vicinities of the targets. Moreover, It can be seen that the performance of using parts of the population as landmarks can be increased when robots share information about the position of the target, by allowing the robots' knowledge to spread through the population in the form of a gradient. Indeed, from the experiments performed so far, it can be seen that the average time spent by the robots in performing the task is divided by two for each incremental step of the test i.e. $\simeq$600 for navigating randomly, $\simeq$300 with static landmarks, $\simeq$150 with dynamic landmarks, and $\simeq$75 when creating a gradient of colours.

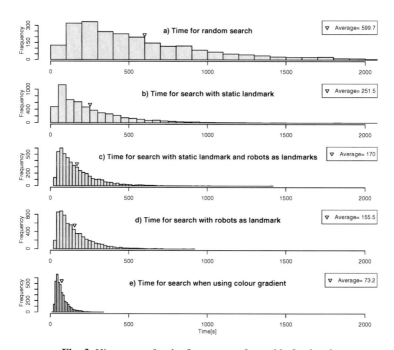

Fig. 3 Histograms for the five tests performed beforehand

4.2 Varying Attraction between Robots

The influence of component **t** (attraction to landmark) can be modulated by a parameter of attraction we called k.

$$S_x = b_x + f_x + k * t_x + nt_x \tag{5}$$
$$S_y = b_y + f_y + k * t_y + nt_y \tag{6}$$

Varying k enables us to control the speed of the robots when approaching other robots, and thus modifying the time a robot spends in finding a target. By changing the speed of the robots, we can also modify the dynamics of the population, making the navigation less or more fluid. We measured the time each robot spent in going from one tag to the other, and after 1000 single trips we obtained the results shown in figure 4. In this case we collect data while changing also the amount of robots performing the task of searching the targets. The population size was always kept constant.

The amount of potential collisions is also affected as a result of changing the attraction between robots. Low values of k yield a smoother navigation at the expense

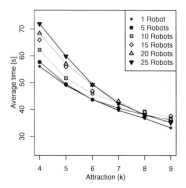

Fig. 4 Average time spent by a robot with respect to the value of attraction in the case of using the information gathered by the population about the location of the targets. The experiment summarizes a set of 1000 single trips performed by the robots

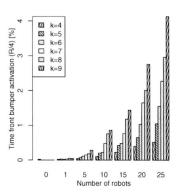

Fig. 5 Percent of time the frontal proximity sensor is activated in the case of using the information gathered by the population about the location of the targets. The experiment summarizes a set of 1000 single trips performed by the robots

of longer travel times. As it can be seen on figure 4, shorter times are achieved by increasing k until asymptotically reaching a minimum. Moreover, robots quickly tend to agglomerate or even collide when k is increased. This fact can be seen on figure 5 which shows the percent of time a robot frontally approaches an obstacle to a distance less than 0.25 times its radius. Even if the total time in front of an obstacle is not very large (4% in the worst case), a straight increment of this time is produced when increasing k in every case.

5 Conclusions

Localizing a common target is an essential task in a population of foragers. Robots exploring an area eventually need to find areas which are important for the operation of each one of the members of the population, or for solving the foraging task. Since these key zones are the same for every robot, we propose to use the knowledge disseminated in the population about the location of the targets, in order to guide the navigation of each member of it. The proposed coordination scheme is distributed and uses state comunication in a instrinsic way, i.e. robots transmit some information about their internal state, but they are not aware of whether other robots receive this information or not. This fact simplifies the communication and makes the system more robust. This "social" approach is tested and compared incrementally against different strategies and, as it has been shown in figure 3, it is proven that the performance in finding a target is improved. Moreover, the proposed robot guidance avoids the use of global positioning systems, that might be difficult to deploy in unknown or hostile environments, and avoids also the use of odometry, which is sensitive to cumulated errors after large running periods. Additionally, the fact of being a distributed scheme makes it very robust and scalable.

Moreover, some tests concerning robustness and scalability were performed. A parameter k was added in order to modulate the attraction between robots when approaching a target. As it can be seen in figure 4, changing k changes the average time a robot spends in finding a target and therefore, the trajectories performed by the robots become more or less smooth. Figure 5 shows the percent of time a robot activates its frontal proximity bumper (which can be considered as collisions), and it can be seen that there is a strong relationship between robot agglomeration and parameter k. The amount of robots looking for targets was also changed during simulations. Figure 4 shows that there is a weak relationship between robot performance and the amount of robots performing the task; and that this relationship is even weaker when parameter k is higher. Additionally, figure 5 shows that the robots tend to agglomerate more, activating more often their frontal sensors, when there are more robots looking for targets at the same time.

Besides the tests shown here, some future work concerning the performance of the strategy has been already envisaged. The proposed approach will be tested in a more complicated arena, including fixed obstacles. Moreover, we will test our social target localization approach with moving targets, and with targets which change its value, simulating limited resources.

References

[1] Arkin, R.C.: Cooperation without communication: Multiagent schema-based robot navigation. Journal of Robotic Systems 9(3), 351–364 (1992)
[2] Arkin, R.C., Bekey, G.A. (eds.): Robot colonies. Kluwer Academic Publishers, Norwell (1997)
[3] Balch, T., Arkin, R.C.: Communication in reactive multiagent robotic systems. Auton Robots 1(1), 27–52 (1994)
[4] Bekey, G.A.: Autonomous Robots: From Biological Inspiration to Implementation and Control (Intelligent Robotics and Autonomous Agents). The MIT Press, Cambridge (2005)
[5] Bonani, M., Baaboura, T., Retornaz, P., Vaussard, F., Magnenat, S., Burnier, D., Longchamp, V., Mondada, F.: The marxbot – a modular all-terrain experimentation robot (2009), http://mobots.epfl.ch/marxbot.html
[6] Braitenberg, V.: Vehicles: Experiments in Synthetic Psychology. The MIT Press, Cambridge (1984)
[7] Camazine, S., Crailsheim, K., Hrassnigg, N., Robinson, G.E., Leonhard, B., Kropiunigg, H.: Protein trophallaxis and the regulation of pollen foraging by honey bees (apis mellifera l.). Apidologie 29(1) (1998)
[8] Camazine, S., Deneubourg, J.L., Franks, N.R., Sneyd, J., Theraulaz, G., Bonabeau, E.: Self-Organization in Biological Systems. Princeton University Press, Princeton (2001)
[9] Deneubourg, J.L., Goss, S.: Collective patterns and decision making. Ethology, Ecology & Evolution 1, 295–311 (1989)
[10] Gadagkar, R.: Survival strategies: cooperation and conflict in animal societies. Harvard University Press, USA (1997)
[11] Ijspeert, A., Martinoli, A., Billard, A., Gambardella, L.M.: Collaboration through the Exploitation of Local Interactions in Autonomous Collective Robotics: The Stick Pulling Experiment. Autonomous Robots 11(2), 149–171 (2001)
[12] Kuniyoshi, Y., Kita, N., Rougeaux, S., Sakane, S., Ishii, M., Kakikua, M.: Cooperation by observation: the framework and basic task patterns. In: Proceedings of IEEE International Conference on Robotics and Automation 1994, vol. 1, pp. 767–774 (1994)
[13] Kuniyoshi, Y., Rickki, J., Ishii, M., Rougeaux, S., Kita, N., Sakane, S., Kakikura, M.: Vision-based behaviors for multi-robot cooperation. In: Proceedings of the IEEE/RSJ/GI International Conference on Intelligent Robots and Systems 1994. Advanced Robotic Systems and the Real World, IROS 1994, vol. 2, pp. 925–932 (1994)
[14] Lambrinos, D., Roggendorf, T., Pfeifer, R.: Insect strategies of visual homing in mobile robots. In: Biorobotics - Methods and Applications, pp. 37–66. AAAI Press, Menlo Park (2001)
[15] Magnenat, S., Waibel, M., Beyeler, A.: Enki – an open source fast 2d robot simulator (2009), http://home.gna.org/enki/
[16] Nouyan, S., Gross, R., Dorigo, M., Bonani, M., Mondada, F.: Group transport along a robot chain in a self-organised robot colony. In: Proc. of the 9th Int. Conf. on Intelligent Autonomous Systems, IOS, pp. 433–442. IOS Press, Amsterdam (2005)
[17] Smith, J.M., Szathmary, E.: The Origins of Life: From the Birth of Life to the Origin of Language. Oxford University Press, USA (2000)
[18] Werger, B., Mataric, M.J.: Robotic"food" chains: Externalization of state and program for minimal-agent foraging. In: Proc. 4th Int. Conf. Simulation of Adaptive Behavior: From Animals to Animats, vol. 4, pp. 625–634. The MIT Press, Cambridge (1996)
[19] Winfield, A.: Towards an engineering science of robot foraging. Distributed Autonomous Robotic Systems 8, 185–192 (2009)

Using Knowledge Discovery in Cooperative Strategies: Two Case Studies

A.D. Masegosa, E. Muñoz, D. Pelta, and J.M. Cadenas

Abstract. In this work we discuss to what extent and in what contexts the use of knowledge discovery techniques can improve the performance of cooperative strategies for optimization. The study is approached over two different cases study that differs in terms of the definition of the initial cooperative strategy, the problem chosen as test bed (Uncapacitated Single Allocation p Hub Median and knapsack problems) and the number of instances available for applying data mining. The results obtained show that this techniques can lead to an improvement of the cooperatives strategies as long as the application context fulfils certain characteristics.

1 Introduction

Although some algorithms have a good performance in a specific problem, there is hardly an algorithm which behaves better than others in a wide set of instances of such problem. This fact corresponds with the No Free Lunch Theorem [21]. In this way, it is very complicated to determine what the best method for a given instance is, specially if there are big differences in performance from one algorithm to another. Formally, this is known as the "Algorithm Selection problem" [20], and was defined by Rice in 1976.

This problem has been treated in various areas. One of them is Machine Learning [5, 11, 12]. These kind of techniques have been used to estimate the execution time required by an algorithm to solve a determined type of instances, so that through this

A.D. Masegosa · D. Pelta
Dept. of Computer Science and Artificial Intelligence
University of Granada, Granada, Spain
e-mail: {admase,dpelta}@decsai.ugr.es

E. Muñoz · J.M. Cadenas
Dept. Ingeniería de la Información y las Comunicaciones
University of Murcia, Murcia, Spain
e-mail: enriquemuba@dif.um.es,jcadenas@um.es

J.R. González et al. (Eds.): NICSO 2010, SCI 284, pp. 25–38, 2010.
springerlink.com   © Springer-Verlag Berlin Heidelberg 2010

information, we can choose the best expected method when we face a new instance. Another technique is associated with the "Algorithm Portfolio" paradigm, where, instead of selecting a single algorithm, a set of methods are executed in parallel until the fastest one solves the problem. An example of this type of strategies can be found in [17]. When the algorithms are allowed to exchange information among them, then cooperative search strategies arise, and this collaboration leads to a dramatically improve in the robustness and the quality of the solutions obtained with respect to the independent version of the strategy [2, 6]. This concept of cooperation is successfully used, explicit or implicitly, in other types of metaheuristics as multi-agent systems (ACO's [8], PSO's[14]), memetic algorithms [15] and hyper-heuristics [3].

In this paper we are going to treat with both areas, cooperative strategies and Machine Learning. Concretely, we will discuss to what extent and in what contexts the use of knowledge discovery techniques can improve the performance of cooperative strategies. For this purpose, a centralised cooperative strategy based on simple elements of Soft Computing, previously presented in [4, 7, 19], will be consider as the baseline case. From this starting point, we will analyse the improvement produced by the use of new control rules and two alternatives for setting the initial parameters of the methods composing the cooperative strategy. These features are obtained using data mining. The study will be conducted on two different scenarios that differ in terms of the baseline implementation and test bed used (Uncapacitated Single Allocation p-Hub Median Problem (USApHMP) and the Knapsack problem). We have chosen these two problems for the following reasons: the USApHMP is a NP-hard problem where only small datasets of solved instances can be found, and for that reason we have little information in order to perform the training phase in the knowledge discovery process. On the other hand, Knapsack Problem is one of the "easiest" NP-hard problems, in which simple resolution algorithms obtain good results, and where we can find big datasets of solved instances for training the system. These test beds are two extreme situations in which we want to check the improvements obtained by the KD.

This work is structured as follows. Firstly, we will describe the centralised cooperative strategy used as base case. In Section 3, the new control rule and the two types of initial parameter tune will be shown. Section 4 is devoted to state the two case studies used to test the cooperative method. After that, we will relate the experimentation done and the results obtained. To finish, in Section 6, the conclusions of this work, will be discussed.

2   A Centralized Cooperative Search Strategy

The cooperative strategy described in [7, 19], consists on a set of solvers/threads, each one implementing the same or a different resolution strategy for the problem at hand. These threads are controlled by a coordinator which processes the information received from them and, making use of a fuzzy rule base, produces subsequent adjustments of solver behaviours by sending "orders". The information exchange process is done through a blackboard architecture [9].

A important part of this strategy is the information flow, that is divided in the three steps: 1) performance information (report) is sent to the coordinator from the solvers, 2) this information is stored and processed by the coordinator and 3) coordinator sends orders to the solvers.

Each report in the first step contains:

- Solver identification
- A time stamp t
- The current solution of the solver at that time s^t
- The best solution reached until that time by this solver s_{best}

The coordinator stores the last two reports from each solver, so in the information processing step, the improvement rate is calculated as $\Delta_f = \frac{f(s^t) - f(s^{t'})}{t - t'}$, where $t - t'$ represents the elapsed time between two consecutive reports, s^t is the current solution sent by the solver in the last report and f is the objective function. The values Δ_f and $f(s^t)$ are then stored in two fixed length ordered "memories", one for improvements and another for costs.

Over those memories, a fuzzy control rule is constructed. This rule allows the coordinator to determine if a solver is working fine or not. It was designed based on expert knowledge following the principle: *If a solver is working well, keep it*; but *if a solver seems to be trapped, do something to alter its behaviour*. From now on, this rule is called *EK* and its definition is the next one:

IF the quality of the current solution reported by *solver$_i$* is *low* **AND** the improvement rate of *solver$_i$* is *low* **THEN** send C_{best} to *solver$_i$*

The label *low* is defined as a fuzzy set whose membership function $\mu(x)$ is shown in Figure 1 (a). The variable x will correspond with the relative position (resembling the notion of percentile rank) of a value (an improvement rate or a cost) in the samples stored in memory of improvements or memory of costs, respectively, and the other parameters are fixed to $a = 80$ and $b = 100$ for the memory of costs, and $a = 0$ and $b = 20$ for the memory of improvements. C_{best} denotes the best solution ever recorded by the coordinator. In short, what the rule says is that if the values reported by a solver are among the worst in the memories, then such a solver should be changed in some way.

By means of sending C_{best}, it is expected that the solvers will concentrate around the most promising regions of the search space, which will be sampled using different schemes (the ones defined by the solver threads themselves). This increases the chances of finding better and better solutions.

Depending on the nature of the solvers (trajectory-based or population-based), the solution C_{best} is sent in a different way. For trajectory based methods, a new solution C'_{best} is obtained from C_{best} using a mutation operator. When the solver receives C'_{best}, then it will restart the search from that new point. Such modification tries to avoid relocating all of the solvers in the same point of the search space.

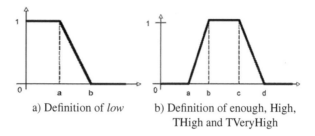

a) Definition of *low*

b) Definition of enough, High,
THigh and TVeryHigh

Fig. 1 Definition of low, enough, High, THigh and TVeryHigh

However, for population based methods, a proportion of the worst individuals of the receiver is substituted by a set of mutated solutions obtained from C_{best} using the same operator as before.

3 Knowledge Discovery for Rule Design and Parameters Setup

Any of the components defining the basic cooperative strategy could be changed. In this work, we will consider new definitions for two relevant components: 1) a new set of control rules and 2) a mechanism to setup the initial parameters of the threads. Both features will be obtained using knowledge discovery techniques that are fully described in [4]. The basic ideas of the process is briefly presented here.

The first step is the data generation process where we have:

- $\{m_0, \ldots, m_k\}$, a set k metaheuristics
- $\{c_{i,0}, \ldots, c_{i,d}\}$, a set of possible parameter combinations for m_i
- $\{p_0, \ldots, p_l\}$ the set of training instances

Then, every m_i is run over each p_t with every possible combination of parameters c_{ij} in order to obtain a performance information database. The second step in the knowledge discovery process is to extract several decision trees. Then, when the cooperative system is presented with a new instance to solve, the trees are traversed and certain weights for the control rules are returned (from the "Weights Tree"), and a list of "good parameter configurations" is constructed (from the "Parameters Tree"). From this list, the system will setup the parameters of the threads. See Figure 2 for a schematic description.

3.1 New Set of Control Rules

The new set of control rules has two parameterized rules: the first one allows to change the position in the search space of a thread (because it may show a bad performance) making it closer to the one of another metaheuristic with a better behavior; the second rule allows to dynamically change the parameters governing the behaviour of a thread.

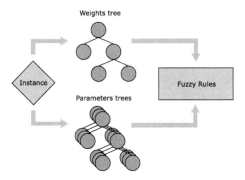

Fig. 2 Given a new instance to solve, new control rules weights and solvers' initial parameters are calculated

The parameterized rules (one for each thread or metaheuristic) are as follows:

- IF [$solver_i$ IS *theWorst*] AND [($wm_1 * d_1$ OR ... OR $wm_n * d_n$) IS *enough*] THEN change the current solution of $solver_i$.
- IF [($wm_1 * d_1$ OR ... OR $wm_n * d_n$) IS *High* AND (*time* IS *THigh* OR *TVeryHigh*)] THEN *changeParameterValues* of $solver_i$.

where:

- n is the number of solvers.
- $solver_i$ is the solver being evaluated by the rule.
- *theWorst* evaluates if the solver being studied now is having the worst performance according to any previously defined measure.
- $d_i = (perf_i - perf_{MH})/maximum(perf_i, perf_{MH})$, where $perf$ is a measure of performance previously defined.
- $wm_{m_i} \in [0,1]$ where $\sum_{i=1}^{n} wm_{m_i} = 1$ and wm_{m_i} represents the weight of solver i (importance of metaheuristic i for solving the current instance). These weights are calculated from the "Weights Tree" obtained from the data mining process.
- *Enough*, *High*, *THigh* and *TVeryHigh* are fuzzy sets with trapezoidal membership functions with support contained in $[0,1]$ defined by a cuadruplet (a,b,c,d). Its representation is shown in Figure 1b).
- *ChangeParameterValues* is a function that changes the values of the parameters of a solver. As stated before, a list of "good parameter configurations" was obtained from the "Parameters Tree". So, when the rule is triggered, the next configuration from the list is selected.

3.2 Parameters Adjustment

The initial parameters of the solvers are calculated from the "Parameters Tree". In fact, there exist to different operational modes. In the first one, the parameters are calculated as a function of the type of the instance, while in the second one, the parameters are independent from the instance being solved. In this last case, the best configuration of parameters is the one that allowed to obtain, on average,

Table 1 Main features of the two case studies proposed

	Case study 1	Case study 2
Communication mode	asynchronous	synchronous
Stop condition and communication frequency	evaluation number	Time
Implemented solvers	VND, Tabu, SA	Tabu, SA, GA
Test problem	USApHMP	knapsack
Number of instances	34	2000
Number of instances for training	34	500
Number of instances per size and type	1	25
Number of instances for test	34	20

the best results over the set of training instances. Both strategies are considered in this work.

4 Case Studies Details

This section is devoted to describe the two case studies designed to assess to what extent and in what contexts the use of knowledge discovery techniques can improve the performance of cooperative strategies. The scenarios differ in the type of basic cooperative strategy used, in the communication model, test problem and information available for the data mining stage. The next two subsections fully describe the two scenarios, while their main features are displayed in Table 1.

4.1 Case Study 1

When implementing multi-threaded cooperative strategies, one can resort to parallel schemes if time is important, or one can simulate the parallelism in a one-processor computer. This is the strategy taken here and the procedure is extremely simple. We construct an array of solvers and we run them using a round-robin schema. This implementation uses a synchronous communication mode that is simulated in this way: solvers are executed during 100 ms each one and after this period of time, information exchanges are performed. These steps are repeated until the stopping condition, given in terms of running time, is fulfilled.

Regarding the fuzzy rule (EK rule), the size of the memory of costs and improvements was set to be double the number of solvers. Three different heuristic searches were chosen as solvers: Genetic Algorithm (GA), Tabu Search (TS) and Simulated Annealing (SA). Their implementation follows the basic guidelines described in [10] and no specific tailoring of operators to problem was done. The description of these methods is omitted due to space constraints.

The test bed used in this case is the well known knapsack problem. The problem is defined as follows: Given a set of items, each with a cost and a benefit, determine a subset such that the total cost is less than a given limit and the total benefit is as large as possible.

From the point of view of the Knowledge Discovery process, and due to the availability of an instance generator, the number of instances available for training the system was 500, with 25 instances per size and type considered. We used four different sizes: 500, 1000, 1500 and 2000 objects and five types of instances, given by Dr D. Pisinger in[13], were taken into account:

- Spanner: These instances are constructed in such a way that all their items are multiple of a small set of items called key. That key was generated using three distributions:

  - Uncorrelated,
  - Weakly correlated,
  - Strongly correlated.

- Profit ceiling: In these instances all the benefits are multiple of a given parameter d.
- Circle: These instances are generated in such a way that the benefits are a function of the weights, having its graph an elliptic representation.

To carry out the tests we solved a database of instances composed of 20 instances (one per type and size). In order to asses the quality of the solutions returned by the strategy, we consider an error as $error = 100 \times \frac{obtained\ value - optimum}{optimum}$.

4.2 Case Study 2 Description

In this case study, the implementation is broadly the same with some slight variations. Firstly, here the communication mode is asynchronous and is not determined by CPU time but by objective function evaluations. Concretely, the solvers are run during a random number of evaluations that varies from 100 to 150. The process is repeated until a maximum number of objective function evaluations has been done.

Other important difference with respect to the former one are the heuristic implemented by the solvers, since now all of them are trajectory based. The three different heuristic searches chosen were: Tabu Search, Simulated Annealing (SA) and Variable Neighborhood Descent search (VND). As before, their implementation follows the basic guidelines described in [10] and no specific tailoring of operators to problem was done.

The test bed is a hub location problem. The aim on this type of problems is composed of two steps: 1) **Hub location:** to determine which and how many nodes should be the hubs, in order to distribute the flow across them, and 2) **Non-hub to hub allocation:** to assign the rest of the nodes to the hubs. Generally, these tasks are performed by minimizing an objective function that describes the exchange flow and its cost. We will focus on a particular case: the Uncapacitated Single Allocation p-Hub Median Problem (USApHMP), which is consider as a NP-hard problem. Its quadratic integer formulation was given by O'Kelly in [18].

The instances chosen for the experimentation were obtained from the resource ORLIB [1]. Concretely, we used the AP (Australian Post) data set derived from a study of a postal delivery system. The data set contains a first group of instances with 10,20,25,40 and 50 nodes (having 2,3,4 hubs), and a second group where the instances have 100 and 200 nodes with 2, 3, 4, 5,10,15 and 20 hubs. The optimum for those instances with a number of nodes less than 50 was provided by the resource ORLIB, and for the other instances we considered the best solution found for one of the state-of-art algorithms for this problem, presented in [16]. The quality of the solutions is measured as in the previous case study. To finish this section, we should remark other significant distinction with respect to the case 1, since this one only have available a total of 34 instances and there is just one instance per type and size.

5   Experiments and Results

The experimentation done in this paper has as target to analyse the benefits contributed by the Knowledge Discovery process seen in Section 3. For this purpose, the baseline for comparison is the strategy seen in Section 2, where there is just one control rule (EK Rule) and the initial parameters for all the threads are those that gave the best results when averaged over all the training instances. In other words, they are independent from the instance being solved. The following combinations will be tested:

- KD rule is used instead of the EK rule.
- The parameters are calculated as a function of the instance being solved.

In this way, from the base case we can obtain the next strategies:

$$\text{basic strategy} \begin{cases} EK\ rule \begin{cases} parameters: instance\,independent\,(EK+IIP) \\ parameters: instance\,dependent\,(EK+IDP) \end{cases} \\ KD\ rule \begin{cases} parameters: instance\,independent\,(KD+IIP) \\ parameters: instance\,dependent\,(KD+IDP) \end{cases} \end{cases}$$

Each one of the four cooperative strategies obtained is run over a set of test instances for every case study. We will first analyse the impact of the KD rule and then, that of the parameter's setting mode.

5.1   On the Impact of KD Rule versus EK Rule

Here, we compare the behaviour of the strategies EK+IIP vs. KD+IIP and EK+IDP vs. KD+IDP on each case study.

We will start the analysis with the first case study. Figure 3 shows two scatter plots where EK and KD rules are compared for both types of parameter's adjustment considered. In the scatter plots, each point represents a test instance and shows the relative deviation from the optimum for the two strategies compared. This is defined as $d = \frac{q-q^*}{q^*}$. Each point is an average over the total of runs. In this type of plots,

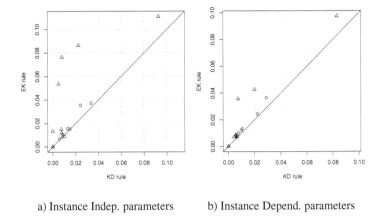

a) Instance Indep. parameters   b) Instance Depend. parameters

Fig. 3 Case Study 1: Comparison of the average relative deviation from optimum (smaller values are better). Triangles represent the instances on which the algorithms being compared perform differently at significance level 0.05 (Mann-Whitney U test(Wilcoxon's unpaired rank sum test))

when a point is below the diagonal line means that the strategy of the X axis has a worse average value than the strategy of the Y axis, and viceversa. When a point is represented by a triangle indicates that the difference is statistically significant (confidence level 0.05 by a Mann-Whitney U-test (Wilcoxon's unpaired rank sum test)) whereas in the opposite case, the point is showed as a circle.

Figure 3 (a) shows an important improvement when the KD rule is used with respect to EK. The cooperative strategy coupled with the KD rules always obtained equal or better average values (except in one instance). Moreover, the differences were statistically significant for 9 cases. When the parameters are set in terms of the type of the instance being solved, Figure 3 (b), the results are very similar. There is no point below the diagonal and the number of significant differences here is 5. In short, we can conclude that for this case study, the basic cooperative strategy can be enhanced with data mining techniques.

For the second case study, we are going to follow the same analysis structure. Figure 4 a) shows the performance of the EK rule vs the KD rule when both strategies use instance independent parameters. The differences in terms of results between the two rules are only statistically significant in 5 instances, three of which are positives for KD and the other two for EK. In the rest of the cases, the results are almost the same.

When the parameters are tuned accordingly with the type of instance, Figure 4b), there seems to be a slight improvement when using KD rules with respect to EK. Now, KD overcomes EK in most of the instances, being three cases statistically significant whereas this condition is only fulfilled in one occasion when such difference has the opposite sign. However, this result should be carefully analysed as the improve is not due to an enhancement of the KD rule, but a deterioration of EK, as we will see in the next subsection.

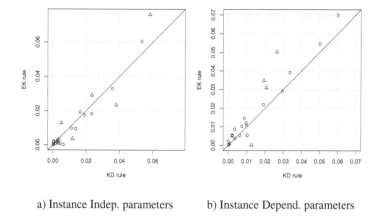

a) Instance Indep. parameters   b) Instance Depend. parameters

Fig. 4 Case Study 2: comparison of the average relative deviation from optimum (smaller values are better). Triangles represent the instances on which the algorithms being compared perform differently at significance level 0.05 (Mann-Whitney U test (Wilcoxon's unpaired rank sum test))

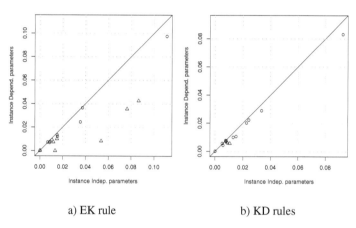

a) EK rule   b) KD rules

Fig. 5 Case Study 1:Comparison of the average relative deviation from optimum (smaller values are better). Triangles represent the instances on which the algorithms being compared perform differently at significance level 0.05 (Mann-Whitney U test (Wilcoxon's unpaired rank sum test))

5.2   On the Impact of the Parameters Setup Method

This part of the result analysis is devoted to study to what extent the strategy improve its performance when the parameters of the heuristic are tuned as a function of the instance characteristics, so we will focus on EK+IIP vs. EK+IDP and KD+IIP vs. KD+IDP for both case studies.

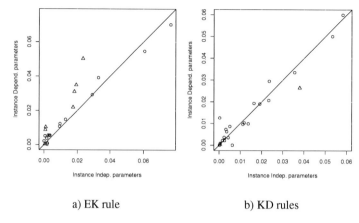

a) EK rule   b) KD rules

Fig. 6 Case Study 2:Comparison of the average relative deviation from optimum (smaller values are better). Triangles represent the instances on which the algorithms being compared perform differently at significance level 0.05 (Mann-Whitney U test(Wilcoxon's unpaired rank sum test))

In the first case study, the new parameter set up mechanism leads to a performance improvement that is very notorious for the EK rule, as we can see in Figure 5 a). The improvement achieved by the instance dependent parameter setting is significant in 7 instances and never drives the search to a deterioration. However, this enhancement is less appreciable for the KD rule. Viewing Figure 5 b), we can check that although the strategy always work better when the parameter are adjusted by this method, now the difference with respect to the other alternatives only statistically significant in one case.

For the second case study, we will start with the EK rule. Viewing the results shown by Figure 6a), we can observe the behaviour we pointed out before. The use of EK+IDP produce a high performance degradation of the basic strategy leading to worse results (the difference is statistically significant) in six instances.

When the control of the strategy is carried out by the KD rules, we can observe in Figure 6b) that the performance of KD+IIP and KD+IDP are almost the same.

6   Discussions

In this work we have seen how and in what contexts, Knowledge Discovery can be used to improve a centralised cooperative strategy. Concretely, the Knowledge Discovery has been incorporated in two different ways:

- By means of new parameterized control rules, where the parameters are determined using Data Mining
- Defining alternatives for setting up the parameters governing the behaviour of the metaheuristics: instance independent and instance dependent parameters that are provided, for each metaheuristic, by a decision tree.

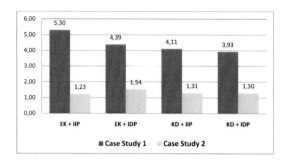

Fig. 7 Average error over all the instances of the corresponding case study, for every strategy evaluated

In order to analyse the suitability of the methodology, we proposed two case studies that differs in terms of the definition of the basic cooperative strategy (implemented heuristics, communication mode, ...), problem type, and amount of information available for doing knowledge extraction.

In the first case study, we observed that these new components led to cooperative strategies (EK+IDP, KD+IIP, KD+IDP) whose performance is better than the basic strategy (EK+IIP). This is clearer if we look at Figure 7 (case study 1) where the average error over all the test instances is shown for every strategy. This nice and clear behaviour is not present in the second case study.

In our opinion, the difference is related with amount of available information to "learn" in each case study. In other words, with the number of available instances to generate the performance information that then, should be mined to extract the weights and parameters that will govern the cooperative system. As we saw formerly, in the second case study we only had 34 instances for training with a unique sample per size and type, very low values to achieve a robust learning, specially if they are compare with such values in the first case study: 500 and 20 respectively.

Nevertheless, some conclusions can be obtained. First one is: if enough information is available to apply Knowledge Discovery techniques, then better cooperative strategies can be obtained. In second place, the benefit of using an instance dependent parameter setting needs to be further analysed because it depends on how well the instances in the training set could be characterized. If not enough information is available, then it will be safer not to use it. In the contrary, the use of KD rules when combined with an instance independent parameter setting leads to cooperative strategies that, at least, are as good as those using an expert designed rule for both case studies.

As future work, we plan to improve the learning process in order to reduce the amount of information needed to obtain meaningful knowledge. Another line of research consist on using online learning instead of the current offline data generation and processing method. In this way, the overhead of the learning process will be reduced and the future comparison against state of the art algorithms for specific problems could be fairly done.

Acknowledgements. A.D. Masegosa is supported by the scholarship program FPI from the Spanish Ministry of Science and Innovation. E. Muñoz is supported by "Fundación Séneca, Agencia de Ciencia y Tecnología de la Región de Murcia", under "Programa Séneca" action. This work has been partially funded by the projects TIN2008-01948 and TIN2008-06872-C04-03 from the Spanish Ministry of Science and Innovation and the "Fondo Europeo de Desarrollo Regional" (FEDER). Support from Andalusian Government through project P07-TIC-02970 is also acknowledged.

References

[1] Beasley, J.: Obtaining test problems via internet. Journal of Global Optimization 8(4), 429–433 (1996)

[2] Bouthillier, A.L., Crainic, T.G.: A cooperative parallel meta-heuristic for the vehicle routing problem with time windows. Comput. Oper. Res. 32(7), 1685–1708 (2005)

[3] Burke, E., Kendall, G., Newall, J., Hart, E., Ross, P., Schulenburg, S.: Hyper-heuristics: an emerging direction in modern search technology. In: Handbook of metaheuristics, pp. 457–474. Kluwer Academic Publishers, Dordrecht (2003)

[4] Cadenas, J., Garrido, M., Hernández, L., Muñoz, E.: Towards a definition of a data mining process based on fuzzy sets for cooperative metaheuristic systems. In: Proceedings of IPMU 2006, pp. 2828–2835 (2006)

[5] Carchrae, T., Beck, J.C.: Applying machine learning to low-knowledge control of optimization algorithms. Computational Intelligence 21(4), 372–387 (2005)

[6] Crainic, T.G., Gendreau, M., Hansen, P., Mladenović, N.: Cooperative parallel variable neighborhood search for the p-median. Journal of Heuristics 10(3), 293–314 (2004)

[7] Cruz, C., Pelta, D.: Soft computing and cooperative strategies for optimization. Applied Soft Computing Journal (2007) (In press) doi:10.1016/j.asoc.2007.12.007

[8] Dorigo, M., Stützle, T.: Ant Colony Optimization. Bradford Book (2004)

[9] Ferber, J.: Multi-Agent Systems: An Introduction to Distributed Artificial Intelligence. Addison-Wesley Longman Publishing Co., Inc, Boston (1999)

[10] Glover, F.W., Kochenberger, G.A. (eds.): Handbook of metaheuristics. Kluwer Academic Publishers, Dordrecht (2003)

[11] Guo, H., Hsu, W.H.: A machine learning approach to algorithm selection for np-hard optimization problems: a case study on the mpe problem. Annals of Operations Research 156(1), 61–82 (2007)

[12] Houstis, E., Catlin, A., Rice, J.R., Verykios, V., Ramakrishnan, N., Houstis, C.: Pythia-ii: a knowledge/database system for managing performance data and recommending scientific software. ACM Transactions on Mathematical Software 26(2), 227–253 (2000)

[13] Kellerer, H., Pferschy, U., Pisinger, D.: Knapsack Problems. Springer, Heidelberg (October 2004)

[14] Kennedy, J., Eberhart, R.C.: Swarm intelligence. Morgan Kaufmann Publishers Inc., San Francisco (2001)

[15] Krasnogor, N., Pelta, D.A.: Fuzzy Memes in Multimeme Algorithms: a Fuzzy-Evolutionary Hybrid. In: Fuzzy Sets based Heuristics for Optimization. Studies in Fuzziness and Soft Computing, vol. 126, pp. 49–66. Springer, Heidelberg (2002)

[16] Kratica, J., Stanimirović, Z., Duščan Tovšić, V.F.: Two genetic algorithms for solving the uncapacitated single allocation p-hub median problem. European Journal of Operational Research 182(1), 15–28 (2007)

[17] Leyton-Brown, K., Nudelman, E., Andrew, G., McFadden, J., Shoham, Y.: Boosting as a metaphor for algorithm design. In: Rossi, F. (ed.) CP 2003. LNCS, vol. 2833, pp. 899–903. Springer, Heidelberg (2003)

[18] O'Kelly, M., Morton, E.: A quadratic integer program for the location of interacting hub facilities. European Journal of Operational Research 32(3), 393–404 (1987)

[19] Pelta, D., Sancho-Royo, A., Cruz, C., Verdegay, J.L.: Using memory and fuzzy rules in a co-operative multi-thread strategy for optimization. Information Sciences 176(13), 1849–1868 (2006)

[20] Rice, J.: The algorithm selection problem. Advances in Computers 15, 65–118 (1976)

[21] Wolpert, D.H., Macready, W.G.: No free lunch theorems for optimization. IEEE Transactions on Evolutionary Computation 1, 67–82 (1997)

Hybrid Cooperation Models for the Tool Switching Problem

Jhon Edgar Amaya, Carlos Cotta, and Antonio J. Fernández Leiva

Abstract. The Tool Switching Problem (ToSP) is a hard combinatorial optimization problem of relevance in the field of flexible manufacturing systems (FMS), that has been tackled in the literature using both complete and heuristic methods, including local-search metaheuristics, population-based methods and hybrids thereof (e.g., memetic algorithms). This work approaches the ToSP using several hybrid cooperative models where spatially-structured agents are endowed with specific local-search/population-based strategies. Issues such as the intervening techniques and the communication topology are analyzed via an extensive empirical evaluation. It is shown that the cooperative models provide better results than their constituent parts. Furthermore, they not only provide solutions of similar quality to those returned by the memetic approach but raise interest prospects with respect to its scalability.

1 Introduction

The uniform tool switching problem (ToSP) is a hard combinatorial optimization problem that appears in Flexible Manufacturing Systems (FMSs), an alternative to rigid production systems that has the capability to be adjusted for generating different products and/or for changing the order of product generation. This problem arises in a single machine that has several slots into which different tools can be loaded. Each slot just admits one tool, and each job executed on that machine requires a particular set of tools to be completed. Jobs are sequentially executed, and

Jhon Edgar Amaya
Universidad Nacional Experimental del Táchira (UNET),
Laboratorio de Computación de Alto Rendimiento (LCAR), San Cristóbal, Venezuela
e-mail: jedgar@unet.edu.ve

Carlos Cotta · Antonio J. Fernández Leiva
Dept. Lenguajes y Ciencias de la Computación, ETSI Informática, University of Málaga,
Campus de Teatinos, 29071 - Málaga, Spain
e-mail: {ccottap,afdez}@lcc.uma.es

J.R. González et al. (Eds.): NICSO 2010, SCI 284, pp. 39–52, 2010.
springerlink.com   © Springer-Verlag Berlin Heidelberg 2010

therefore each time a job is to be processed, the corresponding tools must be loaded in the machine magazine. The ToSP consists of finding an appropriate job sequence in which jobs will be executed, and an associated sequence of tool loading/unloading operations that minimizes the number of tool switches in the magazine. In this context, tool management is a challenging task that directly influences the efficiency of flexible manufacturing systems. Different examples of the problem can be found in diverse areas such as electronics manufacturing, metalworking industry, computer memory management, and aeronautics, among others [3, 4, 30, 32].

Exact methods ranging from integer linear programming (ILP) techniques to heuristic constructive algorithms have been already applied to the problem with moderate success. The reason is clear: the ToSP has been proved to be NP-hard when the magazine capacity is higher than two (which is the usual case) and thus exact methods are inherently limited. In this context the use of alternative techniques that might eventually overcome this limitation has been explored. In particular, the use of metaheuristic techniques can be considered. In this line of work, [2] recently proposed three methods to tackle the ToSP: a simple local search (LS) scheme based on hill climbing, a genetic algorithm (which, as far as we know, constituted the first population-based approach to solve the uniform ToSP[1]), and a memetic algorithm [19, 26] (MA), based on the hybridization of the two latter methods. Related to this latter approach, this work proceeds along the cooperative side of hybridization by considering composite models in which different search techniques cooperate for solving the ToSP. These models can be arranged in a plethora of ways, and as a first step we have focused on the use of local-search metaheuristics as basic search strategies, and more precisely on how they can synergistically interact and the effect of the communication topology.

2 Background

Before proceeding, let us firstly describe more in depth the ToSP. Then, we will review previous related work.

2.1 The Tool Switching Problem

In light of the informal description of the uniform ToSP given before, there are two major elements in the problem: a machine M and a collection of jobs $J = \{J_1, \cdots, J_n\}$ to be processed. Regarding the latter, the relevant information for the optimization process is the tool requirements for each job. We assume that there is a set of tools $T = \{\tau_1, \cdots, \tau_m\}$, and that each job J_i requires a certain subset $T^{(J_i)} \subseteq T$ of tools to be processed. As to the machine, we will just consider one piece of information: the capacity C of the magazine (i.e., the number of available slots). Given the previous elements, we can formalize the ToSP as follows: let a ToSP instance

---

[1] Note that genetic algorithms (GAs) have been applied to other variants of the problem –e.g., [17]– though.

be represented by a pair, $I = \langle C, A \rangle$ where C denotes the magazine capacity, and A is a $m \times n$ binary matrix that defines the tool requirements to execute each job, i.e., $A_{ij} = 1$ if, and only if, tool τ_i is required to execute job J_j.

We assume that $C < m$; otherwise the problem is trivial. The solution to such an instance is a sequence $\langle J_{i_1}, \cdots, J_{i_n} \rangle$ (where $i_1, \ldots, i_n$ is a permutation of numbers $1, \ldots, n$) determining the order in which the jobs are executed, and a sequence $T_1, \cdots, T_n$ of tool configurations ($T_i \subset T$) determining which tools are loaded in the magazine at a certain time. Note that for this sequence of tool configurations to be feasible, it must hold that $T^{(J_{i_j})} \subseteq T_j$.

Let $\mathbb{N}_h^+ = \{1, \cdots, h\}$ henceforth. We will index jobs (resp. tools) with integers from $\mathbb{N}_n^+$ (resp. $\mathbb{N}_m^+$). An ILP formulation for the ToSP is shown below, using two sets of zero-one decision variables – x_{jk} ($j \in \mathbb{N}_n^+, k \in \mathbb{N}_n^+$), and y_{ik} ($i \in \mathbb{N}_m^+, k \in \mathbb{N}_n^+$) – that respectively indicate whether a job j is executed at time k or not, or whether a tool i is in the magazine at time k or not. Notice that since each job makes exclusive use of the machine, time-step k can be assimilated to the time at which the kth job is executed.

Processing each job requires a particular collection of tools loaded in the magazine. It is assumed that no job requires a number of tools higher than the magazine capacity, i.e., $\sum_{i=1}^{m} A_{ij} \leqslant C$ for all $j \in \mathbb{N}_n^+$. Tool requirements are reflected in Eq. (5). Following [3], we assume the initial condition $y_{i0} = 1$ for all $i \in \mathbb{N}_m^+$. This initial condition amounts to the fact that the initial loading of the magazine is not considered as part of the cost of the solution (in fact, no actual switching is required for this initial load). The objective function $F(\cdot)$ counts the number of switches that have to be done for a particular job sequence:

$$\min F(y) = \sum_{k=1}^{n} \sum_{i=1}^{m} y_{ik}(1 - y_{i,k-1}) \qquad (1)$$

$$\forall j \in \mathbb{N}_n^+ : \sum_{k=1}^{n} x_{jk} = 1 \qquad (2)$$

$$\forall k \in \mathbb{N}_n^+ : \sum_{j=1}^{n} x_{jk} = 1 \qquad (3)$$

$$\forall k \in \mathbb{N}_n^+ : \sum_{i=1}^{m} y_{ik} \leqslant C \qquad (4)$$

$$\forall j, k \in \mathbb{N}_n^+ \ \forall i \in \mathbb{N}_m^+ : A_{ij} x_{jk} \leqslant y_{ik} \qquad (5)$$

$$\forall j, k \in \mathbb{N}_n^+ \ \forall i \in \mathbb{N}_m^+ : x_{jk}, y_{ij} \in \{0, 1\} \qquad (6)$$

This general definition shown above corresponds to the *uniform ToSP* in which each tool fits in just one slot. The ToSP can be divided into three subproblems [35]: the first subproblem is *machine loading* and consists of determining the sequence of jobs; the second subproblem is *tool loading*, consisting of determining which tool to switch (if a switch is needed) before processing a job; finally, the third subproblem is *slot loading*, and consists of deciding where (i.e., in which slot) to place each tool. Since we are considering the uniform ToSP, the third subproblem does not apply (all slots are identical, and the order of tools is irrelevant). Moreover, and without loss of generality, the cost of switching a tool is considered constant (the same for all tools) in the uniform ToSP. Under this assumption, the tool loading subproblem can also be obviated because if the job sequence is fixed, the optimal tool switching policy can be determined in polynomial time using a greedy procedure termed *Keep Tool Needed Soonest* (KTNS) [3, 32]. The importance of this policy is that given a job sequence KTNS obtains its optimal number of tool switches. Therefore, we can concentrate on the machine loading subproblem, and use KTNS as a subordinate procedure to solve the subsequent tool loading subproblem.

2.2 Related Work on the ToSP

This paper focuses on the uniform case of the ToSP, in which there is one magazine, no job requires more tools than the magazine capacity, and the slot size is constant. To the best of our knowledge, the first reference to the uniform ToSP can be found in the literature as early as in the 1960's [4]; since then, the uniform ToSP has been tackled via many different techniques. The late 1980's contributed specially to solve the problem [3, 11, 18, 32]. This way, [32] proposed an ILP formulation of the problem, and [3] formulated the ToSP as a non-linear integer program with a dual-based relaxation heuristic. More recently, [20] proposed two exact algorithms: a branch-and-bound approach and a linear programming-based branch-and-cut algorithm.

Despite the moderate success of exact methods, it must be noted that they are inherently limited, since [27] and [8] proved formally that the ToSP is NP-hard for $C > 2$. This limitation was already highlighted by Laporte et al. [20] who reported that their algorithm was capable of dealing with instances with 9 jobs, but provided very low success ratios for instances with more than 10 jobs. Some ad hoc heuristics have been devised in response to this complexity barrier (e.g., [10, 14, 30]).

The use of metaheuristics has been also considered recently. For instance, local search methods such as tabu search (TS) have been proposed [1, 13]. Among these, we find specifically interesting the approach presented by [1], due to the quality of the obtained results; they defined three different versions of TS that arose from the inclusion of different algorithmic mechanisms such as long-term memory and oscillation strategies. We will return later to this approach and describe it in more detail since it has been included in our experimental comparison. A different, and very interesting, approach has been described by [36], who proposed a beam search algorithm. Beam search (BS) is a derivate of branch-and-bound that uses a breadth-first traversal of the search tree, and incorporates a heuristic choice to keep at each level only the best (according to some *quality* measure) β nodes (the so-called *beam width*). This sacrifices completeness, but provides a very effective heuristic search approach. Actually, this method provided good results, e.g., better than those of Bard's heuristics, and will be also included in the experimental comparison.

2.3 Background on Cooperative Models

Different schemes have been proposed for cooperating metaheuristics. For example, Toulouse *et al.* [33] considered using multiple instances of tabu search running in parallel, eventually exchanging some of the attributes stored in tabu memory.

Later on, Toulouse *et al.* [34] proposed a a hierarchical cooperative model based on problem decomposition. Crainic and Gendreau [6] presented a cooperative parallel tabu search method for capacitated network design problem that was shown to outperform independent search strategies. Crainic *et al.* [7] also proposed a method for asynchronous cooperative multi-search using variable neighborhood search with application to the p-median problem. Pelta *et al.* [28] presented a cooperative multi-thread search-based optimization strategy, in which several solvers were controlled by a higher-level coordination algorithm which collected information on their search performance and altered the behavior of the solvers accordingly (see also [9]).

More recently, Lu *et al.* [23] presented a hybrid cooperative version of quantum particle swam optimization aimed to improving the diversity of the swarms. Another approach for the implementation of cooperative mechanisms with metaheuristics is multi-agent systems. Milano and Roli [25] developed a multi-agent system called MAGMA (multiagent metaheuristic architecture) allowing the use of metaheuristics at different levels (creating solutions, improving them, defining the search strategy, and coordinating lower-level agents). Malek [24] introduced a multi-agent system like MAGMA which considered particular metaheuristics implemented by individual agents and the exchange of solutions between these.

To the best of our knowledge, no cooperative scheme has been applied to tackle the ToSP, perhaps with the exception of our memetic proposal described in [2] that can be catalogued as an integrative cooperation according to the classification described in [29] (note at any rate that none of the techniques involved in the MA is a complete algorithm). In any case, no classical cooperation model in the sense of "search algorithms working in parallel with a varying level of communication" [5] has been tried. This paper presents the first cooperative models according to this mentioned schema for solving the ToSP.

3   Hybrid Cooperative Models

We have considered four collaborative architectures. In three of them, agents are attached to a certain spatial structure endowed with a LS mechanism. These architectures are defined on the basis of the particular LS methods used, and on their interaction topology. Therefore, these two aspects are defined separately in Sections 3.1 and 3.2 respectively. A fourth architecture, also described in Section 3.2, is defined on the basis of a model based in heterogeneous techniques for executing search, diversification and intensification.

3.1   Local Searchers

LS metaheuristics are based on exploring the neighborhood of a certain "current" solution. It is thus convenient to address firstly the representation of solutions and the structure of this neighborhood, and subsequently of the underlying search space. A permutational encoding arises as the natural way to represent solutions. Thus, a candidate solution for a specific ToSP instance $I = \langle C, A \rangle$ is simply a permutation

$\pi = \langle \pi_1, \cdots, \pi_n \rangle \in \mathbb{P}_n$ where $\pi_i \in \mathbb{N}_n^+$, and $\mathbb{P}_n$ is the set of all permutations of elements in $\mathbb{N}_n^+$. The KTNS algorithm is used to obtain the actual tool configuration of the machine for the corresponding job sequence.

Having defined the representation, we now turn our attention to the neighborhood structure. In general, we have considered the well-known *swap* neighborhood $\mathcal{N}_{swap}(\cdot)$, in which two permutations are neighbors if they just differ in two positions of the sequence, that is, for a permutation $\pi \in \mathbb{P}_n$, $\mathcal{N}_{swap}(\pi) = \{\pi' \in \mathbb{P}_n \mid H(\pi, \pi') = 2\}$ where $H(\pi, \pi') = n - \sum_{i=1}^{n} [\pi_i = \pi_i']$ is the Hamming distance between sequences π and π' (the number of positions in which the sequences differ), and $[\cdot]$ is Iverson bracket (i.e., $[P] = 1$ if P is true, and $[P] = 0$ otherwise). Given the permutational nature of sequences, this implies that the contents of the two differing positions have been swapped. For some specific applications (named when necessary), we have also considered a specific neighborhood called *block* neighborhood $\mathcal{N}_{block}(\cdot)$. This is a generalization of the swap neighborhood in which two non-overlapping blocks (i.e., subsequences of adjacent positions) of a randomly chosen length $b_l \in \mathbb{N}_{n/2}^+$ are selected at random within a permutation, and swapped.

These neighborhoods are exploited within two different LS frameworks. The first one is steepest-ascent Hill Climbing (HC), in which given a current solution π, its neighborhood $\mathcal{N}(\pi)$ is explored, and the best solution found is taken as the new current solution, provided it is better than the current one (ties are randomly broken). If no such neighboring solution exist, the search is considered stagnated, and can be restarted from a different initial point. The second LS technique considered is a Tabu Search (TS) method along the lines of the proposal in [1]. This TS method is based on a strategic oscillation mechanism which switches between the two neighborhoods defined before. A deterministic criterion based on switching the neighborhood structure after a fixed number of iterations was reported by [1] to perform better than a probabilistic criterion (i.e., choosing the neighborhood structure in each step, according to a certain probability distribution). We implement a long term memory scheme using a frequency based memory structure with a mechanism based in swapping to select new candidate solutions [1]. No aspiration criterion is used in this referred algorithm.

3.2 Interaction Topology

Let R be an architecture with n agents; each agent a_i ($0 \leqslant i \leqslant n - 1$) in R consists of one of the metaheuristics described in Sect. 3.1. These agents engage in periods of isolated exploration followed by synchronous communication. We denote as $cycles_{max}$ the maximum number of such exploration/communication cycles in a certain cooperative model. Also, let S_i be the best solution found by agent a_i, and let $\mathbf{T}_R \subseteq \mathbb{N}_n^+ \times \mathbb{N}_n^+$ be the communication topology over R (i.e., if $(i, j) \in \mathbf{T}_R$ then a_i can send information to agent a_j). The general architecture of the model is then described in Algorithm 1. Firstly all the agents are initialized with random initial solution (lines 1-3). Then, the algorithm is executed for a maximum number of iterations cycles (lines 5-15) where, in each cycle, a local improvement of the solution

---

Algorithm 1. COOPERATIVE-MODEL$_n$

---
1 **for** $i \in \mathbb{N}_n^+$ **do**
2 $\quad\mid\quad S_i \leftarrow GenerateInitialSolution()$;
3 **endfor**
4 $cycles \leftarrow 1$;
5 **while** $cycles \leqslant cycles_{\max}$ **do**
6 $\quad$ **for** $i \in \mathbb{N}_n^+$ **do**
7 $\quad\quad\mid\quad S_i \leftarrow a_i(S_i)$;
8 $\quad$ **endfor**
9 $\quad$ **for** $(i, j) \in \mathbf{T}_R$ **do**
10 $\quad\quad\mid\quad$ **if** $KTNS(S_i) < KTNS(S_j)$ **then**
11 $\quad\quad\quad\mid\quad S_j \leftarrow S_i$;
12 $\quad\quad\mid\quad$ **endif**
13 $\quad$ **endfor**
14 $\quad cycles \leftarrow cycles + 1$
15 **endw**
16 **return** $\max^{-1}\{KTNS(S_i) \mid i \in \mathbb{N}_n^+\}$;

---

kept in each agent is done (lines 6-8), and solutions are fed from an agent to another according to the topology considered (lines 9-13). Note that an agent only accepts an incoming solution if it is better than its incumbent. Observe also that, for a maximum number of evaluations $E_{\max}$ and for a specific number of cycles $cycles_{\max}$, each cycle in our cooperative algorithms spends $E_{cycle} = E_{\max}/cycles_{\max}$ evaluations, and the specific LS method of any agent takes E_{cycle}/n evaluations at most.

Three strategies based on different interaction topologies are considered:

- RING: $\mathbf{T}_R = \{(i, i(n) + 1) \mid i \in \mathbb{N}_n^+ \text{ and } i(n) \text{ denotes } i \text{ modulo } n\}$. Thus, there exists a circular list of agents in which each node only sends (resp. receives) information to its successor (resp. from its predecessor).
- BROADCAST: $\mathbf{T}_R = \mathbb{N}_n^+ \times \mathbb{N}_n^+$, i.e., a *go with the winners*-like topology in which the best overall solution at each synchronization point is transmitted to all agents. This means all agents executes intensification over the same local region of the search space at the beginning of each cycle.
- RANDOM: $\mathbf{T}_R$ is composed by n pairs (i, j) that are randomly sampled from $\mathbb{N}_n^+ \times \mathbb{N}_n^+$. This sampling is done each time communication takes place, and hence any two agents might eventually communicate in any step.

In addition to these strategies we have considered a so-called Ring SDI model, based on an interesting proposal described in [31]. SDI stands for Search, Diversification and Intensification, and hence the SDI architecture consists of three agents dedicated to distinct purposes: the first one to local search, the second one to diversification and the third one to intensification. As described in next section, within this SDI model the intervening techniques are not just local searchers, but other techniques can be used for intensification/diversification purposes.

4 Computational Results

As far as we know, no standard data instance exists for this problem (at least publicly available) so that we have selected a wide set of problem instances that were attacked in [1, 3, 14, 36]; more specifically, 16 instances were chosen with values for the number of jobs, number of tools, and machine capacity ranging in [10,50], [9,60] and [4,25] respectively. Table 1 shows the different problem instances chosen for the experimental evaluation where a specific instance with n jobs, m tools and machine capacity C is labeled as $C\zeta_n^m$.

Table 1 Problem Instances considered in the experimental evaluation. The minimum and maximum of tools required for all the jobs is indicated in second and third rows respectively. Fourth row shows the work from which the problem instance was obtained

	$4\zeta_{10}^{10}$	$4\zeta_{10}^{9}$	$6\zeta_{10}^{15}$	$6\zeta_{15}^{12}$	$6\zeta_{15}^{20}$	$8\zeta_{20}^{15}$	$8\zeta_{20}^{16}$	$10\zeta_{20}^{20}$	$24\zeta_{20}^{30}$	$24\zeta_{20}^{36}$	$30\zeta_{20}^{40}$	$10\zeta_{30}^{25}$	$15\zeta_{40}^{40}$	$15\zeta_{40}^{30}$	$20\zeta_{40}^{60}$	$25\zeta_{50}^{40}$
Min.	2	2	3	3	3	3	3	4	9	9	11	4	6	6	7	9
Max.	4	4	6	6	6	8	8	10	24	24	30	10	15	15	20	20
Source	[14]	[3]		[3]			[3]	[3]	[3]	[3]						
	[1]	[36]	[36]	[36]	[14]	[1]	[36]	[36]	[36]	[36]	[36]	[1]	[14]	[1]	[14]	[1]

Five different datasets[2] (i.e., incident matrixes or relations among tools and jobs) were generated randomly per instance. Each dataset was generated with the restriction, already imposed in previous works such as [14], that no job is *covered* by any other job in the sense that $\forall i, j \in \mathbb{N}_n^+, i \neq j, T^{(J_i)} \not\subseteq T^{(J_j)}$. The reason to enforce this constraint is to avoid the simplification of the problem by preprocessing techniques as done for instance in [3] and [36].

The experiments have been performed using a wide set of different algorithms: the beam search (BS) presented in [36], three LS methods, a GA, the memetic approach (denoted as MaHC) presented in [2], and the four cooperative algorithms described in this paper. From these, a wide number of algorithms were devised and tested. For instance, in the case of BS, five different values $\beta \in \mathbb{N}_5^+$ were considered for the beam width. Regarding LS methods, we consider the TS proposed in [1], and HC as described previously. Moreover, we have taken into account also LS versions in which a partial exploration of the neighborhood is done by obtaining a fixed-size random sample; in particular, the size of this sample has been chosen to be αn, i.e., proportional to the number of jobs (the value $\alpha = 4$ has been used). The notation HCP and HCF (resp. TSP and TSF) is used to indicate the HC variant (resp. TS variant) in which the neighborhood is partially or fully explored respectively. Also, in the case of HC, the search is restarted from a different initial point if stagnation takes place before consuming the allotted number of evaluations. Regarding TS, the tabu tenure is 5, and the number of iterations on each neighborhood for performing strategic oscillation is 3. In both cases, this corresponds to the setting used by [1].

---

[2] All datasets are available at *http://www.unet.edu.ve/~jedgar/ToSP/ToSP.htm*

The GA is a steady-state genetic algorithm whose parameters are exactly as those described in [2], that is to say, $popsize = 30$, $p_X = 1.0$, and $p_M = 1/n$ where n is the number of jobs, with binary tournament selection; alternating position crossover (APX) is used [21], and mutation is done by applying the random block swap as operator. The MaHC consists of a combination of this GA with HCP where HCP was always applied to each offspring generated after the mutation step. The election of the parameter values (including the value for α) was done after an extensive phase of experimentation with many different values. The best combinations of the values were finally selected.

Regarding the cooperative models, we have used $cycles_{max} \in \{3, 4, 5\}$, and have focused on models with 3 agents to make easier the comparison with the SDI model. In this latter RINGSDI model we connect HCP for LS, GA for diversification, and for intensification we plug in the *KickOperator* that was also used in [31]. In our rendition of this operator it acts as a first-ascent HC on the swap neighborhood. As to the basic RING, BROADCAST and RANDOM topologies, their three agents were loaded with HCF, HCP and TSP techniques respectively.

All algorithms were run 10 times (per instance and dataset) and a maximum of $E_{max} = \varphi n(m - C)$ evaluations[3] per run (with $\varphi > 0$). Preliminary experiments on the value of φ proved that $\varphi = 100$ is an appropriate value that allows to keep an acceptable relation between solution quality and computational cost. Regarding the BS algorithm, because of its deterministic nature, just one execution per dataset (and per value of beam width) was run and the algorithm was allowed to be executed until exhaustion (i.e., until completing the search).

Due to space limitations we will not present all the obtained results for each of the instances and for all the algorithms involved in the comparison, and will use a rank-based approach in order to analyze the significance of the results. To do so, we have computed the rank r_j^i of each algorithm j on each instance i (rank 1 for the best, and rank k for the worst, where $k = 23$ is the number of algorithms; in case of ties, an average rank is awarded). The distribution of these ranks is shown in Fig. 1. Here one can extract important conclusions: the most important is that in general, the cooperative models behaves better than its constituent parts. This is an important fact as the cooperative models have not been optimized exhaustively (due to the high number of possible metaheuristics combinations to be loaded in the agents). Also, the fact that RINGSDI is better than RING might indicate the need for a diversification algorithm to increase the area of exploration in the search landscape.

Next, we have used two well-known non-parametric statistical tests [22] to compare ranks, namely Friedman test [12] and Iman-Davenport test [16]. The results are shown in Table 2. As seen in the first row, the statistic values obtained are clearly

---

[3] Observe that the number of evaluations depends directly on the number of jobs although it seems evident that the problem difficulty lies in the relation between number of tools and magazine capacity. In this sense, the number of evaluations increases with the number of tools (assumed to be directly related with problem difficulty) and decreases when the magazine capacity increases (that, in some sense, it is also inversely related to the problem difficulty).

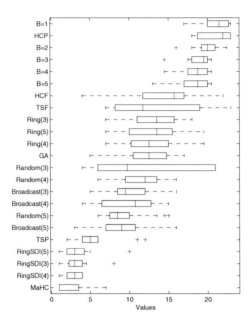

Fig. 1 Rank distribution of each algorithm across all instances. As usual, each box comprises the second and third quartiles of the distribution, the median is marked with a vertical line, whiskers span 1.5 times the inter-quartile range, and outliers are indicated with a plus sign. The numbers in parentheses indicate the number of cycles of execution (i.e., $cycles_{max}$)

Table 2 Results of Friedman and Iman-Davenport tests

	Friedman value	critical χ^2 value	Iman-Davenport value	critical F_F value
all	269.80	33.92	49.23	1.57
top 5	19.04	9.49	6.35	2.53

higher than the critical values, and therefore the null hypothesis, namely that all algorithms are equivalent, can be rejected. Since there are algorithms with markedly poor performance, we have repeated the test with the top 5 algorithms (i.e., the MaHC, all RINGSDI and TS), whose performance places them in a separate cluster from the remaining algorithms. Again, it can be seen that the statistical test is passed, thus indicating significant differences in their ranks at the standard $\alpha = 0.05$ level.

Subsequently, we have focused in these top 5 algorithms, and performed Holm's test [15] in order to determine whether there exists significant differences with respect to a control algorithm (in this case MaHC, the algorithm with the best mean rank). The results are shown in Table 3. Notice that the test is passed for all algorithms with respect to MaHC and that there is no statistical difference between MaHC, RINGSDI(4) and RINGSDI(3).

Table 3 Results of Holm's test using *MaHC* as control algorithm

i	algorithm	z-statistic	p-value	$\alpha/(k-i)$
1	RingSDI(4)	1.286	0.09926	0.013
2	RingSDI(3)	1.957	0.02520	0.016
3	RingSDI(5)	2.012	0.02209	0.025
4	TSP	4.249	< 0.00001	0.050

Table 4 Computational results. Best results (in terms of the best solution average) are underlined and in boldface

	4^{10}_{10}	6^{15}_{10}	4^{9}_{10}	6^{12}_{15}	6^{20}_{15}	8^{15}_{20}	8^{16}_{20}	10^{20}_{20}	24^{30}_{20}	24^{36}_{20}	30^{40}_{20}	10^{25}_{30}	15^{40}_{30}	15^{30}_{40}	20^{60}_{40}	25^{40}_{50}
TSP mean	8.8	13.68	8.08	16.46	23.02	23.62	27.92	30.72	25.04	45.9	42.12	67.72	101.72	101.9	213.74	153.58
σ	1.61	2.1	0.74	1.93	2.0	3.63	2.13	2.5	3.02	8.98	4.34	1.52	13.07	8.14	8.38	12.89
best	7	11	7	13	20	18	23	26	21	34	33	65	82	89	199	130
MaHC mean	**8.68**	13.7	**7.86**	**15.5**	**22.38**	**22.36**	**26.66**	**29.92**	24.9	46.54	**41.04**	**64.92**	**100.86**	**97.96**	211.88	153.36
σ	1.62	2.09	0.721	1.982	1.938	3.576	1.986	2.357	3.28	8.81	4.54	1.573	12.9	7.887	7.812	13.52
best	7	11	7	12	20	17	23	26	20	36	31	62	81	86	201	132
RingSDI (3) mean	8.72	13.74	7.9	16.14	23.0	23.34	27.6	30.72	24.76	**45.26**	41.3	66.98	102.66	99.52	210.68	148.14
σ	1.64	2.04	0.68	1.85	2.21	3.36	2.21	2.31	3.17	8.46	4.43	2.64	13.03	7.71	7.79	11.75
best	7	11	7	12	20	17	23	26	20	36	31	62	81	87	197	129
RingSDI (4) mean	8.72	**13.66**	7.9	16.16	22.96	23.06	27.44	30.72	**24.46**	45.84	41.8	67.32	102.28	99.16	211.08	**146.34**
σ	1.65	2.12	0.73	1.9	2.14	3.34	2.05	2.41	3.51	7.91	4.79	2.69	12.37	7.64	9.34	12.28
best	7	11	7	13	20	17	23	25	19	36	32	61	79	88	198	126
RingSDI (5) mean	8.7	13.74	7.92	16.1	23.1	23.2	27.28	30.74	**24.46**	45.64	41.58	68.1	102.2	100.24	**210.44**	146.98
σ	1.62	2.07	0.69	2.02	2.08	3.42	2.3	2.38	3.13	8.55	4.5	2.76	12.78	8.07	9.17	12.12
best	7	11	7	12	20	18	22	25	20	33	32	63	81	87	195	124

Also, Table 4 shows the obtained results, grouped by problem instances, for these top 5 algorithms. One can observe that all RINGSDI algorithms perform better than MaHC in several instances, particularly in the largest one (i.e., last column), in which Wilcoxon's ranksum test indicates that RingSDI[4] significantly outperforms (at the standard 0.05 level) the MA in all five datasets generated for this parameter combination. This raises interest prospects for the scalability of these models, thus hinting the need for experiments at a larger scale to confirm this.

5 Conclusions

Collaborative optimization models constitute a very appropriate framework for integrating different search techniques. Each of these techniques has a different view of the search landscape, and by combining the corresponding different exploration patterns, the search can benefit from an increased capability for escaping from local optima. Of course, this capability is more useful whenever the problem tackled poses a challenging optimization task to the individual search algorithms. Otherwise, computational power is diversified in unproductive explorations.

We have tackled here the tool switching problem and have proposed four cooperative methods to attack it. An empirical evaluation was executed in order to prove the validity and performance of the proposed techniques. One topology based on

---

[4] All tables are available in *http://www.unet.edu.ve/~jedgar/ToSP/Wilcoxon.htm*

heterogenous intervening techniques (RINGSDI) provides better computational results than well-known algorithms for solving the ToSP, i.e., beam search and tabu search, and does not perform worse than a memetic algorithm. Indeed, some results with larger instances lead us to hypothesize that this model might have better scalability properties than the MA. This issue will be analyzed in future work.

We believe that there is room for improvement. For instance, it would be interesting to test other alternatives to LS. More precisely, the MA is a killer approach for the ToSP, so it may be interesting to include this technique in the cooperative model. In this case, it would be necessary to re-balance the intensification/exploration ratio, since MAs perform a much more intensified search than other techniques, and thus may require a more explorative counterweight. This line of research is in progress.

Acknowledgements

The authors wish to thank the anonymous reviewers for their constructive comments and suggestions, which have improved the readability of the paper. The second and third authors were partially supported by Spanish MICINN project under contract TIN2008-05941 (NEMESIS project).

References

[1] Al-Fawzan, M., Al-Sultan, K.: A tabu search based algorithm for minimizing the number of tool switches on a flexible machine. Computers & Industrial Engineering 44(1), 35–47 (2003)
[2] Amaya, J., Cotta, C., Fernández, A.: A memetic algorithm for the tool switching problem. In: Blesa, M.J., Blum, C., Cotta, C., Fernández, A.J., Gallardo, J.E., Roli, A., Sampels, M. (eds.) HM 2008. LNCS, vol. 5296, pp. 190–202. Springer, Heidelberg (2008)
[3] Bard, J.F.: A heuristic for minimizing the number of tool switches on a flexible machine. IIE Transactions 20(4), 382–391 (1988)
[4] Belady, L.: A study of replacement algorithms for virtual storage computers. IBM Systems Journal 5, 78–101 (1966)
[5] Blum, C., Roli, A.: Metaheuristics in combinatorial optimization: Overview and conceptual comparison. ACM Comput. Surv. 35(3), 268–308 (2003)
[6] Crainic, T.G., Gendreau, M.: Cooperative parallel tabu search for capacitated network design. Journal of Heuristics 8(6), 601–627 (2002)
[7] Crainic, T.G., Gendreau, M., Hansen, P., Mladenović, N.: Cooperative parallel variable neighborhood search for the p-median. Journal of Heuristics 10(3), 293–314 (2004)
[8] Crama, Y., Kolen, A., Oerlemans, A., Spieksma, F.: Minimizing the number of tool switches on a flexible machine. International Journal of Flexible Manufacturing Systems 6, 33–54 (1994)
[9] Cruz, C., Pelta, D.A.: Soft computing and cooperative strategies for optimization. Applied Soft Computing 9(1), 30–38 (2009)
[10] Djellab, H., Djellab, K., Gourgand, M.: A new heuristic based on a hypergraph representation for the tool switching problem. International Journal of Production Economics 64(1-3), 165–176 (2000)

[11] ElMaraghy, H.: Automated tool management in flexible manufacturing. Journal of Manufacturing Systems 4(1), 1–14 (1985)
[12] Friedman, M.: The use of ranks to avoid the assumption of normality implicit in the analysis of variance. Journal of the American Statistical Association 32(200), 675–701 (1937)
[13] Hertz, A., Widmer, M.: An improved tabu search approach for solving the job shop scheduling problem with tooling constraints. Discrete Applied Mathematics 65, 319–345 (1993)
[14] Hertz, A., Laporte, G., Mittaz, M., Stecke, K.: Heuristics for minimizing tool switches when scheduling part types on a flexible machine. IIE Transactions 30, 689–694 (1998)
[15] Holm, S.: A simple sequentially rejective multiple test procedure. Scandinavian Journal of Statistics 6, 65–70 (1979)
[16] Iman, R., Davenport, J.: Approximations of the critical region of the Friedman statistic. Communications in Statistics 9, 571–595 (1980)
[17] Keung, K.W., Ip, W.H., Lee, T.C.: A genetic algorithm approach to the multiple machine tool selection problem. Journal of Intelligent Manufacturing 12(4), 331–342 (2001)
[18] Kiran, A., Krason, R.: Automated tooling in a flexible manufacturing system. Industrial Engineering 20, 52–57 (1988)
[19] Krasnogor, N., Smith, J.: A tutorial for competent memetic algorithms: model, taxonomy, and design issues. IEEE Transactions on Evolutionary Computation 9(5), 474–488 (2005)
[20] Laporte, G., Salazar-González, J., Semet, F.: Exact algorithms for the job sequencing and tool switching problem. IIE Transactions 36(1), 37–45 (2004)
[21] Larrañaga, P., Kuijpers, C., Murga, R., Inza, I., Dizdarevic, S.: Genetic algorithms for the travelling salesman problem: A review of representations and operators. Articial Intelligence Review 13, 129–170 (1999)
[22] Lehmann, E., D'Abrera, H.: Nonparametrics: statistical methods based on ranks. Prentice-Hall, Englewood Cliffs (1998)
[23] Lu, S., Sun, C.: Coevolutionary quantum-behaved particle swarm optimization with hybrid cooperative search. In: Proceedings of the Pacific-Asia Workshop on Computational Intelligence and Industrial Applications PACIIA 2008, vol. 1, pp. 109–113 (2008)
[24] Malek, R.: Collaboration of metaheuristic algorithms through a multi-agent system. In: Mařík, V., Strasser, T., Zoitl, A. (eds.) Holonic and Multi-Agent Systems for Manufacturing. LNCS, vol. 5696, pp. 72–81. Springer, Heidelberg (2009)
[25] Milano, M., Roli, A.: Magma: a multiagent architecture for metaheuristics. IEEE Transactions on Systems, Man, and Cybernetics, Part B 34(2), 925–941 (2004)
[26] Moscato, P., Cotta, C.: Memetic algorithms. In: González, T. (ed.) Handbook of Approximation Algorithms and Metaheuristics, ch. 27. Chapman & Hall/CRC Press (2007)
[27] Oerlemans, A.: Production planning for flexible manufacturing systems. Ph.d. dissertation, University of Limburg, Maastricht, Limburg, Netherlands (1992)
[28] Pelta, D., Cruz, C., Sancho-Royo, A., Verdegay, J.: Using memory and fuzzy rules in a co-operative multi-thread strategy for optimization. Information Sciences 176, 1849–1868 (2006)
[29] Puchinger, J., Raidl, G.R.: Combining metaheuristics and exact algorithms in combinatorial optimization: A survey and classification. In: Mira, J., Álvarez, J.R. (eds.) IWINAC 2005. LNCS, vol. 3562, pp. 41–53. Springer, Heidelberg (2005)
[30] Shirazi, R., Frizelle, G.: Minimizing the number of tool switches on a flexible machine: an empirical study. International Journal of Production Research 39(15), 3547–3560 (2001)

[31] Talbi, E.G., Bachelet, V.: Cosearch: A parallel cooperative metaheuristic. Journal of Mathematical Modelling and Algorithms 5(1), 5–22 (2006)

[32] Tang, C., Denardo, E.: Models arising from a flexible manufacturing machine, part I: minimization of the number of tool switches. Operations Research 36(5), 767–777 (1988)

[33] Toulouse, M., Crainic, T.G., Sanso, B., Thulasiraman, K.: Self-organization in cooperative tabu search algorithms. In: Proceedings of the IEEE International Conference on Systems, Man, and Cybernetics, vol. 3, pp. 2379–2384 (1998)

[34] Toulouse, M., Thulasiraman, K., Glover, F.: Multi-level cooperative search: A new paradigm for combinatorial optimization and an application to graph partitioning. In: Amestoy, P.R., Berger, P., Daydé, M., Duff, I.S., Fraysse, V., Giraud, L., Ruiz, D. (eds.) Euro-Par 1999. LNCS, vol. 1685, pp. 533–542. Springer, Heidelberg (1999)

[35] Tzur, M., Altman, A.: Minimization of tool switches for a flexible manufacturing machine with slot assignment of different tool sizes. IIE Transactions 36(2), 95–110 (2004)

[36] Zhou, B.H., Xi, L.F., Cao, Y.S.: A beam-search-based algorithm for the tool switching problem on a flexible machine. The International Journal of Advanced Manufacturing Technology 25(9-10), 876–882 (2005)

Fault Diagnosis in Industrial Systems Using Bioinspired Cooperative Strategies

Lídice Camps Echevarría, Orestes Llanes-Santiago, and Antônio José da Silva Neto

Abstract. This paper explores the application of bioinspired cooperative strategies for optimization on Fault Diagnosis in industrial systems. As a first step, the
Differential Evolution and Ant Colony Optimization algorithms are considered.
Both algorithms have been applied to a benchmark problem, the two tanks system.
The experiments have considered noisy data in order to compare the robustness
of the diagnosis. The preliminary results indicate that the proposed approach, basically the combination of the two algorithms, characterizes a promising methodology
for the Fault Detection and Isolation problem.

1 Introduction

The increases on the complexity of the industrial systems implies that the probability of fault occurrence is more significant. The faults change the characteristic
properties of the system and produce its incapacity to fulfill the intended purpose,
[6]. Therefore, an automatic supervisor should be used to detect and isolate, (FDI),
the faults as early as possible. This is a reason for which in the last three decades a
wide variety of FDI methods have been developed.

Lídice Camps Echevarría
Departamento de Matemáticas, Facultad de Ingeniería Mecánica,
Instituto Superior Politécnico José Antonio Echeverría (ISPJAE), Ciudad de La Habana, Cuba
e-mail: lidice@mecanica.cujae.edu.cu

Orestes Llanes-Santiago
Departamento de Automática y Computación, Facultad de Ingeniería Eléctrica,
Instituto Superior Politécnico José Antonio Echeverría (ISPJAE), Ciudad de La Habana, Cuba
e-mail: orestes@electrica.cujae.edu.cu

Antônio José da Silva Neto
Department of Mechanical Engineering and Energy, Instituto Politécnico (IPRJ),
Universidade do Estado do Rio de Janeiro, UERJ, Nova Friburgo, RJ, Brazil
e-mail: ajsneto@iprj.uerj.br

J.R. González et al. (Eds.): NICSO 2010, SCI 284, pp. 53–63, 2010.
springerlink.com   © Springer-Verlag Berlin Heidelberg 2010

The FDI methods are divided in two general groups, those which use a model of the process and those which do not use it. Although many approaches have been developed, [2, 6, 11], robust FDI is still considered as a problem open to further research, [14], due to the the unavoidable process disturbances and the modeling errors which make almost unfeasible the use of many FDI methods in practical application.[11].

The FDI problem approach by the model-based methods has the following structure: based on some observations and the direct model, it is necessary to establish the causes of this observed behavior. In some cases, the identification of model parameter with fault of the system allows the FDI via the parameters estimation.

Recently some articles have reported applications of meta heuristics to the FDI problems via parameters estimation, [16–18]. In this sense, the FDI via parameter estimation based on the approach of the meta heuristic algorithms seems to be an adequate alternative. The simple structure of these algorithms and their robustness reported in the solution of many parameters estimation inverse problems, [12], [1, 8, 9, 13], indicate that they are a promising alternative for FDI methods which need to be fast and simple (for online process) and robust to external perturbations. Moreover, estimations based on heuristic algorithms are absolutely viable when a nonlinear model is considered, making perfectly feasible the use of non linear models in order to prevent some modeling errors when linearizing the nonlinear process.

This work presents the application of two bioinspired algorithms, Differential Evolution (DE), [15], and Ant Colony Optimization (ACO), [4], to the FDI problem in order to study and compare the capabilities of both algorithms and their combination for the FDI problems . As a case of study it has been simulated the problem of the two tanks system. This system is a simplified version of the three tanks system, which was adopted as a benchmark problem for FDI and reconfigurable control [10]. In order to verify and compare the robustness of the diagnosis, several simulations were made and different fault situations were considered. In all cases noisy data were considered. The results are presented using comparative tables and figures.

The structure of the paper is the following: in the next section the basis of DE and the ACO are described. The third section shows the benchmark problem of the two tanks system. The section number 4 shows the simulations, the experimental results and the analysis of these results. Finally, section 5 summarizes the contributions and achievements of the paper.

2   Differential Evolution and Ant Colony Optimization

This section describes the basis of the two algorithms that are used during this paper.

2.1   Differential Evolution

The Differential Evolution (DE) was proposed around 1995, for optimization problems, [15]. DE is an improved version of the Goldberg's Genetic Algorithm (GA), [5], taking the basis of Simulated Annealing (SA), [7]. Some of the most important advantages of DE are: simple structure, simple computational implementation, speed and robustness, [15].

Basically, DE generates new parameter vectors by adding the weighted difference between a pair population vectors to a certain vector (the number of pair can be changed). This configuration is summarized by the notation $DE/X/\alpha/\beta$ where X denotes the vector to disturb, α the number of pair of vectors for disturbing X and β indicates the type of crossover to be used. In this case was considered $DE/X_j^{best}/1/bin$. The key parameters of control in DE are the population size, N, the crossover constant, C_R, and the weight applied to random differential or scaling factor, F_s. In [15] some simple rules for choosing the parameters of DE for any application are given: usually, N should be about 5 to 10 times the dimension of the variable of the problem, D and F_s lie in the range 0.4 to 1.0. Initially, $D = 0.5$ can be tried, and then can be increased if the population converges prematurely.

2.2 Ant Colony Optimization

ACO was initially proposed, [4], for integer programming problems but it has been extended to continuous optimization problems. This algorithm is inspired on the behavior of ants seeking a path between their colony and a source of food. This behavior is due to the deposit and evaporation of pheromone.

For the continuous case the idea of the ACO is to mimic this behavior with simulated ants which are identified with a feasible solution. The first step is to discretize the feasible interval of each variable of the problem in n possible values. On each iteration of the algorithm a family of N new ants are generated based on the information obtained from the previous ants. This information is saved on the pheromone probability matrix Pf (dimensions $m \times n$ where m is the number of variables in the problem) which is updated at each iteration based on a evaporation factor C_{evap} and an incremental factor C_{inc}:

$$pf_{ij}(t) = \frac{\sum_{l=1}^{j} [f_{il}(t)]^{\alpha}}{\sum_{l=1}^{n} [f_{il}(t)]^{\alpha}} \tag{1}$$

where $\alpha = 1$ and f_{ij} is the element of the pheromone matrix which expresses the pheromone level of the discrete value $j - esimo$ of the variable i, and it is updated on each iteration:

$$f_{ij}(t+1) = (1 - C_{evap})f_{ij}(t) + \delta_{ij,best}C_{inc}f_{ij}(t) \tag{2}$$

3 The Two Tanks System

The two tank system considered for study is represented in Fig.1.

The system consists of two liquid tanks that can be filled with two similar and independent pumps acting on the tank 1 and tank 2, which have the same cross section $S_1 = S_2$. The pumps deliver the flow rates q_1 in tank 1 and q_2 in tank 2. The tanks are interconnected to each other through lower pipes. All the pipes have the same cross section S_p. The liquid levels L_1 and L_2 in each tank are the controlled variables and they are measured with continuous valued level sensors. The variables q_1 and q_2 are chosen as manipulated variables to control the levels of tank 1 and tank 2.

The system has two faults to be detected and isolated:

- Fault 1 : Leak at the tank 1, an outflow with magnitude q_{f_1}.
- Fault 2 : Leak at the tank 2, an outflow with magnitude q_{f_2}.

The differential equations that describe the system, under the presence of faults, is derived from conservation of mass in the system of the two tanks:

$$\dot{L}_1 = \frac{q_1}{S_1} - \frac{q_{10}}{S_1} - \frac{q_{12}}{S_1} - \frac{q_{f_1}}{S_1} \tag{3}$$

$$\dot{L}_2 = \frac{q_2}{S_2} - \frac{q_{20}}{S_2} + \frac{q_{12}}{S_2} - \frac{q_{f_2}}{S_2}, \tag{4}$$

and by the application of the Torricelli's law:

$$q_{i0} = \mu_i S_p \sqrt{2gL_i} \tag{5}$$

$$q_{ij} = \mu_i S_p \sqrt{2g\,|L_i - L_j|}\,sign\,(L_i - L_j), \tag{6}$$

where μ_i are flow coefficients and considering

$$C_i = \mu_i S_p \sqrt{2g}, \tag{7}$$

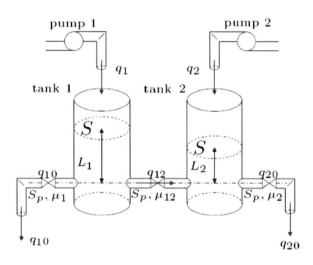

Fig. 1 Two tanks system

Table 1 Values of the constants of the two tanks system

C_1, C_2	$0.3028\ m^2$
S_1, S_2	$2.54\ m^2$
S_p	$0.1\ m^2$
Acceleration due to gravity, g	$9.8\ m/s^2$

the following system of equations is obtained:

$$\dot{L}_1 = \frac{q_1}{S_1} - \frac{C_1}{S_1}\sqrt{L_1} - \frac{C_1}{S_1}\sqrt{|L_1 - L_2|}\,sign\,(L_1 - L_2) - \frac{q_{f_1}}{S_1}$$
$$\dot{L}_2 = \frac{q_2}{S_2} - \frac{C_2}{S_2}\sqrt{L_2} + \frac{C_1}{S_2}\sqrt{|L_1 - L_2|}\,sign\,(L_1 - L_2) - \frac{q_{f_2}}{S_2} \qquad (8)$$
$$y_1 = L_1$$
$$y_2 = L_2$$

For more details see [3].

The goal here is to diagnosis the presence of the faults 1 or 2, even more, their magnitude. As a first approach it has been supposed that the leak at both tanks do not change in time and it is assumed that the magnitude of the leaks is less than $1000\ ml/s$. In other words, the following restrictions for the parameters $q_{f_i}, i = 1,2$ have been established:

$$q_{f_1}, q_{f_2} \in \Re : 0 \le q_{f_1}, q_{f_2} \le 1\ ml/s$$

Estimation of the parameters q_{f_1} and q_{f_2} permit to diagnosis the system. In order to estimate these parameters, the following problem is formulated:

$$\min F\,(\mathbf{v}) = \min \sum_{n=1}^{N} \left[\bar{L}_n^{exp} - \bar{L}_n^{cal}\,(\mathbf{v}) \right]^2 \qquad (9)$$

where $\mathbf{v} = \left(q_{f_1}, q_{f_2}\right)^t$, $\bar{L}_n^{exp} = (L_1^n, L_2^n)^t$ are the observations of the liquid levels at different instants of time, $\bar{L}_n^{cal} = \left(L_{cal(1)}^n, L_{cal(2)}^n\right)^t$ are the liquid levels computed by the model (9) using Runge Kuta 4.

The table 1 shows the values of the constants considered in the model of the two tanks system.

4 Results and Discussion

The closed loop behavior of the process was simulated when no faults are present. This behavior is shown in Fig. 2.

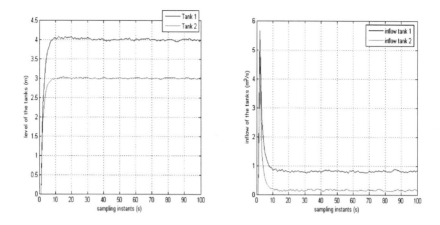

Fig. 2 Closed loop behavior of the process when no faults are present, noise data 2-5 %

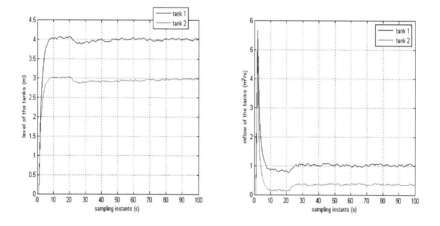

Fig. 3 Closed loop behavior of the process when leaks of 200 ml/s are present in both tanks, noise data 2-5 %

The closed loop behavior of the process when a leak of magnitude 200 ml/s in each tank ($q_{f_1} = q_{f_2} = 0.2$) is introduced at time $t = 20\,s$ is shown in Fig. 3.

The closed loop behavior of the process when a leak of magnitude 50 ml/s in tank 2 ($q_{f_1} = 0$, $q_{f_2} = 0.05$) is introduced at time $t = 20\,s$ is shown in Fig. 4. The effect of this leak in tank 2 is graphically imperceptible.

In order to diagnosis the faults, the minimization of the objective function $F(\mathbf{v})$ was implemented, in the first case based on the DE algorithm. The population was considered to be 10 and the mutation mechanism is $(DE/x_j^{best}/1/bin)$. The second case considered the minimization of the objective function by means of ACO, described on 2.2, with 10 ants. Both algorithms stop when 100 iterations are achieved.

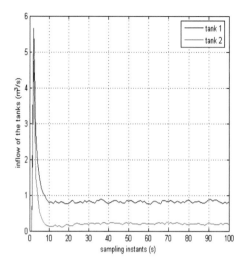

Fig. 4 Closed loop behavior of the process when leaks of 50 ml/s is present in tank 2, noise data 2-5 %

Table 2 Diagnosis obtained in five runs of leaks of 200 ml/s in each tank: noisy data between 2 and 5 % error

alg	fault	$\bar{q}_{f_1}$	$\bar{q}_{f_2}$	iter	t(s)
DE	$q_{f_1} = q_{f_2} = 0.2$	0.1827	0.1901	93	56.5681
DE	$q_{f_1} = q_{f_2} = 0.2$	0.1985	0.2034	67	29.5475
DE	$q_{f_1} = q_{f_2} = 0.2$	0.1988	0.2062	65	28.3845
DE	$q_{f_1} = q_{f_2} = 0.2$	0.2036	0.2044	60	25.5477
DE	$q_{f_1} = q_{f_2} = 0.2$	0.1993	0.1969	56	23.8943
ACO	$q_{f_1} = q_{f_2} = 0.2$	0.1738	0.2414	75	75.0830
ACO	$q_{f_1} = q_{f_2} = 0.2$	0.1775	0.2287	57	48.0647
ACO	$q_{f_1} = q_{f_2} = 0.2$	0.1832	0.2198	54	45.9054
ACO	$q_{f_1} = q_{f_2} = 0.2$	0.2087	0.2401	44	34.9798
ACO	$q_{f_1} = q_{f_2} = 0.2$	0.1860	0.1829	41	33.9888

The tables 2 and 3 shows the results of the diagnosis of different faulty situation by both algorithms. All cases considered data with 2-5 % of noise. The abbreviations used in the tables are alg for algorithm and iter for number of iterations. The notation introduced is t for the computing time, in seconds, of the algorithm. Both algorithms detected the presence of faults but the DE algorithm is more accurate in the determination of the leak magnitudes. Both algorithms are fast, which is good for the online diagnosis, but DE is faster.

In Fig. 5 are shown the evolution of both algorithms for two situation described in table 2. The figures suggest a way of combination of ACO and DE in order to

Table 3 Diagnosis obtained in five runs of a leaks of 50 ml/s in tank 1: noisy data between 2 and 5 % error

alg	fault		$\bar{q}_{f_1}$	$\bar{q}_{f_2}$	iter	t(s)
DE	$q_{f_1} = 0.05$	$q_{f_2} = 0$	0.0498	0.0007	79	46.8123
DE	$q_{f_1} = 0.05$	$q_{f_2} = 0$	0.0515	0.0000	64	35.9895
DE	$q_{f_1} = 0.05$	$q_{f_2} = 0$	0.0508	0.0000	64	35.4267
DE	$q_{f_1} = 0.05$	$q_{f_2} = 0$	0.0531	0.0000	43	28.7774
DE	$q_{f_1} = 0.05$	$q_{f_2} = 0$	0.0489	0.0001	42	18.0390
ACO	$q_{f_1} = 0.05$	$q_{f_2} = 0$	0.0622	0.0000	49	55.1990
ACO	$q_{f_1} = 0.05$	$q_{f_2} = 0$	0.0595	0.0000	31	37.0283
ACO	$q_{f_1} = 0.05$	$q_{f_2} = 0$	0.0624	0.0008	30	36.8984
ACO	$q_{f_1} = 0.05$	$q_{f_2} = 0$	0.0683	0.0007	30	36.7891
ACO	$q_{f_1} = 0.05$	$q_{f_2} = 0$	0.0814	0.0001	26	34.2564

Table 4 Comparison of the diagnosis obtained in runs of leaks of different magnitudes: noisy data between 2 and 5 % error

alg	fault		mean $\bar{q}_{f_1}$	mean $\bar{q}_{f_2}$	mean iter	mean t(s)
ACO-DE	$q_{f_1} = 0.6$	$q_{f_2} = 0$	0.6081	0.0000	53	38.2838
DE	$q_{f_1} = 0.6$	$q_{f_2} = 0$	0.5459	0.0000	76	44.9995
ACO	$q_{f_1} = 0.6$	$q_{f_2} = 0$	0.5038	0.0009	51	57.3317
ACO-DE	$q_{f_1} = 0.6$	$q_{f_2} = 0.6$	0.5901	0.6013	42	19.0167
DE	$q_{f_1} = 0.6$	$q_{f_2} = 0.6$	0.6068	0.6109	50	20.0031
ACO	$q_{f_1} = 0.6$	$q_{f_2} = 0.6$	0.6001	0.4683	51	45.1023

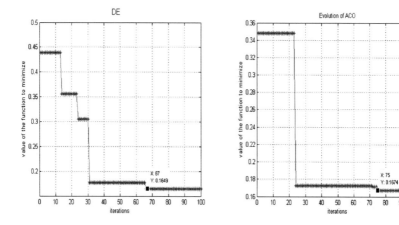

Fig. 5 Evolution of the DE and ACO for a case of the table 2

obtain better and faster diagnosis: start the minimization with ACO a few number of iterations, (no more than 30 taking in count the experimental results), and then use this solution as initial solution for DE. The table 4 shows the comparison between the hybrid algorithm ACO-DE and the diagnosis when using pure algorithms for two faults situations. Each algorithm was executed 30 times (each one starting from a different initial solution) for each fault situation and the table 4 shows the mean $\bar{q}_{f_1}, \bar{q}_{f_2}$ obtained.

In order to analyze the robustness of the diagnosis to unavoidable process disturbances, some numerical experiments were made with very noisy data (15- 20 % of noise). The Fig. 6 shows a simulation of the process behavior under disturbances which causes observations with noise between 15 and 20 %.

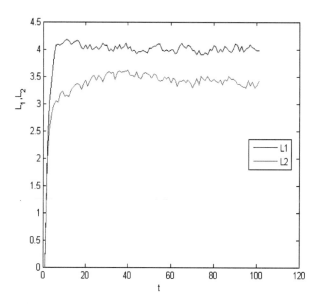

Fig. 6 Closed loop behavior of the process, noisy data 15-20 %

Table 5 Comparison of the diagnosis obtained in runs of leaks of different magnitude: noisy data between 15 and 20 % error

alg	$\bar{q}_{f_1}$	$\bar{q}_{f_2}$	iter	t
ACO-DE(best)	0	0	14	13.9289
DE(best)	0	0	52	16.3422
ACO(best)	0.0600	0.0900	25	19.9327
ACO-DE(worst)	0	0	71	71.0147
DE(worst)	0.1191	0	100	67.5625
ACO(worst)	0.1900	0.0300	100	111.0318

The table 5 shows the best and the worst diagnosis obtained for each algorithm when no faults are present but the system is under disturbances that are represented by noise on the measurable variables. The diagnosis based on the parameter estimation via DE and ACO seems to be robust.

5 Conclusions

This preliminary study indicates that the application of bioinspired algorithms and their cooperative use characterize a promising methodology for the fault diagnosis problem based on a model which does not need to be linear.

There are some advantages observed in the application of the two algorithms to the FDI problem: correct and fast diagnosis, easy structure, robustness to disturbances and less modeling errors due to the use of no linear model. For the detection problem some iterations of the ACO are enough, but for a correct diagnosis the DE algorithm showed better results. In general the cooperative algorithm ACO-DE shows faster diagnosis than pure DE or pure ACO.

In this sense the study of a real cooperative strategies between this two nature inspired algorithms will be done: considering the influence of the parameter α in a more exploration version of the ACO algorithm and the parameter D of the DE algorithm in order to obtain a more exploitation version of DE.

References

[1] Campos Knupp, D., Silva Neto, A.J., Figueiredo Sacco, W.: Estimation of radiactive properties with the particle collision algorithm. In: Inverse Problems, Design and Optimization Symposium, Miami, Florida, USA (2007)

[2] Chen, J., Patton, R.J.: Robust model-based fault diagnosis for dynamic systems. Kluwer Academic Publishers, Dordrecht (1999)

[3] Dolanc, G., Juricic, D., Rakar, A., Petrovcic, J., Vrancic, D.: Three-tank benchmark test. Tech. rep., Copernicus Project Report CT94-02337. J. Stefan Institute (1997)

[4] Dorigo, M.: Ottimizzazione, apprendimento automático, ed algoritmi basati su metafora naturale. PhD thesis, Politécnico di Milano, Italia (1992)

[5] Goldberg, D.E.: Genetic Algorithms in Search, Optimization, and Machine Learning. Addison-Wesley, MA (1989)

[6] Isermann, R.: Process fault detection based on modelling and estimation methods– a survey. Automatica 30(4), 387–404 (1984)

[7] Kirkpatrick, S., Gelatt Jr., C.D., Vecchi, M.P.: Optimization by simulated annealing. Science 220, 671–680 (1983)

[8] Lobato, F.S., Steffen, V., Silva Neto, A.J.: Solution of inverse radiative transfer problems in two-layer participating media with differential evolution. Inverse Problems in Science and Engineering (15), 1–12 (2009)

[9] Lobato, F.S., Steffen, V., Silva Neto, A.J.: Solution of the coupled inverse conduction-radiation problem using multi-objective optimization differential evolution. In: 8th World Congress on Structural and Multidisciplinary Optimization, Lisboa, Portugal (2009)

[10] Lunze, J.: Laboratory three tanks system -benchmark for the reconfiguration problem. Tech. rep., Tech. Univ. of Hamburg-Harburg, Inst. of Control. Eng., Germany (1998)

[11] Patton, R.J., Frank, P.M., Clark, R.N.: Issues of fault diagnosis for dynamic systems. Springer, London (2000)

[12] Sacco, W.F., Oliveira, C.R.E.: A new stochastic optimization algorithm based on particle collisions. In: 2005 ANS Annual Meeting, Transactions of the American Nuclear Society (2005)

[13] Silva Neto, A.J., Moura Neto, F.D.: Problemas Inversos - Conceitos Fundamentais e Aplicações. EdUERJ (2005)

[14] Simani, S., Patton, R.J.: Fault diagnosis of an industrial gas turbine prototype using a system identification approach. Control Engineering Practice 16, 769–786 (2008)

[15] Storn, R., Price, K.: Differential evolution: A simple and efficient adaptive scheme for global optimization over continuous spaces. International Computer Science Institute (1995)

[16] Wang, L., Niu, Q., Fei, M.: A novel quantum ant colony optimization algorithm and its application to fault diagnosis. Transactions of the Institute of Measurement and Control 30(3/4), 313–329 (2008)

[17] Witczak, M.: Advances in model based fault diagnosis with evolutionary algorithms and neural networks. Int. J. Appl. Math. Comput. Sci. 16(1), 85–99 (2006)

[18] Yang, E., Xiang, H., Gu, D., Zhang, Z.: A comparative study of genetic algorithm parameters for the inverse problem-based fault diagnosis of liquid rocket propulsion systems. International Journal of Automation and Computing 04(3), 255–261 (2007)

A New Metaheuristic Bat-Inspired Algorithm

Xin-She Yang

Abstract. Metaheuristic algorithms such as particle swarm optimization, firefly algorithm and harmony search are now becoming powerful methods for solving many tough optimization problems. In this paper, we propose a new metaheuristic method, the Bat Algorithm, based on the echolocation behaviour of bats. We also intend to combine the advantages of existing algorithms into the new bat algorithm. After a detailed formulation and explanation of its implementation, we will then compare the proposed algorithm with other existing algorithms, including genetic algorithms and particle swarm optimization. Simulations show that the proposed algorithm seems much superior to other algorithms, and further studies are also discussed.

1 Introduction

Metaheuristic algorithms such as particle swarm optimization and simulated annealing are now becoming powerful methods for solving many tough optimization problems [3-7,11]. The vast majority of heuristic and metaheuristic algorithms have been derived from the behaviour of biological systems and/or physical systems in nature. For example, particle swarm optimization was developed based on the swarm behaviour of birds and fish [6, 7], while simulated annealing was based on the annealing process of metals [8].

New algorithms are also emerging recently, including harmony search and the firefly algorithm. The former was inspired by the improvising process of composing a piece of music [4], while the latter was formulated based on the flashing behaviour of fireflies [13]. Each of these algorithms has certain advantages and disadvantages.

Xin-She Yang
Department of Engineering, University of Cambridge,
Trumpington Street, Cambridge CB2 1PZ, UK
e-mail: xy227@cam.ac.uk

J.R. González et al. (Eds.): NICSO 2010, SCI 284, pp. 65–74, 2010.
springerlink.com

For example, simulating annealing can almost guarantee to find the optimal solution if the cooling process is slow enough and the simulation is running long enough; however, the fine adjustment in parameters does affect the convergence rate of the optimization process. A natural question is whether it is possible to combine major advantages of these algorithms and try to develop a potentially better algorithm? This paper is such an attempt to address this issue.

In this paper, we intend to propose a new metaheuristic method, namely, the Bat Algorithm (BA), based on the echolocation behaviour of bats. The capability of echolocation of microbats is fascinating as these bats can find their prey and discriminate different types of insects even in complete darkness. We will first formulate the bat algorithm by idealizing the echolocation behaviour of bats. We then describe how it works and make comparison with other existing algorithms. Finally, we will discuss some implications for further studies.

2 Echolocation of Bats

2.1 Behaviour of Microbats

Bats are fascinating animals. They are the only mammals with wings and they also have advanced capability of echolocation. It is estimated that there are about 996 different species which account for up to 20% of all mammal species [1, 2]. Their size ranges from the tiny bumblebee bat (of about 1.5 to 2g) to the giant bats with wingspan of about 2 m and weight up to about 1 kg. Microbats typically have forearm length of about 2.2 to 11cm. Most bats uses echolocation to a certain degree; among all the species, microbats are a famous example as microbats use echolocation extensively while megabats do not [11, 12].

Most microbats are insectivores. Microbats use a type of sonar, called, echolocation, to detect prey, avoid obstacles, and locate their roosting crevices in the dark. These bats emit a very loud sound pulse and listen for the echo that bounces back from the surrounding objects. Their pulses vary in properties and can be correlated with their hunting strategies, depending on the species. Most bats use short, frequency-modulated signals to sweep through about an octave, while others more often use constant-frequency signals for echolocation. Their signal bandwidth varies depends on the species, and often increased by using more harmonics.

2.2 Acoustics of Echolocation

Though each pulse only lasts a few thousandths of a second (up to about 8 to 10 ms), however, it has a constant frequency which is usually in the region of 25kHz to 150 kHz. The typical range of frequencies for most bat species are in the region between 25kHz and 100kHz, though some species can emit higher frequencies up to 150 kHz. Each ultrasonic burst may last typically 5 to 20 ms, and microbats emit about 10 to 20 such sound bursts every second. When hunting for prey, the rate of pulse emission can be sped up to about 200 pulses per second when they fly near their prey. Such short sound bursts imply the fantastic ability of the signal processing power of bats. In fact, studies shows the integration time of the bat ear is typically about 300 to 400 μs.

As the speed of sound in air is typically $v = 340$ m/s, the wavelength λ of the ultrasonic sound bursts with a constant frequency f is given by

$$\lambda = \frac{v}{f}, \tag{1}$$

which is in the range of 2mm to 14mm for the typical frequency range from 25kHz to 150 kHz. Such wavelengths are in the same order of their prey sizes.

Amazingly, the emitted pulse could be as loud as 110 dB, and, fortunately, they are in the ultrasonic region. The loudness also varies from the loudest when searching for prey and to a quieter base when homing towards the prey. The travelling range of such short pulses are typically a few metres, depending on the actual frequencies [11]. Microbats can manage to avoid obstacles as small as thin human hairs.

Studies show that microbats use the time delay from the emission and detection of the echo, the time difference between their two ears, and the loudness variations of the echoes to build up three dimensional scenario of the surrounding. They can detect the distance and orientation of the target, the type of prey, and even the moving speed of the prey such as small insects. Indeed, studies suggested that bats seem to be able to discriminate targets by the variations of the Doppler effect induced by the wing-flutter rates of the target insects [1].

Obviously, some bats have good eyesight, and most bats also have very sensitive smell sense. In reality, they will use all the senses as a combination to maximize the efficient detection of prey and smooth navigation. However, here we are only interested in the echolocation and the associated behaviour.

Such echolocation behaviour of microbats can be formulated in such a way that it can be associated with the objective function to be optimized, and this make it possible to formulate new optimization algorithms. In the rest of this paper, we will first outline the basic formulation of the Bat Algorithm (BA) and then discuss the implementation and comparison in detail.

3 Bat Algorithm

If we idealize some of the echolocation characteristics of microbats, we can develop various bat-inspired algorithms or bat algorithms. For simplicity, we now use the following approximate or idealized rules:

1. All bats use echolocation to sense distance, and they also 'know' the difference between food/prey and background barriers in some magical way;
2. Bats fly randomly with velocity v_i at position x_i with a fixed frequency f_{min}, varying wavelength λ and loudness A_0 to search for prey. They can automatically

adjust the wavelength (or frequency) of their emitted pulses and adjust the rate of pulse emission $r \in [0, 1]$, depending on the proximity of their target;

3. Although the loudness can vary in many ways, we assume that the loudness varies from a large (positive) A_0 to a minimum constant value A_{min}.

Another obvious simplification is that no ray tracing is used in estimating the time delay and three dimensional topography. Though this might be a good feature for the application in computational geometry, however, we will not use this as it is more computationally extensive in multidimensional cases.

In addition to these simplified assumptions, we also use the following approximations, for simplicity. In general the frequency f in a range $[f_{min}, f_{max}]$ corresponds to a range of wavelengths $[\lambda_{min}, \lambda_{max}]$. For example a frequency range of [20kHz, 500kHz] corresponds to a range of wavelengths from 0.7mm to 17mm.

For a given problem, we can also use any wavelength for the ease of implementation. In the actual implementation, we can adjust the range by adjusting the wavelengths (or frequencies), and the detectable range (or the largest wavelength) should be chosen such that it is comparable to the size of the domain of interest, and then toning down to smaller ranges. Furthermore, we do not necessarily have to use the wavelengths themselves, instead, we can also vary the frequency while fixing the wavelength λ. This is because λ and f are related due to the fact λf is constant. We will use this later approach in our implementation.

For simplicity, we can assume $f \in [0, f_{max}]$. We know that higher frequencies have short wavelengths and travel a shorter distance. For bats, the typical ranges are a few metres. The rate of pulse can simply be in the range of $[0, 1]$ where 0 means no pulses at all, and 1 means the maximum rate of pulse emission.

Based on these approximations and idealization, the basic steps of the Bat Algorithm (BA) can be summarized as the pseudo code shown in Fig. 1.

3.1 Movement of Virtual Bats

In simulations, we use virtual bats naturally. We have to define the rules how their positions $\mathbf{x}_i$ and velocities $\mathbf{v}_i$ in a d-dimensional search space are updated. The new solutions $\mathbf{x}_i^t$ and velocities $\mathbf{v}_i^t$ at time step t are given by

$$f_i = f_{min} + (f_{max} - f_{min})\beta, \tag{2}$$

$$\mathbf{v}_i^t = \mathbf{v}_i^{t-1} + (\mathbf{x}_i^t - \mathbf{x}_*)f_i, \tag{3}$$

$$\mathbf{x}_i^t = \mathbf{x}_i^{t-1} + \mathbf{v}_i^t, \tag{4}$$

where $\beta \in [0, 1]$ is a random vector drawn from a uniform distribution. Here $\mathbf{x}_*$ is the current global best location (solution) which is located after comparing all the solutions among all the n bats. As the product $\lambda_i f_i$ is the velocity increment, we can use either f_i (or λ_i) to adjust the velocity change while fixing the other factor λ_i (or

Bat Algorithm

Objective function $f(\mathbf{x})$, $\mathbf{x} = (x_1,...,x_d)^T$
Initialize the bat population $\mathbf{x}_i$ $(i = 1,2,...,n)$ and $\mathbf{v}_i$
Define pulse frequency f_i at $\mathbf{x}_i$
Initialize pulse rates r_i and the loudness A_i
while *(t <Max number of iterations)*
Generate new solutions by adjusting frequency,
and updating velocities and locations/solutions [equations (2) to (4)]
    if *(rand > r_i)*
    Select a solution among the best solutions
    Generate a local solution around the selected best solution
    end if
    Generate a new solution by flying randomly
    if *(rand < A_i & $f(\mathbf{x}_i) < f(\mathbf{x}_*)$)*
    Accept the new solutions
    Increase r_i and reduce A_i
    end if
Rank the bats and find the current best $\mathbf{x}_*$
end while
Postprocess results and visualization

Fig. 1 Pseudo code of the bat algorithm (BA)

f_i), depending on the type of the problem of interest. In our implementation, we will use $f_{min} = 0$ and $f_{max} = 100$, depending the domain size of the problem of interest. Initially, each bat is randomly assigned a frequency which is drawn uniformly from $[f_{min}, f_{max}]$.

For the local search part, once a solution is selected among the current best solutions, a new solution for each bat is generated locally using random walk

$$\mathbf{x}_{new} = \mathbf{x}_{old} + \varepsilon A^t, \qquad (5)$$

where $\varepsilon \in [-1,1]$ is a random number, while $A^t = <A_i^t>$ is the average loudness of all the bats at this time step.

The update of the velocities and positions of bats have some similarity to the procedure in the standard particle swarm optimization [6] as f_i essentially controls the pace and range of the movement of the swarming particles. To a degree, BA can be considered as a balanced combination of the standard particle swarm optimization and the intensive local search controlled by the loudness and pulse rate.

3.2 Loudness and Pulse Emission

Furthermore, the loudness A_i and the rate r_i of pulse emission have to be updated accordingly as the iterations proceed. As the loudness usually decreases once a bat has found its prey, while the rate of pulse emission increases, the loudness can be chosen as any value of convenience. For example, we can use $A_0 = 100$ and $A_{min} = 1$. For simplicity, we can also use $A_0 = 1$ and $A_{min} = 0$, assuming $A_{min} = 0$ means that a bat has just found the prey and temporarily stop emitting any sound. Now we have

$$A_i^{t+1} = \alpha A_i^t, \quad r_i^{t+1} = r_i^0[1 - \exp(-\gamma t)], \tag{6}$$

where α and γ are constants. In fact, α is similar to the cooling factor of a cooling schedule in the simulated annealing [8]. For any $0 < \alpha < 1$ and $\gamma > 0$, we have

$$A_i^t \to 0, \quad r_i^t \to r_i^0, \quad \text{as } t \to \infty. \tag{7}$$

In the simplicity case, we can use $\alpha = \gamma$, and we have used $\alpha = \gamma = 0.9$ in our simulations. The choice of parameters requires some experimenting. Initially, each bat should have different values of loudness and pulse emission rate, and this can be achieved by randomization. For example, the initial loudness A_i^0 can typically be $[1, 2]$, while the initial emission rate r_i^0 can be around zero, or any value $r_i^0 \in [0, 1]$ if using (6). Their loudness and emission rates will be updated only if the new solutions are improved, which means that these bats are moving towards the optimal solution.

4 Validation and Comparison

From the pseudo code, it is relatively straightforward to implement the Bat Algorithm in any programming language. For the ease of visualization, we have implemented it using Matlab for various test functions.

4.1 Benchmark Functions

There are many standard test functions for validating new algorithms. In the current benchmark validation, we have chosen the well-known Rosenbrock's function

$$f(\mathbf{x}) = \sum_{i=1}^{d-1} (1 - x_i^2)^2 + 100(x_{i+1} - x_i^2)^2, \quad -2.048 \le x_i \le 2.048, \tag{8}$$

and the eggcrate function

$$g(x, y) = x^2 + y^2 + 25(\sin^2 x + \sin^2 y), \quad (x, y) \in [-2\pi, 2\pi] \times [-2\pi, 2\pi]. \tag{9}$$

We know that $f(\mathbf{x})$ has a global minimum $f_{min} = 0$ at $(1, 1)$ in 2D, while $g(x, y)$ has a global minimum $g_{min} = 0$ at $(0, 0)$. De Jong's standard sphere function

$$h(\mathbf{x}) = \sum_{i=1}^{d} x_i^2, \quad -10 \le x_i \le 10, \tag{10}$$

has also been used. Its minimum is $h_{\min} = 0$ at $(0,0,...,0)$ for any $d \ge 3$.

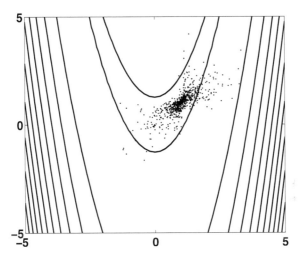

Fig. 2 The paths of 25 virtual bats during 20 consecutive iterations. They converge into $(1,1)$

In addition, we have also used other standard test functions for numerical global optimization [9] such as Ackley's function

$$s(\mathbf{x}) = 20 + e - 20\exp\left[-0.2\sqrt{\frac{1}{d}\sum_{i=1}^{d} x_i^2}\right] - \exp\left[\frac{1}{d}\sum_{i=1}^{d}\cos(2\pi x_i)\right], \tag{11}$$

where $-30 \le x_i \le 30$. It has the global minimum $s_{\min} = 0$ at $(0,0,...,0)$.
  Michalewicz's test function

$$f(\mathbf{x}) = -\sum_{i=1}^{d} \sin(x_i)\left[\sin(\frac{ix_i^2}{\pi})\right]^{2m}, \quad (m=10), \tag{12}$$

has $d!$ local optima in the the domain $0 \le x_i \le \pi$ where $i = 1,2,...,d$. The global minimum is $f_* \approx -1.801$ for $d = 2$, while $f_* \approx -4.6877$ for $d = 5$.
  In our implementation, we use $n = 25$ to 50 virtual bats, and $\alpha = 0.9$. For Rosenbrock's 2-D banana function, the paths of 25 virtual bats during the consecutive 20 time steps are shown in Fig. 2 where we can see that the bats converge at the global optimum $(1,1)$. For the multimodal eggcrate function, a snapshot of the last 10 iterations is shown in Fig. 3. Again, all bats move towards the global best $(0,0)$.

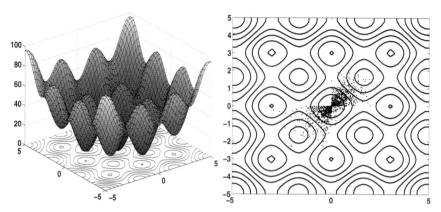

Fig. 3 The eggcrate function (left) and the locations of 40 bats in the last ten iterations (right)

4.2 Comparison with Other Algorithms

In order to compare the performance of the new algorithm, we have tested it against other heuristic algorithms, including genetic algorithms (GA) [5, 10], and particle swarm optimization (PSO) [6, 7]. There are many variants of PSO, and some variants such as the mean PSO could perform better than the standard PSO [3]; however, the standard PSO is by far the most popularly used. Therefore, we will also use the standard PSO in our comparison.

There are many ways to carry out the comparison of algorithm performance, and two obvious approaches are: to compare the numbers of function evaluations for a given tolerance or accuracy, or to compare their accuracies for a fixed number of function evaluations. Here we will use the first approach. In our simulations, we use a fixed tolerance $\varepsilon \leq 10^{-5}$, and we run each algorithm for 100 times so that we can do meaningful statistical analysis.

For genetic algorithms, we have used the standard version with no elitism with the mutation probability of $p_m = 0.05$ and crossover probability of 0.95. For particle swarm optimization, we have also used the standard version with learning parameters $\alpha = \beta = 2$ and the inertia function $I = 1$ [6, 7]. The simulations have been carried out using Matlab on a standard 3GHz desktop computer. Each run with about 10,000 function evaluations typically takes less than 5 seconds. Furthermore, we have tried to use different population sizes from $n = 10$ to 250, and we found that for most problems, $n = 15$ to 50 is sufficient. Therefore, we use a fixed population $n = 40$ for all simulations. Table 1 shows the number of function evaluations in the form of mean $\pm$ the standard deviation (success rate of the algorithm in finding the global optima).

From Table 1, we can see that PSO performs much better than genetic algorithms, while the Bat Algorithm is much superior to other algorithms in terms of accuracy and efficiency. This is no surprising as the aim of developing the new algorithm was to try to use the advantages of existing algorithms and other interesting feature inspired by the fantastic behaviour of echolocation of microbats.

Table 1 Comparison of BA with GA, and PSO

Functions/Algorithms	GA	PSO	BA
Multiple peaks	$52124 \pm 3277(98\%)$	$3719 \pm 205(97\%)$	$1152 \pm 245(100\%)$
Michalewicz's $(d=16)$	$89325 \pm 7914(95\%)$	$6922 \pm 537(98\%)$	$4752 \pm 753(100\%)$
Rosenbrock's $(d=16)$	$55723 \pm 8901(90\%)$	$32756 \pm 5325(98\%)$	$7923 \pm 3293(100\%)$
De Jong's $(d=256)$	$25412 \pm 1237(100\%)$	$17040 \pm 1123(100\%)$	$5273 \pm 490(100\%)$
Schwefel's $(d=128)$	$227329 \pm 7572(95\%)$	$14522 \pm 1275(97\%)$	$8929 \pm 729(99\%)$
Ackley's $(d=128)$	$32720 \pm 3327(90\%)$	$23407 \pm 4325(92\%)$	$6933 \pm 2317(100\%)$
Rastrigin's	$110523 \pm 5199(77\%)$	$79491 \pm 3715(90\%)$	$12573 \pm 3372(100\%)$
Easom's	$19239 \pm 3307(92\%)$	$17273 \pm 2929(90\%)$	$7532 \pm 1702(99\%)$
Griewangk's	$70925 \pm 7652(90\%)$	$55970 \pm 4223(92\%)$	$9792 \pm 4732(100\%)$
Shubert's (18 minima)	$54077 \pm 4997(89\%)$	$23992 \pm 3755(92\%)$	$11925 \pm 4049(100\%)$

If we replace the variations of the frequency f_i by a random parameter and setting $A_i = 0$ and $r_i = 1$, the bat algorithm essentially becomes the standard Particle Swarm Optimization (PSO). Similarly, if we do not use the velocities, but we use fixed loudness and rate: A_i and r_i. For example, $A_i = r_i = 0.7$, this algorithm is virtually reduced to a simple Harmony Search (HS) as the frequency/wavelength change is essentially the pitch adjustment, while the rate of pulse emission is similar to the harmonic acceptance rate (here with a twist) in the harmony search algorithm [4, 14]. The current studies implies that the proposed new algorithm is potentially more powerful and thus should be investigated further in many applications of engineering and industrial optimization problems.

5 Discussions

In this paper, we have successfully formulated a new Bat Algorithm for continuous constrained optimization problems. From the formulation of the Bat Algorithm and its implementation and comparison, we can see that it is a very promising algorithm. It is potentially more powerful than particle swarm optimization and genetic algorithms as well as Harmony Search. The primary reason is that BA uses a good combination of major advantages of these algorithms in some way. Moreover, PSO and harmony search are the special cases of the Bat Algorithm under appropriate simplifications.

In addition, the fine adjustment of the parameters α and γ can affect the convergence rate of the bat algorithm. In fact, parameter α acts in a similar role as the cooling schedule in the simulated annealing. Though the implementation is more complicated than many other metaheuristic algorithms; however, it does show that it utilizes a balanced combination of the advantages of existing successful algorithms with innovative feature based on the echolocation behaviour of bats. New solutions are generated by adjusting frequencies, loudness and pulse emission rates, while the proposed solution is accepted or not depends on the quality of the solutions controlled or characterized by loudness and pulse rate which are in turn related to the closeness or the fitness of the locations/solution to the global optimal solution.

The exciting results suggest that more studies will highly be needed to carry out the sensitivity analysis, to analyze the rate of algorithm convergence, and to improve the convergence rate even further. More extensive comparison studies with a more wide range of existing algorithms using much tough test functions in higher dimensions will pose more challenges to the algorithms, and thus such comparisons will potentially reveal the virtues and weakness of all the algorithms of interest.

An interesting extension will be to use different schemes of wavelength or frequency variations instead of the current linear implementation. In addition, the rates of pulse emission and loudness can also be varied in a more sophisticated manner. Another extension for discrete problems is to use the time delay between pulse emission and the echo bounced back. For example, in the travelling salesman problem, the distance between two adjacent nodes/cities can easily be coded as time delay. As microbats use time difference between their two ears to obtain three-dimensional information, they can identify the type of prey and the velocity of a flying insect. Therefore, a further natural extension to the current bat algorithm would be to use the directional echolocation and Doppler effect, which may lead to even more interesting variants and new algorithms.

References

[1] Altringham, J.D.: Bats: Biology and Behaviour. Oxford Univesity Press, Oxford (1996)
[2] Colin, T.: The Variety of Life. Oxford University Press, Oxford (2000)
[3] Deep, K., Bansal, J.C.: Mean particle swarm optimisation for function optimisation. Int. J. Comput. Intel. Studies 1, 72–92 (2009)
[4] Geem, Z.W., Kim, J.H., Loganathan, G.V.: A new heuristic optimization algorithm: Harmony search. Simulation 76, 60–68 (2001)
[5] Holland, J.H.: Adapation in Natural and Artificial Systems. University of Michigan Press, Ann Arbor (1975)
[6] Kennedy, J., Eberhart, R.: Particle swarm optimization. In: Proc. IEEE Int. Conf. Neural Networks, Perth, Australia, pp. 1942–1945 (1995)
[7] Kennedy, J., Eberhart, R.: Swarm Intelligence. Academic Press, London (2001)
[8] Kirkpatrick, S., Gelatt, C.D., Vecchi, M.P.: Optimization by simulated annealing. Science 220, 671–680 (1983)
[9] Liang, J.J., Suganthan, P.N., Deb, K.: Novel composition test functions for numerical global optimization. In: Proc. IEEE Int. Swarm Intel. Symp., pp. 68–75 (2005)
[10] Mitchell, M.: An Introduction to Genetic Algorithms. MIT Press, Cambridge (1998)
[11] Richardson, P.: Bats. Natural History Museum, London (2008)
[12] Richardson, P.: The secrete life of bats, http://www.nhm.ac.uk
[13] Yang, X.-S.: Nature-inspired Metaheuristic Algorithms. Luniver Press (2008)
[14] Yang, X.-S.: Harmony search as a metaheuristic algorithm. In: Geem, Z.W. (ed.) Music-Inspired Harmony Search Algorithm: Theory and Applications, pp. 1–14. Springer, Heidelberg (2009)

Evaluation of a Catalytic Search Algorithm

Lidia Yamamoto

Abstract. We investigate the search properties of pre-evolutionary random catalytic reaction networks, where reactions might be reversible, and replication is not taken for granted. Since it counts only on slow growth rates and weak selective pressure to steer the search process, catalytic search is an inherently slow process. However it presents interesting properties worth exploring, such as the potential to steer the computation flow towards good solutions, and to prevent premature convergence. We have designed a simple catalytic search algorithm, in order to assess its beamed search ability. In this paper we report preliminary results that show that although weak, the search strength achieved with catalytic search is sufficient to solve simple problems, and to find good approximations for more complex problems, while keeping a diversity of solutions and their building blocks in the population.

1 Introduction

Artificial Chemistries have the ability not only to *model* evolutionary behavior but also to *create* it, or to cause it to *emerge* spontaneously [3, 9–11]. However, the exact conditions upon which such evolutionary behaviour could emerge are not entirely clear, and are deeply linked to the conditions for the transition from inanimate to living matter. Another aspect that remains still unclear so far is how to harness the *emergent computation* [7, 12] properties of such chemistries for the construction of beamed search schemes able to optimize solutions to user-defined problems.

A number of chemically-inspired approaches to optimization towards user-defined goals have been proposed [5, 6, 14, 20, 22]. The reaction networks created by such chemistries may exhibit complex dynamics, hence the general problem of searching with a chemistry remains poorly understood.

Lidia Yamamoto
Computer Science Department, University of Basel,
Bernoullistrasse 16, CH-4056, Basel, Switzerland
e-mail: Lidia.Yamamoto@unibas.ch

J.R. González et al. (Eds.): NICSO 2010, SCI 284, pp. 75–87, 2010.
springerlink.com   © Springer-Verlag Berlin Heidelberg 2010

Some chemistries take evolution elements for granted, such as replication, therefore the problem of how to get evolutionary behavior is not an issue for them. In contrast, in this paper we look at the particular case of chemistries that do not assume replication, and that must comply to some physical laws such as mass and energy conservation. The behavior of such chemical search can be classified as *pre-evolutionary* [18]. Chemical reactions consume educts to produce new molecules, in a mass-conserving way, and most reactions are reversible. Catalysts may be present to enhance the rate of some reactions. The resulting mechanism is a *Catalytic Search*, that relies only on slow growth rates and weak selective pressure to steer the search process.

Catalytic search is an inherently slow process in general, but after some time it could reach an *autocatalytic* stage where some elements of the network become able to replicate directly or collectively via cooperative interactions. The question that remains unanswered is whether such a slow process is sufficient to ignite a faster, more efficient search process exhibiting full Darwinian evolutionary dynamics within feasible runtimes, and if yes, how this could be achieved.

Although typically slow, catalytic search has a useful potential as a "soft search" mechanism, which remains under-explored so far. As pointed out in [22], catalytic search presents interesting properties worth exploring, such as the potential to undo wrong computations through reversible reactions, to steer the flow of the system towards the production of good products by shifting the equilibrium distribution of molecules, a certain robustness to noisy fitness feedback, and the prevention of premature convergence. Moreover catalytic search is inherently *cooperative*: since molecules cannot self-replicate in principle, they need the help of other molecules in order to grow. Hence they are forced to self-organize into a network of positive interactions that construct and deconstruct solutions dynamically, according to the objective function to be computed.

We have designed a simple catalytic search algorithm, in order to assess the ability of a catalytic artificial chemistry subject to pre-evolutionary dynamics to exhibit a beamed search behavior. The chemical search scheme is built on top of a thermodynamic model which steers the candidate solutions not only towards better fitness but also towards lower computation costs. We compare Catalytic Search with a pure random search (for the sake of sanity check only), and with a variation of a Steady-State Genetic Algorithm (SSGA) [16] implemented with an artificial chemistry. As expected, the performance of the catalytic search scheme lies between that of a pure random search and that of plain evolutionary search represented by the genetic algorithm. However catalytic search presents other interesting properties, such as the preservation of diversity and of partial solution building blocks in the population.

The paper is organized as follows: Section 2 surveys the related literature on search schemes based on chemistry. Section 3 describes our catalytic search algorithm, and report some experimental results on the simple problem of finding a hidden sentence. Section 4 concludes with an outlook on the many interesting avenues to explore.

2 Background

A survey of optimization schemes based on artificial chemistries can be found in Section 4.3 of [10]. Here we summarize and update it.

Chemical approaches to optimization towards a user-defined goal have been proposed in [5, 6, 14, 20, 22]. These approaches can be divided into two categories: In the first category we find search algorithms inspired by chemistry, but for which the actual solutions searched are not encoded as chemical computing programs but as parameters to be optimized [5, 14], as partial solutions to the problem [22], or as conventional program trees [20]. In the second category we find centralized evolutionary algorithms used to evolve chemical reaction networks by manipulating their graphs [8], or to evolve chemical computing programs by genetic programming [6]. The work reported in this paper falls into the first category. However our long term goal is to combine both categories in one, obtaining a search process based on chemistry, for searching solutions encoded as chemical programs.

An optimization method inspired by chemistry is presented in [5]. Candidate solutions are encoded as strings analogous to macromolecules in prebiotic evolution. These strings carry a quality signal (fitness value). Machines (analogous to enzymes) operate on the strings to change and select them according to a fitness function embedded within the machine's knowledge.

The Chemical Casting Model (CCM) [14] is inspired by the process of entropy reduction which is behind many self-organization phenomena in chemistry. Candidate solutions are encoded as molecules; reaction rules modify and select molecules, driving the system towards a more ordered state (with lower entropy) in which molecules encode better solutions. CCM has been successfully applied to many different problems ranging from constraint satisfaction to graph coloring and the traveling salesperson.

Chemical Genetic Programming [20] takes inspiration from gene expression and chemistry in order to construct program trees in Genetic Programming.

ARMS (Abstract Rewriting Systems on Multisets [21]) is a chemical evolution system based on Membrane Computing or P Systems [19]. Membrane Computing allows hierarchies of multiset compartments to be constructed recursively. ARMS makes use of this feature to evolve populations of artificial cells by a process of cell growth and division. The resulting cells may exhibit a rich internal structure, sometimes resembling protocells models such as the chemoton [13]. The ARMS system has been applied to the evolution of artificial cells both for biological and for computational purposes.

Our work builds upon the Artificial Catalysed Reaction Networks from [22]. That scheme takes inspiration from Kaufffman's autocatalytic networks [4, 15]. Each molecule is a partial solution. The algorithm starts with a population of small molecules and builds larger ones via polymerization reactions. Fitter products are rewarded by catalyzing their own production. Each molecule is therefore an autocatalyst. Compared with [22], in our work we do not assume that molecules are autocatalysts, and we use a crossover operator that includes polymerization as a special case.

3   Catalytic Search Algorithm

The Catalytic Search Algorithm works as follows: initially, a random soup of molecules is generated. Each molecule is a candidate solution represented as a string of symbols from an alphabet Σ. At every time step, two molecules are chosen for collision. They react with a probability k_f, which maps to the kinetic coefficient of the reaction. If they react, a crossover of the two molecules is produced, and the two resulting molecules are injected into the soup. The collision is elastic with probability $(1 - k_f)$, in which case the molecules are put back into the soup and no products are generated.

A crossover reaction can be written as follows:

$$A + B \underset{k_r}{\overset{k_f}{\rightleftharpoons}} C + D \tag{1}$$

Here is an example, for strings from an alphabet $\Sigma = \{a, b, c\}$:

$$abba + ccb \underset{k_r}{\overset{k_f}{\rightleftharpoons}} abbc + acb \tag{2}$$

Crossover is a mass-conserving operation, i.e. it conserves the total number of symbols before and after the reaction.

The initial population is always a soup of monomers (strings of length one). Solutions are then built by concatenating these monomers. This is a special case of a crossover operation, where the crossover point on one of the strings is the end of the string, and the crossover point on the other string is at the beginning. More complex solutions can then be constructed out of these basic building blocks.

We choose the coefficients k_f and k_r to be a function of the fitness and the computation cost associated with the solution, in order to steer the search by differentiated reaction rates. This mapping will be explained below.

Once the molecules have collided, the reaction only occurs if the molecules have sufficient kinetic energy in order to overcome the *activation energy barrier* (E_a), which corresponds to the minimum amount of kinetic energy that is needed for the reaction to occur.

Figure 1 shows a typical plot of the potential energy changes during a chemical reaction. The horizontal axis is called *reaction coordinate*, and shows the progression of the (forward) reaction from reactants X on the left side to products Y on the right (symmetrically, the corresponding reverse reaction can be read from right to left). The vertical axis shows the corresponding potential energy. The height of the peaks with respect to the initial state corresponds to the *activation energy* of the reaction. A *catalyst* is a substance that participates in a chemical reaction by accelerating it without being consumed in the process. Its effect is to lower the reaction's activation energy peak, thereby accelerating the reaction, while leaving the initial and final states unchanged. The difference in potential energy before and after the reaction is given by ΔG:

$$\Delta G = G_p - G_e \tag{3}$$

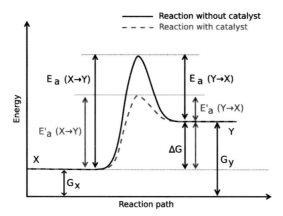

Fig. 1 Potential energy changes during catalysed and uncatalysed chemical reactions. Figure adapted from [1]

If $\Delta G > 0$ then the reaction is *endergonic*, i.e. it absorbs energy from its surroundings, while if $\Delta G < 0$ it is *exergonic*, releasing energy. Endergonic reactions are typically non-spontaneous, i.e. their equilibrium is shifted towards the educts, while exergonic reactions occur typically spontaneously, resulting in larger quantities of products.

In order to steer the system towards the production of fitter solutions, we map the fitness of the solution to the potential energy of its molecule. A lower value of the fitness function is often associated with a better fitness, for instance, a shorter distance to the optimum. In this case, we can associate fitness with the the *potential energy* of the molecule directly. The total potential energy of the educts G_e (resp. products G_p) is the sum of potential energies of each educt (resp. product) involved in the reaction, i.e. the sum of their fitness values. In this way, the production of fitter solutions (i.e. with lower potential energy) is spontaneous, while the production of poor solutions is non-spontaneous.

In order to provide the system with an incentive for efficient computations, we further map the activation energy for a reaction to the estimated computation cost of producing a solution. For instance, let us take a simple case in which the cost is a linear function of the length of the candidate solutions. Since we only consider mass-conserving (i.e. symbol-conserving) operations, the total number of atoms is the same on both educt and product sides. An increase in activation energy ΔE_a is then added on top of the highest potential energy G. ΔE_a corresponds to the portion $E_a(Y \rightarrow X)$ in Figure 1. As a result, the side of the reaction with the lowest potential energy (the X side to the left of Figure 1) will see an activation energy of $E_a = \Delta E_a + |\Delta G|$, while the other side ($Y$, on the right) will see $E_a = \Delta E_a$. The portion ΔE_a of the total activation energy is set to the average length of the educts (or products):

$$\Delta E_a = \frac{|A| + |B|}{2} = \frac{|C| + |D|}{2} \qquad (4)$$

The activation energies of the forward and reverse reactions, E_{af} and E_{ar} respectively, are:

$$\text{if} \quad \Delta G \leq 0 \quad \begin{cases} E_{af} = \Delta E_a \\ E_{ar} = \Delta E_a - \Delta G \end{cases} \tag{5}$$

$$\text{if} \quad \Delta G > 0 \quad \begin{cases} E_{af} = \Delta E_a + \Delta G \\ E_{ar} = \Delta E_a \end{cases} \tag{6}$$

The coefficient k_f (resp. k_r) is determined as a function of the activation energy, following the *Arrhenius equation* from chemistry [2]:

$$k = A e^{-\frac{E_a}{RT}} \tag{7}$$

where A is the so-called *pre-exponential factor* of the reaction, E_a is its *activation energy*, and RT are constants. In our case, we set $A = 1$ and $\beta = \frac{1}{RT}$ is a configuration parameter of our algorithm (currently set to $\beta = 1$).

The constants k_f (resp. k_r) determine the probability that the reaction is successful once the reactants collide. According to the Arrhenius equation (Eq. 7), these coefficients decrease exponentially with the activation energy barrier E_a seen by the reactants. Since E_a increases with ΔE_a, which is mapped to the computational cost of the operation, the probability of the reaction to occur decreases with its cost, as desired. Similarly, since the height of the E_a barrier observed is higher on the side of the reaction with lower G, it is more difficult to "cross the barrier" from this side, therefore it is more difficult to move from a lower (closer to the optimum) to a higher (farther from the optimum) fitness value, which is also the behavior that we are seeking. While lower, there is still some probability to move towards worse solutions, since that may help creating new solutions which might be useful for the search.

In this way, this scheme is able to steer the flow of production of candidate solutions towards better ones. There is no explicit replication, and no memory of which molecules produced good solutions. The search process is guided by the differences in reaction rates to move from one pair of candidate solutions to another.

3.1 Catalysts

The above scheme is able to steer the search process, but in a weak way. In order to improve steering and to make the search more beamed, enzymes that catalyse the reactions can be included. Enzymes decrease the activation energy necessary for the reaction, as depicted in Figure 1. They do so on both forward and reverse sides of the reaction, therefore the equilibrium concentrations do not change. However, as shown in [4], under some conditions, catalysts can focus the reaction network into a few species, creating a selection pressure towards a metabolic core. One of the conditions for obtaining such *catalytic focusing* is that the system is kept out of equilibrium by an inflow of food material.

Algorithm 2. Catalytic Search Algorithm

```
 1:  T: maximum number of iterations
 2:  0 ≤ t < T: current iteration
 3:  S: multiset of candidate solutions
 4:  C: pool of enzymes (catalysts) of maximum capacity C_max
 5:  initialization:
 6:  t = 0
 7:  S = random soup of N monomers m ∈ Σ
 8:  C = ∅
 9:  while t < T and solution not found do
10:      expel two random molecules e₁ and e₂ out of S
11:      (i₁, i₂) = random crossover points within e₁ and e₂
12:      (p₁, p₂) ← crossover(e₁, e₂, i₁, i₂)
13:      G_e = fitness(e₁) + fitness(e₂)
14:      G_p = fitness(p₁) + fitness(p₂)
15:      ΔG = G_p − G_e
16:      E_a = (|e₁| + |e₂|)/2
17:      if ΔG > 0 then
18:          E_a ← E_a + ΔG
19:      else if ΔG < 0 then
20:          c = "crossover(e₁, e₂, i₁, i₂)": the enzyme that catalyses this reaction
21:          n_c = multiplicity of c in C
22:          if n_c > 0 then
23:              E_a ← E_a/n_c
24:          end if
25:          p_c = |ΔG|/|ΔG_max|
26:          add another instance of c to C with probability p_c
27:          while |C| > C_max do
28:              destroy a random catalyst from C
29:          end while
30:      end if
31:      k_f = e^{−βE_a}
32:      if dice(k_f) then
33:          inject new products p₁ and p₂ into S
34:      else
35:          inject educts e₁ and e₂ back to S
36:      end if
37:      t ← t + 1
38:  end while
```

Here we introduce a simpler kind of catalyst which is not entirely faithful to chemistry, as it will work to reduce the activation energy barrier, but only in the direction of fitness improvement. We have temporarily adopted such annoying violation of the chemical laws because our first experiments have shown that maintaining the system out of equilibrium for such an optimization purpose is not such an easy task: in order to keep the system within a reasonable mass balance, an inflow of material (e.g. monomers) requires a corresponding outflow of other, potentially more complex solution molecules. If we remove such molecules at random, we might lose important partial solutions. Since the system is slow to replenish them, the optimization process is hindered. If we remove worse fit molecules with a higher probability, then the equilibrium is shifted towards the production of more of such bad molecules. If we do the opposite, i.e. remove the fitter molecules, then the system will tend to replenish them, but too slowly. Similar problems are reported in [22]. A good method for keeping the system out of equilibrium without disrupting the search process is still lacking. This topic deserves further investigation.

Our catalysts are strings of the form: "$op(s_1, s_2, p_1, p_2)$", where op is an operator (currently only the crossover operator is supported), s_1 and s_2 are the educt strings, $p1$ and p_2 are parameters indicating the crossover points in s_1 and s_2 respectively.

When two molecules collide, it is checked whether they have one or more matching catalysts. If matching catalysts are found, they will be used to increase the reaction probability, as explained below. Currently only exact match is supported. In the future, enzymes could bind to their substrates with a certain affinity, proportional to how well their strings match, for instance by using a distance metric such as the Hamming distance or a string alignment algorithm such as the edit distance.

The complete algorithm is shown in Algorithm 2. Enzymes are kept in a separate pool. When two molecules collide, if the reaction results in $\Delta G < 0$, i.e. in better fit products, then an enzyme might be created for this reaction, with a probability p_c proportional to the amount of improvement $|\Delta G|$. The next time similar molecules collide, the enzyme will facilitate their reaction, by lowering the corresponding ΔE_a, which then becomes:

$$\Delta E_a' = \frac{\Delta E_a}{n_c} \tag{8}$$

where ΔE_a is calculated according to Equation (4), and n_c is the concentration (multiplicity) of the corresponding catalyst in the catalyst pool.

3.2 Find the Hidden Sentence

We compare Catalytic Search with a pure random search (for the sake of sanity check only), and a variation of a Steady-State Genetic Algorithm (SSGA) based on a tournament selection mechanism implemented using an artificial chemistry. SSGA [16] is a non-generational evolutionary algorithm in which at each time step, individuals are selected for evaluation and reproduction, without a synchronized generational loop.

The three algorithms have been applied to the simple problem of finding a hidden string. In [22] Catalytic Search is applied to the OneMax problem, which consists in maximizing the number of ones in a binary string. This is a special case of finding a hidden string, i.e. a sentence Σ^+ made of a sequence of letters from an alphabet Σ. The length of the sentence can be variable, and the algorithm does not know anything about the nature of the solution. It is guided only by the fitness, given as the distance from the optimum. This problem has a smooth fitness landscape with a unique peak, and is therefore easy to optimize.

The fitness function for this problem is simply:

$$f(i) = d(i, i^*) \tag{9}$$

where $d(i, i^*)$ is the distance between the candidate solution i and the target sentence i^*. The function $d(i, j)$ is taken as the edit distance between the two strings, i.e. the minimum number of edit operations (add, delete, replace symbol) necessary to convert one string into the other. The best fitness value is thus the smallest distance, i.e. $d(i, j) = 0$.

3.3 Experimental Results

We have simulated the three algorithms on simple cases, and show a few preliminary results in this section. For each of the three algorithms three test cases have been performed, according to Table 1, where L is the length of the target solution i^*, and s is the size of the search space for each case, for sentences of length up to L.

Table 1 Test cases

case n.	alphabet Σ	target solution i^*	length $L = |i^*|$	search space s
0	ABCD	AABBCCDD	6	87380
1	01	1111111111111111	16	131070
2	a-z	thisisatest	11	3.81716e+15

For each algorithm, 100 runs were performed. The genetic algorithm was run with a tournament of size $r = 4$, a mutation probability of $p_m = 0.1$ and a crossover probability of $p_c = 0.9$. The population size was $N = 100$ molecules for all the algorithms and cases, and the maximum number of iterations was $T = 10000$.

Table 2 shows the number of exact solutions found, per test case and per algorithm.

Table 2 Number of exact solutions found, out of 100 runs per algorithm per test case

case n.	random search	catalytic search	genetic algorithm
0	0	38	75
1	1	0	100
2	0	0	3

As expected, the genetic algorithm is able to find a higher number of exact solutions due to its stronger replication and selective pressure. Also as expected, random search performed very poorly. Catalytic search was only able to find a significant amount of exact solutions on the first, easier case. The best solutions found in other cases were approaching the optimum but very slowly. We can see this in Fig. 2 (left), which shows the average best fitness for case 2, per algorithm. In Fig. 2 (top left) we can see that random search not only does not make progress, but diverges to worse solutions. The catalytic search (Fig. 2 (middle left)), although not entirely optimal, displays a qualitative behavior that is similar to the genetic algorithm, showing steady progress towards the optimum.

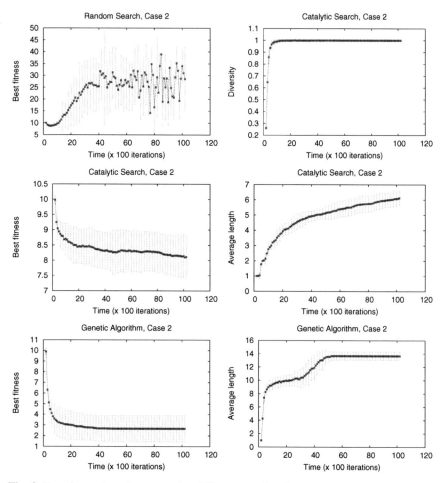

Fig. 2 Experimental results comparing different search schemes. Average values over 100 runs, with errorbars indicating the standard deviation. Left: Average best fitness for each algorithm. Top right: diversity of the population for the catalytic search. Middle right: Average length of the solutions in catalytic search. Bottom right: Average length for the GA

The diversity of the population has been measured using a multiset diversity metric [17]. It measures the fraction of unique elements (molecules) over the total size of the multiset (population size). Although it rises almost to the maximum for the catalytic search scheme (Fig. 2 top right), the system does not get "lost", and still displays a well-behaved search towards the solution. In comparison, the diversity in the genetic algorithm goes up at the beginning, and then drops to a middle level, as the system approaches the optimum (not shown).

The solutions in catalytic search show modest elongation towards the optimum length ($L = 11$ for case 2) (Fig. 2 middle right), while for the same case the genetic algorithm quickly moves towards solutions that are longer than the optimum (Fig. 2 bottom right). Catalytic search conserves the number of atoms, while the genetic algorithm produces an increasing number of atoms at the beginning, and this number then slowly drops and then stabilizes as the optimum is approached.

Similar qualitative behaviors have been observed for the other test cases from Table 1 (not shown).

4   Conclusions and Future Work

Catalytic search illustrates that optimization is possible even in the absence of explicit Darwinian selection. The selection force here is much weaker, progress is slower, and the systems not always converges to the optimum. Such a search method is inherently suboptimal, and not intended as a replacement for evolutionary algorithms or other successfully established heuristic search methods. As shown in models of pre-evolutionary dynamics [18], a prelife model with no established Darwinian evolution properties can be invaded as soon as self-replicants cross an efficiency threshold. In the optimization domain, catalytic search relates to prelife as genetic algorithms relate to Darwinian evolution. However, in the same way as prelife played a crucial role towards life, catalytic search can play a role as a "soft" search method, in a more exploratory phase of the search. It might prove useful in dynamic or noisy environments, to let a variety of solutions survive, to dampen temporary fluctuations in input parameters, and to undo or revert to past states when needed. We believe that there is a potential that remains to be explored in such soft search schemes, although we are not able to show this entire potential here to its full extent. We were able to show some properties such as an apparent ability to keep a higher diversity of solutions in the population without any explicit diversity maintenance mechanism. Other properties described in [22] remain to be demonstrated. We are particularly interested in the potential to undo wrong computations via reversible reactions, and to steer the flow of computation using an open system driven out of equilibrium as in [4]. This is difficult to achieve in a search algorithm, due to the risk of flushing out good solutions or their building blocks.

Many points remain to be improved in our current implementation: The catalysis model should support affinity matching. Catalysts should be inserted in the same pool together with the candidate solutions, and the reaction algorithm should model collisions involving catalysts and substrates explicitly. Furthermore, the saturation of enzymes must be considered, moving from mass action to enzyme kinetics. A more accurate diversity metric should be considered, taking into account the distance between strings. An analysis of the topology of catalytic networks should be undertaken, in order to detect potential autocatalytic sets, and search for emergent feedback loops and collective replicators. The main unsolved issue so far is to find a good way to keep the system out of equilibrium and yet in a focused optimizing mode.

Acknowledgements. This work has been supported by the European Union through FET Project BIONETS. The author would also like to thank Thomas Meyer, Wolfgang Banzhaf, and the anonymous reviewers for their helpful comments and encouragement.

References

[1] Activation Energy, Wikipedia (2006),
    http://en.wikipedia.org/wiki/Activation_energy
[2] Atkins, P., de Paula, J.: Physical Chemistry. Oxford University Press, Oxford (2002)
[3] Bagley, R., Farmer, J., Fontana, W.: Evolution of a Metabolism. In: Artificial Life II, pp. 141–158. Addison-Wesley, Reading (1991)
[4] Bagley, R.J., Farmer, J.: Spontaneous Emergence of a Metabolism. In: Artificial Life II, pp. 93–140. Addison-Wesley, Reading (1991)
[5] Banzhaf, W.: The "Molecular" Traveling Salesman. Biological Cybernetics 64, 7–14 (1990)
[6] Banzhaf, W., Lasarczyk, C.: Genetic Programming of an Algorithmic Chemistry. In: O'Reilly, et al. (eds.) Genetic Programming Theory and Practice II, vol. 8, ch.11. pp. 175–190. Kluwer/Springer (2004)
[7] Banzhaf, W., Dittrich, P., Rauhe, H.: Emergent Computation by Catalytic Reactions. Nanotechnology 7, 307–314 (1996)
[8] Deckard, A., Sauro, H.M.: Preliminary Studies on the In Silico Evolution of Biochemical Networks. ChemBioChem. 5(10), 1423–1431 (2004)
[9] Dittrich, P., Banzhaf, W.: Self-Evolution in a Constructive Binary String System. Artificial Life 4, 203–220 (1909)
[10] Dittrich, P., Ziegler, J., Banzhaf, W.: Artificial Chemistries – A Review. Artificial Life 7(3), 225–275 (2001)
[11] Fontana, W., Buss, L.W.: The Arrival of the Fittest: Toward a Theory of Biological Organization. Bulletin of Mathematical Biology 56, 1–64 (1994)
[12] Forrest, S.: Emergent Computation: Self-organizing, Collective, and Cooperative Phenomena in Natural and Artificial Computing Networks. Physica D 42(1-3), 1–11 (1990)
[13] Gánti, T.: Chemoton Theory, Volume 1: Theoretical Foundations of Fluid Machineries. Kluwer Academic, Dordrecht (2003)
[14] Kanada, Y.: Combinatorial Problem Solving Using Randomized Dynamic Composition of Production Rules. In: IEEE International Conference on Evolutionary Computation, pp. 467–472 (1995)
[15] Kauffman, S.A.: The Origins of Order: Self-Organization and Selection in Evolution. Oxford University Press, Oxford (1993)
[16] Lozano, M., Herrera, F., Cano, J.R.: Replacement Strategies to Preserve Useful Diversity in Steady-State Genetic Algorithms. Information Sciences 178(23), 4421–4433 (2008)

[17] Mattiussi, C., Waibel, M., Floreano, D.: Measures of Diversity for Populations and Distances Between Individuals with Highly Reorganizable Genomes. Evolutionary Computation 12(4), 495–515 (2004)
[18] Nowak, M.A., Ohtsuki, H.: Prevolutionary Dynamics and the Origin of Evolution. PNAS 105(39) (2008)
[19] Paun, G.: Computing with Membranes. Journal of Computer and System Sciences 61(1), 108–143 (2000)
[20] Piaseczny, W., Suzuki, H., Sawai, H.: Chemical Genetic Programming - Evolution of Amino Acid Rewriting Rules Used for Genotype-Phenotype Translation. In: Congress on Evolutionary Computation (CEC), vol. 2, pp. 1639–1646 (2004)
[21] Suzuki, Y., Fujiwara, Y., Takabayashi, J., Tanaka, H.: Artificial Life Applications of a Class of P Systems: Abstract Rewriting Systems on Multisets. In: Workshop on Multiset Processing (WMP), pp. 299–346. Springer, London (2001)
[22] Weeks, A., Stepney, S.: Artificial Catalysed Reaction Networks for Search. In: ECAL Workshop on Artificial Chemistry (2005)

Discovering Beneficial Cooperative Structures for the Automated Construction of Heuristics

Germán Terrazas, Dario Landa-Silva, and Natalio Krasnogor

Abstract. The current research trends on hyper-heuristics design have sprung up in two different flavours: heuristics that choose heuristics and heuristics that generate heuristics. In the latter, the goal is to develop a problem-domain independent strategy to automatically generate a good performing heuristic for specific problems, that is, the input to the algorithm are problems and the output are problem-tailored heuristics. This can be done, for example, by automatically selecting and combining different low-level heuristics into a problem specific and effective strategy. Thus, hyper-heuristics raise the level of generality on automated problem solving by attempting to select and/or generate tailored heuristics for the problem in hand. Some approaches like genetic programming have been proposed for this. In this paper, we report on an alternative methodology that sheds light on simple methodologies that efficiently cooperate by means of local interactions. These entities are seen as building blocks, the combination of which is employed for the automated manufacture of good performing heuristic search strategies. We present proof-of-concept results of applying this methodology to instances of the well-known symmetric TSP. The goal here is to demonstrate *feasibility* rather than compete with state of the art TSP solvers. This TSP is chosen only because it is an easy to state and well known problem.

1 Introduction

A *hyper-heuristic* is a search methodology that selects and combines heuristics to generate good solutions for a given problem. To investigate on the design of hyper-heuristics is important because they provide a problem independent level of abstraction for the automatic generation of good performing algorithms. Given a

Germán Terrazas · Dario Landa-Silva · Natalio Krasnogor
ASAP Group, School of Computer Science
University of Nottingham, UK
e-mail: {gzt,jds,nxk}@cs.nott.ac.uk

J.R. González et al. (Eds.): NICSO 2010, SCI 284, pp. 89–100, 2010.
springerlink.com   © Springer-Verlag Berlin Heidelberg 2010

computational search problem and a set of simpler heuristics, hyper-heuristics contribute with a methodology for the manufacture of heuristic capable of producing high quality solutions when applied to the problem in hand. We consider that developing a systematic procedure in which beneficial entities are identified and combined for the automated manufacture of good performing heuristics is a suitable approach. The purpose of this paper is then to propose a method for the automated construction of heuristic search strategies in terms of simpler heuristic building blocks which cooperate efficiently. Our methodology has three main stages: pattern-based heuristics generation, cross validation and template-based heuristics distilling. In the following, Section 2 gives a brief introduction to hyper-heuristics and the context of our investigation. Section 3 expands on the proposed approach giving details of the model components and the methodology. After that, experiments and results are presented and discussed in Section 4. Finally, conclusions and further work are the subject of Section 5.

2 Heuristics Design

Hyper-heuristics are defined as search methodologies that select and combine low-level heuristics to solve hard computational search problems [6, 16]. The general aim of a hyper-heuristic is to manufacture unknown heuristics which are fast, well performing and widely applicable to a range of problems. During the process of fabrication, hyper-heuristics receive feedback from the problem domain which indicates how good the chosen heuristics for solving the problem in hand, hence driving the search process. Hyper-heuristics do not violate the no-free-lunch theory which indicates that over all problems, no algorithm performs better than another. Studying novel approaches for the development of hyper-heuristics is important since they are domain-independent problem strategies that operate on a space of heuristics, rather than on a space of solutions, and rise the level of generality on automated problem solving. Hyper-heuristics have been employed for solving search and optimisation problems such as bin-packing [4, 17], timetabling [14], scheduling [8, 9] and satisfiability [2] among others. For detailed reviews of hyper-heuristics and their applications, please refer to [7, 13, 16].

The automated manufacture of heuristic search strategies by means of hyper-heuristics has received increasing attention in the last ten years or so. Recent investigations have sprung up in two main different directions of hyper-heuristics: 1) heuristics that choose heuristics and 2) heuristics that generate heuristics. In the first case, a learning mechanism assists the selection of low-level heuristics according to their historical performance during the search process, e.g. [8]. In the second case, the focus is on searching components that once combined generate a new heuristic suitable for the problem in hand. For example, approaches based on genetic algorithms [9] and genetic programming have been proposed for the automated generation of heuristics [5, 11]. From an engineering point of view, the already existent approaches are defined in terms of the architecture established by the underlying meta-heuristic which sometimes brings unsuspected difficulties such as the correct modelling of solutions or parameters tunning. Hence, the construction of well performing heuristics in terms of low-level heuristics which efficiently cooperate by means of local interactions is an interesting route for developing a new alternative within the second flavour of hyper-heuristics. Our interest lays on the identification of beneficial cooperative structures, the combination of which give rise to a specification for the automated manufacture of good performing heuristic strategies for a given combinatorial optimisation problem.

3 Proposed Approach

Given a set of instances of a combinatorial optimisation problem Π, we propose a methodology composed of pattern-based heuristics generation, cross validation and template-based heuristics distilling. Each stage is associated to a dataset generated from the optimisation problem in hand whilst the output of the methodology is a template to be employed for the manufacture of good performing heuristics. Fig. 1 depicts the methodology and its components.

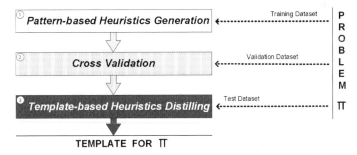

Fig. 1 Schematic representation of the proposed methodology with its three stages, their associated datasets and the achieved template for the problem in hand

In the *pattern-based heuristics generation*, an input dataset is employed to train randomly generated sequences of low-level heuristics (high-level heuristics). This training aims at generating proficient high-level heuristics, the common constituents of which are expected to produce high quality solutions when applied to a given instance of the problem in hand. The research question in this stage is:

Given a set of high-level heuristics, is it possible to generate common combinations of low-level heuristics ? If yes, how do they look like and how reliable are these combinations ?

In order to address the first question, a process that spots common combinations of low-level heuristics (patterns) and constructs pattern-based heuristic is employed. The goal of the *cross validation* is then to assess the performance of the constructed pattern-based heuristic over a validation dataset comprising similar instances of the combinatorial optimisation problem in hand. Thus, the question in this stage is:

What is the performance of a pattern-based heuristic when applied to a set of different problem instances ?

The goal of the *template-based heuristics distilling* stage is to discover cooperative and efficient low-level heuristics (building blocks) among several pattern-based heuristics. These building blocks are expected to give rise to a template from where better than average heuristics could be drawn. Here, an extra dataset is employed to test the performance of the constructed heuristics. The question in this stage is:

Is it possible to distill a template in terms of building blocks of heuristics ? If yes, how is the performance of the template-based heuristics when applied to a set of different problem instances ?

The above methodology is expected to deliver a procedure for the automated construction of effective and efficient heuristic search strategies.

4   Methods and Results

This section presents the findings obtained by the above methodology. The chosen combinatorial optimisation problem is the widely known symmetric Traveling Salesman Problem (TSP). The TSP instance considered here is kroA100 which comprises 100 cities distributed in the Euclidean space. The objective value corresponding to the known optimum solution (shortest tour) for this instance is 21282 (see TSPLIB[1]). For each stage of our methodology, we generated five sets in the following systematic way. Each set is initialised with ten copies of the known optimum solution for kroA100. Each of this initial solutions is then 'disturbed' with n consecutive city swaps. In this way, setting n to 5, 25, 50, 75 and 100, a total of ten independently 'disturbed' tours per set are obtained.

We consider a high-level heuristic as a sequence made of low-level heuristics. The low-level heuristics for the TSP used here can be divided in two types: stochastic low-level heuristics and deterministic low-level heuristics. A low-level heuristic is stochastic if different or the same output tours are returned when applied to the same input tour. Contrary to this, a low-level heuristic is deterministic if the same output tour is returned when applied to the same input tour. In our case, *1-city insertion*, *2-exchange*, *arbitrary insertion* and *inver-over* are the stochastic low-level heuristics, whilst *2-opt*, *3-opt*, *OR-opt* and *node insertion* are the deterministic ones. These eight low-level heuristics were implemented as defined in [1, 3, 10, 15, 19] and operating in a hill climber style [12].

---

[1] http://elib.zib.de/pub/mp-testdata/tsp/tsplib/tsplib.html

4.1   Pattern-Based Heuristics Generation

4.1.1   Training Datasets

In this stage, each of the perturbed tours, labeled as $tkroA100_i^n$, $i = 0...9$, $n = 5, 25, 50, 75, 100$, is independently considered for training. A sample of the training data, grouped by set (n), is listed in Table 1 where the values indicate the percentage distance to the optimum from each perturbed tours.

Table 1 Three sample perturbed tours for each of the five training sets

Tour	$n = 5$	$n = 25$	$n = 50$	$n = 75$	$n = 100$
$tkroA100_0^n$	1.42669	4.25805	6.39869	7.01362	6.80147
$tkroA100_1^n$	1.27600	4.60262	6.46067	6.38215	6.59012
$tkroA100_2^n$	1.79926	4.13631	5.76585	6.75190	6.93252

$tkroA100_i^n$ is the i-th disturbed tour after applying n random swaps to $kroA100$ optimal tour.

4.1.2   Method

For a given disturbed tour ($tkroA100_i^n$), a set containing 500 high-level heuristics generated at random was created. Then, each of the 500 high-level heuristics was independently applied to the associated perturbed tour. In this context, an application is seen as a pipeline process in which the chain of processing elements is given by the sequence of low-level heuristics and the information to be processed is the disturbed tour. Thus, the low-level heuristics are applied one after another in the order in which they appear in the sequence and producing better or equal solutions at each step. To illustrate this process, Fig. 2 depicts how a high-level heuristic comprising 1-city insertion and 2-exchange is applied to a TSP instance.

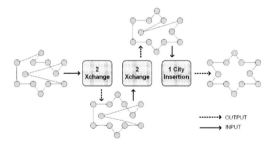

Fig. 2 A high-level heuristic in which successive applications of 1-city insertion and 2-exchange find the optimum solution for the Star of David tour

In order to identify common combinations of low-level heuristics, the 500 high-level heuristics are then sorted according to the distance between the solution that their applications produce and the known optimum solution. The top five high-level heuristics are then selected and encoded as sequences of characters using 'A' to represent 1-city insertion, 'C' to represent 2-opt, 'D' to represent 3-opt, 'E' to represent OR-opt, 'T' to represent 2-exchange, 'F' to represent node insertion, 'G' to represent arbitrary insertion and 'H' to represent inver-over. Hence, in order to identify common combinations of low-level heuristics among the filtered sequence, we employ a multiple sequence alignment (MSA) method [18] over the encodings. For instance, Fig. 3 highlights in gray the common combinations found among the best five high-level heuristics generated for $tkroA100_2^{75}$.

```
1-HLH⁷⁵₃   ehhG--thHG------HHtG---------thhD---DhDh
2-HLH⁷⁵₃   ----aD---GdadtdtH---------------DddCD-D-
3-HLH⁷⁵₃   -----D--HG------H--Gaccchccaca--D--C----
4-HLH⁷⁵₃   ---G----HG------HH-G------------D-------
5-HLH⁷⁵₃   ---Ge----G------H--G------------D---D---
           |  |   ||        ||  |           |   || |
           |  |   ||        ||  |           |   || |
PBH⁷⁵₃     ---G-D--HG------HH-G------------D--CD-D-
```

Fig. 3 Multiple sequence alignment of the top five heuristics. Capitals highlighted in gray indicate the common sequences of heuristics

The results obtained by the MSA method reveal that there are indeed occurrences of common combinations, i.e. patterns of low-level heuristics, among the best ranked high-level heuristics. Thus, these findings give a positive answer to the research question stated for the first part of our methodology in Section 3.

From the resulting alignment, we construct a consensus sequence capturing and representing regions of similarity. We define this consensus sequence as a pattern-based heuristic (PBH_i^n) associated to a perturbed tour ($tkroA100_i^n$). The constructing procedure consists in copying the matching characters between two or more encodings into a new sequence from left to right and following the position in which they appear. For instance, Fig. 3 shows that PBH_3^{75} is the resulting pattern-based heuristic encoded as GDHGHHGDCDD, after combining the common patterns from the high-level heuristics 1-HLH_3^{75} to 5-HLH_3^{75}. Given that this new heuristic is built in terms of common combinations of low-level heuristics, its performance is then expected to be as good as (or better than) any of the top ranked. Notice that the length of the constructed heuristic varies according to the number of matches. Since this is related to the way in which the construction procedure is defined, alternative methodologies to obtain the optimal common sequence are open to further investigation.

In order to assess the reliability of the spotted patterns, we then proceed to evaluate the performance of PBH_i^n against a set of high-level heuristics (different than the initial ones) with the hope that, on average, the best tour improvements are obtained by the former. In order to do this, 300 copies of PBH_i^n are obtained and for each of them a new high-level heuristic equal in length is created. Each of these heuristics is then independently applied to $tkroA100_i^n$ a total of 10 times and the average percentage distance between the lengths of the resulting tours and the known optimum

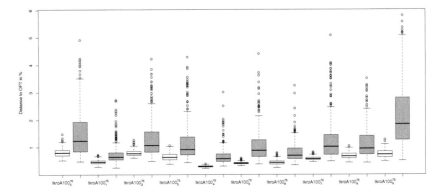

Fig. 4 Assessment of ten pattern-based heuristics resulting from independent sequence alignments. Each pair of boxplots summarises a vis-a-vis comparison between the performance of 300 copies of PBH_i^{75} and the performance of other 300 high-level heuristics when applied to $tkroA100_i^{75}$ for $i = 0 \ldots 9$

is considered as the measure of their performance. As an example, Fig. 4 shows the assessment of the 10 pattern-based heuristics obtained from the data set generated with $n = 75$.

According to the results, it is clear that the performance of pattern-based heuristics (white boxplots) is better in average than the performance of the non-pattern-based high-level heuristics (gray boxplots). These findings constitute a positive answer to the second research question stated in the first stage of the presented methodology, i.e. the identified common-sequences of heuristics are indeed reliable.

4.2 Cross Validation

4.2.1 Validation Dataset

The cross validation data are given in sets of ten perturbed tours $vkroA100_i^n$, $i = 0 \ldots 9$. A sample of the data, grouped by set (n), is listed in Table 2 where the values indicate the percentage distance to the optimum from each perturbed tours.

Table 2 Three sample perturbed tours for each of the five validation sets

Tour	$n = 5$	$n = 25$	$n = 50$	$n = 75$	$n = 100$
$vkroA100_0^n$	1.86490	5.38403	6.85800	6.92453	7.58471
$vkroA100_1^n$	1.72394	5.42246	6.13800	6.57452	6.69500
$vkroA100_2^n$	1.41001	3.76134	6.66469	6.85969	6.90264

$vkroA100_i^n$ is the i-th disturbed tour after applying n random swaps to $kroA100$ optimal tour.

4.2.2   Method

The goal of this stage is to perform a cross validation analysis in order to assess
the performance of the pattern-based heuristics over a set of disturbed tours. Thus,
for each combination of PBH_j^n and $vkroA100_i^n$, $i, j = 0 \ldots 9$, a total of 300 copies of
PBH_j^n were obtained and, for each of the copies, a new high-level heuristic equal
in length was created. Then, the heuristics are independently applied to the given
$vkroA100_i^n$ a total of 10 independent times and the average percentage distance between the lengths of the resulting tours and the known optimum is considered as the
measure of their performance. Fig. 5 shows the resulting assessment of a pattern-
based heuristic, encoded as GDHGHHGDCDD, over the 10 perturbed tours belonging to the data set generated with $n = 75$.

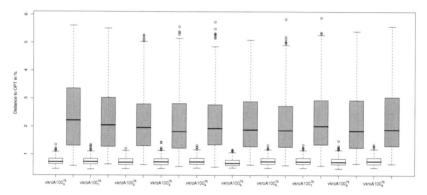

Fig. 5 Performance evaluation of a pattern-based heuristic across the perturbed tours belonging to the data set generated with $n = 75$. Each pair of boxplots summarises a vis-a-vis comparison between the performance of 300 copies of GDHGHHGDCDD and the performance
of other 300 high-level heuristics when applied to $vkroA100_i^{75}$

Clearly, the performances of pattern-based heuristics (white boxplots) are better in average than the performance of the ones generated for assessment (gray
boxplots). These findings answer the research question estated in the second part
of Section 3, revealing that a pattern-based heuristic is in general well performing
when applied to a set of different problem instances. In addition, the similar level of
performance observed among the white boxplots gives an indication that common
low-level heuristics could be acting as building blocks among the PBH_j^n, $j = 1 \ldots 10$.

4.3   Template-Based Heuristics Distilling

4.3.1   Test Dataset

The data used in this last stage comprise five sets, see Table 3 for a sample. Thus, for
a given experiment, each of the ten perturbed tours $dkroA100_i^n$, $i = 0 \ldots 9$, belonging
to a given set is independently employed for testing.

Table 3 Three sample perturbed tours for each of the five test sets

Tour	$n = 5$	$n = 25$	$n = 50$	$n = 75$	$n = 100$
$dkroA100_0^n$	1.43750	4.00032	5.61831	6.34000	6.86100
$dkroA100_1^n$	1.12729	4.70731	6.44469	6.28794	6.69199
$dkroA100_2^n$	0.80584	4.01320	5.96786	6.57973	7.10008

$dkroA100_i^n$ is the i-th disturbed tour after applying n random swaps to $kroA100$ optimal tour.

4.3.2 Method

The purpose of this stage is to identify common building blocks of low-level heuristics among the PBH_j^n assessed in the second part of our methodology. These building blocks are employed to construct templates of heuristics, the instances of which are expected to show similar or better performance when solving any $dkroA100_i^n$. Hence, for each data set, we applied the MSA method over the encodings of PBH_j^n, $i, j = 0 \ldots 9$. For example, Fig. 6 highlights in gray building blocks among the ten pattern-based heuristics found for the data set of perturbed tours generated with $n = 75$.

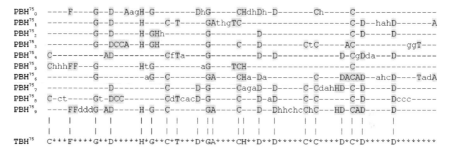

Fig. 6 Multiple sequence alignment of the pattern-based heuristics found for the data set generated with $n = 75$. Capitals highlighted in gray indicate the spotted common building blocks

The resulting alignment reveals that there are common structures among the pattern-based heuristics. A template (TBH^n) is then constructed in terms of building blocks. This procedure consists in copying the matching characters between three or more encodings into a new sequence from left to right and following the position in which they appear. In case no matchings are found or matchings occur only between two encodings, a wildcard character is placed in that position of the sequence. For instance, Fig. 6 shows TBH^{75} as the resulting template after combining the building blocks from the input pattern-based heuristics PBE_0^{75} to PBE_9^{75}.

For each $dkroA100_i^n$, a total of 300 different instances are drawn from the constructed template. During the instantiation process, building blocks are preserved and each of the wildcard characters is either removed or replaced with one of the

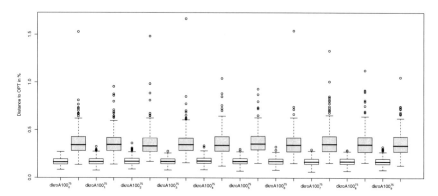

Fig. 7 Assessment of a template-based heuristic across a set of perturbed tours belonging to the data set generated with $n = 75$. Each pair of boxplots summarises a vis-a-vis comparison between the performance of 300 instances drawn from TBH^{75} and the performance of other 300 high-level heuristics when applied to $dkroA100_i^{75}$ for $i = 0 \ldots 9$

eight low-level heuristics chosen at random. In order to assess the reliability of the building blocks, we compared the performance of the 300 instances against new 300 high-level heuristics expecting that, on average, the best tour improvements are obtained by the former. In this way, each of the heuristics is applied to the same perturbed tour a total of 10 independent times and the average distance between the lengths of the resulting tours and the known optimum is considered as the measure of performance. A representative outcome of the assessment is shown in Fig. 7 where the resulting assessment of the instances drawn from TBH^{75} when applied to the data set created with $n = 75$ is depicted.

The results of this stage demonstrate that it is certainly possible to define a template of building blocks of heuristics in terms of common structures identified among a set of pattern-based heuristics. This fact constitutes a positive answer for the first question established in the third part of our methodology. In addition, it is also shown that the performance of template-based heuristics (white boxplots) is on average better than the performance of the randomly generated high-level heuristics (gray boxplots), even though some of the high-level heuristics generated for comparison have outperformed the ones drawn from the template (see Fig. 7). Naturally, one of the reasons for this is that during the random generation, appropriate combinations of low-level heuristics with more efficient local interactions could be generated (by chance). However, the template-based specification still brings a more robust and convenient way for the automated manufacture of good performing heuristic strategies to solve the problem in hand. All in all, the outcome of this assessment answers the last question of the proposed methodology. That is, the instances of such templates are always well performing when applied to any disturbed tour of a given data set.

5 Conclusions

In this paper, we proposed a novel approach for the automated design of heuristics following the rationale of hyper-heuristics which are heuristic methods to generate tailored heuristics for the problem in hand.

The proposed methodology consists of pattern-based heuristics construction, cross validation and template-based heuristics distilling. As a proof of concept, we applied the methodology to instances of the symmetric TSP. On the one hand, our initial findings confirm that there are indeed common patterns of low-level heuristics among the top ranked high-level heuristics. These emergent recurrent structures were subject to a cross validation, the results of which proved them to be local search strategies beneficial to achieve good solutions when solving a symmetric TSP instance. On the other hand, the outcome achieved in the last part of our approach has resulted in a specification to automatically generate a family of heuristics capable of producing high quality solutions when applied to perturbed tours. In particular, these high performing heuristics are made of emergent building blocks extracted from the patterns seen in the first stage.

From a functional point of view, the building blocks achieved in the last stage are beneficial structures needed for the manufacturing of high quality solutions. When these key elements appear in combination with randomly chosen low-level heuristics, they seem to guide the search across the space of solutions. In other words, the local interactions contributed by the building blocks can be seen as artifacts that drive the optimisation process when applied to the combinatorial optimisation problem in hand. Likewise, the local interactions contributed by the randomly created low-level heuristics placed in an instance can be seen as artifacts that contribute with a variety of alternative paths for exploring the space of solutions during the optimisation process. Hence, both types of contributions seem to be properly orchestrated into an instance of a template-based heuristic.

To continue with our methodology, future work involves the extension of our approach to other instances of TSP as well as to different combinatorial optimisation problems. In addition, we also consider to explore alternative ways to generate the family of good performing heuristics in order to get a faster and less human-dependent way. This could be done for instance by means of grammatical inference where the encodings of the pattern-based heuristics would be the input to the grammatical inference algorithm and the resulting grammar would be employed to generate a family of words encoding sequences of low-level heuristics.

References

[1] Babin, G., Deneault, S., Laporte, G.: Improvements to the or-opt heuristic for the symmetric traveling salesman problem. Journal of the Operational Research Society (58), 402–407 (2007)
[2] Bader-El-Den, M., Poli, R.: A gp-based hyper-heuristic framework for evolving 3-sat heuristics. In: Genetic and Evolutionary Computation Conference, p. 1749. ACM, New York (2007)

[3] Brest, J., Zerovnik, J.: A heuristic for the asymmetric traveling salesman problem. In: 6th Metaheuristics International Conference, pp. 145–150 (2005)

[4] Burke, E., Hyde, M., Kendall, G.: Evolving bin packing heuristics with genetic. In: Runarsson, T.P., Beyer, H.-G., Burke, E.K., Merelo-Guervós, J.J., Whitley, L.D., Yao, X. (eds.) PPSN 2006. LNCS, vol. 4193, pp. 860–869. Springer, Heidelberg (2006)

[5] Burke, E., Hyde, M., Kendall, G., Woodward, J.: Automatic heuristic generation with genetic programming: evolving a jack-of-all-trades or a master of one. In: Genetic and Evolutionary Computation Conference, pp. 1559–1565. ACM, New York (2007)

[6] Burke, E.K., Hart, E., Kendall, G.N., Newall, J., Ross, P., Schulenburg, S.: Handbook of Meta-Heuristics. In: chap Hyper-Heuristics: An Emerging Direction in Modern Search Technology, pp. 457–474. Kluwer, Dordrecht (2003)

[7] Chakhlevitch, K., Cowling, P.: Hyperheuristics: Recent developments. In: Adaptive and Multilevel Metaheuristics, vol. 136, pp. 3–29. Springer, Heidelberg (2008)

[8] Cowling, P., Chakhlevitch, K.: Hyperheuristics for managing a large collection of low level heuristics to schedule personnel. In: IEEE Congress on Evolutionary Computation, pp. 1214–1221. IEEE Computer Society, Los Alamitos (2003)

[9] Cowling, P., Kendall, G., Han, L.: An investigation of a hyperheuristic genetic algorithm applied to a trainer scheduling problem. In: IEEE Congress on Evolutionary Computation, pp. 1185–1190. IEEE Computer Society, Los Alamitos (2002)

[10] Krasnogor, N., Smith, J.: Memetic algorithms: The polynomial local search complexity theory perspective. Journal of Mathematical Modelling and Algorithms 7, 3–24 (2008)

[11] Oltean, M., Dumitrescu, D.: Evolving tsp heuristics using multi expression programming. In: Bubak, M., van Albada, G.D., Sloot, P.M.A., Dongarra, J. (eds.) ICCS 2004. LNCS, vol. 3037, pp. 670–673. Springer, Heidelberg (2004)

[12] Özcan, E., Bilgin, B., Korkmaz, E.: Hill climbers and mutational heuristics in hyperheuristics. In: 9th International Conference on PPSN, pp. 202–211 (2006)

[13] Özcan, E., Bilgin, B., Korkmaz, E.: A comprehensive analysis of hyper-heuristics. Intell. Data Anal. 12(1), 3–23 (2008)

[14] Pillay, N., Banzhaf, W.: A study of heuristic combinations for hyper-heuristic systems for the uncapacitated examination timetabling problem. European Journal of Operational Research 197(2), 482–491 (2009)

[15] Reinelt, G.: The traveling salesman: Computational solutions for TSP applications. Springer, Heidelberg (1994)

[16] Ross, P.: Hyper-heuristics. In: Search Methodologies: Introductory Tutorials in Optimization and Decision Support, pp. 529–556. Springer, Heidelberg (2005)

[17] Ross, P., Schulenburg, S., Marín-Blázquez, J., Hart, E.: Hyper-heuristics: Learning to combine simple heuristics in bin-packing problems. In: Genetic and Evolutionary Computation Conference, pp. 942–948. Morgan Kaufmann Publishers Inc., San Francisco (2002)

[18] Setubal, J., Meidanis, J.: Introduction to Computational Molecular Biology. PWS Publishing (1997)

[19] Tao, G., Michalewicz, Z.: Inver-over operator for the tsp. In: Eiben, A.E., Bäck, T., Schoenauer, M., Schwefel, H.-P. (eds.) PPSN 1998. LNCS, vol. 1498, pp. 803–812. Springer, Heidelberg (1998)

Eagle Strategy Using Lévy Walk and Firefly Algorithms for Stochastic Optimization

Xin-She Yang* and Suash Deb

Abstract. Most global optimization problems are nonlinear and thus difficult to solve, and they become even more challenging when uncertainties are present in objective functions and constraints. This paper provides a new two-stage hybrid search method, called Eagle Strategy, for stochastic optimization. This strategy intends to combine the random search using Lévy walk with the firefly algorithm in an iterative manner. Numerical studies and results suggest that the proposed Eagle Strategy is very efficient for stochastic optimization. Finally practical implications and potential topics for further research will be discussed.

1 Introduction

To find the solutions to any optimization problems, we can use either conventional optimization algorithms such as the Hill-climbing and simplex method, or heuristic methods such as genetic algorithms, or their proper combinations. Modern metaheuristic algorithms are becoming powerful in solving global optimization problems [4, 6, 7, 9, 13, 20, 21], especially for the NP-hard problems such as the travelling salesman problem. For example, particle swarm optimization (PSO) was developed by Kennedy and Eberhart in 1995 [8, 9], based on the swarm behaviour such as fish and bird schooling in nature. It has now been applied to find solutions for many optimization applications. Another example is the Firefly Algorithm developed by the first author [13, 20] which has demonstrated promising superiority

Xin-She Yang
Department of Engineering, University of Cambridge, Trumpinton Street,
Cambridge CB2 1PZ, UK
e-mail: xy227@cam.ac.uk

Suash Deb
Department of Computer Science & Engineering, C.V. Raman College of Engineering,
Bidyanagar, Mahura, Janla, Bhubaneswar 752054, India
e-mail: suashdeb@gmail.com

* Corresponding author.

J.R. González et al. (Eds.): NICSO 2010, SCI 284, pp. 101–111, 2010.
springerlink.com   © Springer-Verlag Berlin Heidelberg 2010

over many other algorithms. The search strategies in these multi-agent algorithms are controlled randomization and exploitation of the best solutions. However, such randomization typically uses a uniform distribution or Gaussian distribution. In fact, since the development of PSO, quite a few algorithms have been developed and they can outperform PSO in different ways [20, 22].

On the other hand, there is always some uncertainty and noise associated with all real-world optimization problems. Subsequently, objective functions may have noise and constraints may also have random noise. In this case, a standard optimization problem becomes a stochastic optimization problem. Methods that work well for standard optimization problems cannot directly be applied to stochastic optimization; otherwise, the obtained results are incorrect or even meaningless. Either the optimization problems have to be reformulated properly or the optimization algorithms should be modified accordingly, though in most cases we have to do both [3, 10, 19].

In this paper, we intend to formulate a new metaheuristic search method, called Eagle Stategy (ES), which combines the Lévy walk search with the Firefly Algorithm (FA). We will provide the comparison study of the ES with PSO and other relevant algorithms. We will first outline the basic ideas of the Eagle Strategy, then outline the essence of the firefly algorithm, and finally carry out the comparison about the performance of these algorithms.

2 Stochastic Multiobjective Optimization

An ordinary optimization problem, without noise or uncertainty, can be written as

$$\min_{\mathbf{x} \in \Re^d} f_i(\mathbf{x}), \ (i = 1, 2, ..., N) \tag{1}$$

subject to $\phi_j(\mathbf{x}) = 0, \ (j = 1, 2, ..., J),$

$$\psi_k(\mathbf{x}) \leq 0, \ (k = 1, 2, ..., K), \tag{2}$$

where $\mathbf{x} = (x_1, x_2, ..., x_d)^T$ is the vector of design variables.

For stochastic optimization problems, the effect of uncertainty or noise on the design variable x_i can be described by a random variable ξ_i with a distribution Q_i [10, 19]. That is

$$x_i \mapsto \xi_i(x_i), \tag{3}$$

and

$$\xi_i \sim Q_i. \tag{4}$$

The most widely used distribution is the Gaussian or normal distribution $N(x_i, \sigma_i)$ with a mean x_i and a known standard deviation σ_i. Consequently, the objective functions f_i become random variables $f_i(\mathbf{x}, \xi)$.

Now we have to reformulate the optimization problem as the minimization of the mean of the objective function $f_i(\mathbf{x})$ or μ_{f_i}

$$\min_{\mathbf{x}\in\Re^d}\{\mu_{f_1},...,\mu_{f_N}\}. \tag{5}$$

Here $\mu_{f_i} = E(f_i)$ is the mean or expectation of $f_i(\xi(\mathbf{x}))$ where $i = 1,2,...,N$. More generally, we can also include their uncertainties, which leads to the minimization of

$$\min_{\mathbf{x}\in\Re^d}\{\mu_{f_1}+\lambda\sigma_1,...,\mu_{f_N}+\lambda\sigma_N\}, \tag{6}$$

where $\lambda \geq 0$ is a constant. In addition, the constraints with uncertainty should be modified accordingly.

In order to estimate μ_{f_i}, we have to use some sampling techniques such as the Monte Carlo method. Once we have randomly drawn the samples, we have

$$\mu_{f_i} \approx \frac{1}{N_i}\sum_{p=1}^{N_i} f_i(\mathbf{x},\xi^{(p)}), \tag{7}$$

where N_i is the number of samples.

3 Eagle Strategy

The foraging behaviour of eagles such as golden eagles or *Aquila Chrysaetos* is inspiring. An eagle forages in its own territory by flying freely in a random manner much like the Lévy flights. Once the prey is sighted, the eagle will change its search strategy to an intensive chasing tactics so as to catch the prey as efficiently as possible. There are two important components to an eagle's hunting strategy: random search by Lévy flight (or walk) and intensive chase by locking its aim on the target.

Furthermore, various studies have shown that flight behaviour of many animals and insects has demonstrated the typical characteristics of Lévy flights [5, 12–14]. A recent study by Reynolds and Frye shows that fruit flies or *Drosophila melanogaster*, explore their landscape using a series of straight flight paths punctuated by a sudden 90^0 turn, leading to a Lévy-flight-style intermittent scale-free search pattern. Studies on human behaviour such as the Ju/'hoansi hunter-gatherer foraging patterns also show the typical feature of Lévy flights. Even light can be related to Lévy flights [2]. Subsequently, such behaviour has been applied to optimization and optimal search, and preliminary results show its promising capability [12, 14, 16, 17].

3.1 Eagle Strategy

Now let us idealize the two-stage strategy of an eagle's foraging behaviour. Firstly, we assume that an eagle will perform the Lévy walk in the whole domain. Once it finds a prey it changes to a chase strategy. Secondly, the chase strategy can be considered as an intensive local search using any optimization technique such as the steepest descent method, or the downhill simplex or Nelder-Mead method [11]. Obviously, we can also use any efficient metaheuristic algorithms such as the particle swarm optimization (PSO) and the Firefly Algorithm (FA) to do concentrated local search. The pseudo code of the proposed eagle strategy is outlined in Fig. 1.

The size of the hypersphere depends on the landscape of the objective functions. If the objective functions are unimodal, then the size of the hypersphere can be about the same size of the domain. The global optimum can in principle be found from any initial guess. If the objective are multimodal, then the size of the hypersphere should be the typical size of the local modes. In reality, we do not know much about the landscape of the objective functions before we do the optimization, and we can either start from a larger domain and shrink it down or use a smaller size and then gradually expand it.

On the surface, the eagle strategy has some similarity with the random-restart hill climbing method, but there are two important differences. Firstly, ES is a two-stage strategy rather than a simple iterative method, and thus ES intends to combine a good randomization (diversification) technique of global search with an intensive and efficient local search method. Secondly, ES uses Lévy walk rather than simple randomization, which means that the global search space can be explored more efficiently. In fact, studies show that Lévy walk is far more efficient than simple random-walk exploration.

Eagle Strategy

Objective functions $f_1(\mathbf{x}), ..., f_N(\mathbf{x})$
   Initial guess $\mathbf{x}^{t=0}$
   while *($||\mathbf{x}^{t+1} - \mathbf{x}^t|| > tolerance$)*
   Random search by performing Lévy walk
   Evaluate the objective functions
   Intensive local search with a hypersphere
      via Nelder-Mead or the Firefly Algorithm
     if *(a better solution is found)*
      Update the current best
     end if
     Update $t = t + 1$
     Calculate means and standard deviations
   end while
Postprocess results and visualization

Fig. 1 Pseudo code of the Eagle Strategy (ES)

The Lévy walk has a random step length being drawn from a Lévy distribution

$$\text{Lévy} \sim u = t^{-\lambda}, \quad (1 < \lambda \leq 3), \tag{8}$$

which has an infinite variance with an infinite mean. Here the steps of the eagle motion is essentially a random walk process with a power-law step-length distribution with a heavy tail. The special case $\lambda = 3$ corresponds to Brownian motion, while

$\lambda = 1$ has a characteristics of stochastic tunneling, which may be more efficient in avoiding being trapped in local optima.

For the local search, we can use any efficient optimization algorithm such as the downhill simplex (Nelder-Mead) or metaheuristic algorithms such as PSO and the firefly algorithm. In this paper, we used the firefly algorithm to do the local search, since the firefly algorithm was designed to solve multimodal global optimization problems [20].

3.2 Firefly Algorithm

We now briefly outline the main components of the Firefly Algorithm developed by the first author [13], inspired by the flash pattern and characteristics of fireflies. For simplicity in describing the algorithm, we now use the following three idealized rules: 1) all fireflies are unisex so that one firefly will be attracted to other fireflies regardless of their sex; 2) Attractiveness is proportional to their brightness, thus for any two flashing fireflies, the less brighter one will move towards the brighter one. The attractiveness is proportional to the brightness and they both decrease as their distance increases. If there is no brighter one than a particular firefly, it will move randomly; 3) The brightness of a firefly is affected or determined by the landscape of the objective function. For a maximization problem, the brightness can simply be proportional to the value of the objective functions.

Firefly Algorithm

Objective function $f_p(\mathbf{x})$, $\mathbf{x} = (x_1, ..., x_d)^T$
Initial population of fireflies $\mathbf{x}_i$ $(i = 1, ..., n)$
Light intensity I_i at $\mathbf{x}_i$ is determined by $f_p(\mathbf{x}_i)$
Define light absorption coefficient γ
while *(t <MaxGeneration)*
  for *i = 1 : n all n fireflies*
    for *j = 1 : i all n fireflies*
    if *($I_j > I_i$)*
       Move firefly i towards j (d-dimension)
    end if
    Vary β via $\exp[-\gamma r]$
    Evaluate new solutions and update
    end for *j*
  end for *i*
  Rank the fireflies and find the current best
end while
Postprocess results and visualization

Fig. 2 Pseudo code of the firefly algorithm (FA)

In the firefly algorithm, there are two important issues: the variation of light intensity and formulation of the attractiveness. For simplicity, we can always assume that the attractiveness of a firefly is determined by its brightness which in turn is associated with the encoded objective function.

In the simplest case for maximum optimization problems, the brightness I of a firefly at a particular location $\mathbf{x}$ can be chosen as $I(\mathbf{x}) \propto f(\mathbf{x})$. However, the attractiveness β is relative, it should be seen in the eyes of the beholder or judged by the other fireflies. Thus, it will vary with the distance r_{ij} between firefly i and firefly j. In addition, light intensity decreases with the distance from its source, and light is also absorbed in the media, so we should allow the attractiveness to vary with the degree of absorption. In the simplest form, the light intensity $I(r)$ varies according to the inverse square law $I(r) = \frac{I_s}{r^2}$ where I_s is the intensity at the source. For a given medium with a fixed light absorption coefficient γ, the light intensity I varies with the distance r. That is

$$I = I_0 e^{-\gamma r}, \tag{9}$$

where I_0 is the original light intensity.

As a firefly's attractiveness is proportional to the light intensity seen by adjacent fireflies, we can now define the attractiveness β of a firefly by

$$\beta = \beta_0 e^{-\gamma r^2}, \tag{10}$$

where β_0 is the attractiveness at $r = 0$.

The distance between any two fireflies i and j at $\mathbf{x}_i$ and $\mathbf{x}_j$, respectively, is the Cartesian distance

$$r_{ij} = ||\mathbf{x}_i - \mathbf{x}_j|| = \sqrt{\sum_{k=1}^{d} (x_{i,k} - x_{j,k})^2}, \tag{11}$$

where $x_{i,k}$ is the kth component of the spatial coordinate $\mathbf{x}_i$ of ith firefly. In the 2-D case, we have

$$r_{ij} = \sqrt{(x_i - x_j)^2 + (y_i - y_j)^2}. \tag{12}$$

The movement of a firefly i is attracted to another more attractive (brighter) firefly j is determined by

$$\mathbf{x}_i = \mathbf{x}_i + \beta_0 e^{-\gamma r_{ij}^2}(\mathbf{x}_j - \mathbf{x}_i) + \alpha \left(\text{rand} - \frac{1}{2}\right), \tag{13}$$

where the second term is due to the attraction. The third term is randomization with a control parameter α, which makes the exploration of the search space more efficient.

We have tried to use different values of the parameters α, β_0, γ [13, 20], after some simulations, we concluded that we can use $\beta_0 = 1$, $\alpha \in [0,1]$, $\gamma = 1$, and $\lambda = 1$ for most applications. In addition, if the scales vary significantly in different dimensions such as -10^5 to 10^5 in one dimension while, say, -0.001 to 0.01 along the other, it is a good idea to replace α by αS_k where the scaling parameters

$S_k (k = 1, ..., d)$ in the d dimensions should be determined by the actual scales of the problem of interest.

There are two important limiting cases when $\gamma \to 0$ and $\gamma \to \infty$. For $\gamma \to 0$, the attractiveness is constant $\beta = \beta_0$ and the length scale $\Gamma = 1/\sqrt{\gamma} \to \infty$, this is equivalent to say that the light intensity does not decrease in an idealized sky. Thus, a flashing firefly can be seen anywhere in the domain. Thus, a single (usually global) optimum can easily be reached. This corresponds to a special case of particle swarm optimization (PSO) discussed earlier. Subsequently, the efficiency of this special case could be about the same as that of PSO.

On the other hand, the limiting case $\gamma \to \infty$ leads to $\Gamma \to 0$ and $\beta(r) \to \delta(r)$ (the Dirac delta function), which means that the attractiveness is almost zero in the sight of other fireflies or the fireflies are short-sighted. This is equivalent to the case where the fireflies roam in a very foggy region randomly. No other fireflies can be seen, and each firefly roams in a completely random way. Therefore, this corresponds to the completely random search method. As the firefly algorithm is usually in somewhere between these two extremes, it is possible to adjust the parameter γ and α so that it can outperform both the random search and PSO.

4 Simulations and Comparison

4.1 Validation

In order to validate the proposed algorithm, we have implemented it in Matlab. In our simulations, the values of the parameters are $\alpha = 0.2$, $\gamma = 1$, $\lambda = 1$, and $\beta_0 = 1$. As an example, we now use the ES to find the global optimum of the Ackley function

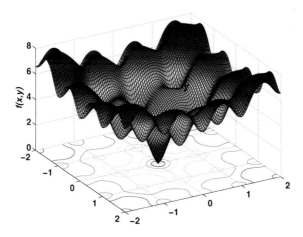

Fig. 3 Ackley's function for two independent variables with a global minimum $f_* = 0$ at $(0,0)$

$$f(\mathbf{x}) = -20\exp[-\frac{1}{5}\sqrt{\frac{1}{d}\sum_{i=1}^{d}x_i^2}] - \exp[\frac{1}{d}\sum_{i=1}^{d}\cos(2\pi x_i)] + 20 + e, \qquad (14)$$

where $(d = 1, 2, ...)$ [1]. The global minimum $f_* = 0$ occurs at $(0, 0, ..., 0)$ in the domain of $-32.768 \leq x_i \leq 32.768$ where $i = 1, 2, ..., d$. The landscape of the 2D Ackley function is shown in Fig. 3, while the landscape of this function with 2.5% noise is shown in Fig. 4

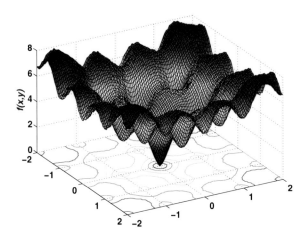

Fig. 4 Ackley's 2D function with Gaussian noise

The global minimum in 2D for a given noise level of 2.5% can be found after about 300 function evaluations (for 20 fireflies after 15 iterations, see Fig. 5).

4.2 Comparison of ES with PSO

Various studies show that PSO algorithms can outperform genetic algorithms (GA) [7] and other conventional algorithms for solving many optimization problems. This is partially due to that fact that the broadcasting ability of the current best estimates gives better and quicker convergence towards the optimality. A general framework for evaluating statistical performance of evolutionary algorithms has been discussed in detail by Shilane et al. [15].

Now we will compare the Eagle Strategy with PSO for various standard test functions. After implementing these algorithms using Matlab, we have carried out extensive simulations and each algorithm has been run at least 100 times so as to carry out meaningful statistical analysis. The algorithms stop when the variations of function values are less than a given tolerance $\varepsilon \leq 10^{-5}$. The results are summarized in the following table (see Table 1) where the global optima are reached. The numbers are in the format: average number of evaluations (success rate), so

$12.7 \pm 1.15(100)$ means that the average number (mean) of function evaluations is $12.7 \times 10^3 = 12700$ with a standard deviation of $1.15 \times 10^3 = 1150$. The success rate of finding the global optima for this algorithm is 100%. Here we have used the following abbreviations: MWZ for Michalewicz's function with $d = 16$, RBK for Rosenbrock with $d = 16$, De Jong for De Jong's sphere function with $d = 256$, Schwefel for Schwefel with $d = 128$, Ackley for Ackley's function with $d = 128$, and Shubert for Shubert's function with 18 minima. In addition, all these test functions have a 2.5% of Gaussian noise, or $\sigma = 0.025$. In addition, we have used the population size $n = 20$ in all our simulations.

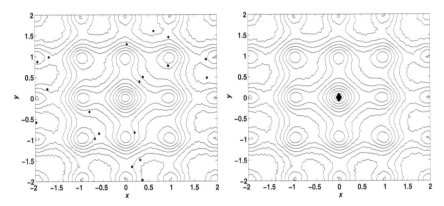

Fig. 5 The initial locations of the 20 fireflies (left) and their locations after 15 iterations (right). We have used $\gamma = 1$

Table 1 Comparison of algorithm performance

	PSO ($\times 10^3$)	ES ($\times 10^3$)
Easom	$185.9 \pm 3.1(97)$	$12.7 \pm 1.15(100)$
MWZ	$346.1 \pm 8.0(98)$	$36.1 \pm 3.5(100)$
Rosenbrock	$1637 \pm 79(98)$	$75 \pm 6.4(100)$
De Jong	$852 \pm 16(100)$	$70.7 \pm 7.3(100)$
Schwefel	$726.1 \pm 25(97)$	$99 \pm 6.7(100)$
Ackley	$1170 \pm 19(92)$	$54 \pm 5.2(100)$
Rastrigin	$3973 \pm 64(90)$	$151 \pm 14(100)$
Easom	$863.7 \pm 55(90)$	$76 \pm 11(100)$
Griewank	$2798 \pm 63(92)$	$134 \pm 9.1(100)$
Shubert	$1197 \pm 56(92)$	$32 \pm 2.5(100)$

We can see that the ES is noticeably more efficient in finding the global optima with the success rates of 100%. Each function evaluation is virtually instantaneous on a modern personal computer. For example, the computing time for 10,000 evaluations on a 3GHz desktop is about 5 seconds. Even with graphics for displaying the locations of the particles and fireflies, it usually takes less than a few minutes. Furthermore, we have used various values of the population size n or the number of fireflies. We found that for most problems $n = 15$ to 50 would be sufficient. For tougher problems, larger n such as $n = 100$ or 250 can be used, though excessively large n should not be used unless there is no better alternative, as it is more computationally extensive.

5 Conclusions

By combining Lévy walk with the firefly algorithm, we have successfully formulated a hybrid optimization algorithm, called Eagle Strategy, for stochastic optimization. After briefly outlining the basic procedure and its similarities and differences with particle swarm optimization, we then implemented and compared these algorithms. Our simulation results for finding the global optima of various test functions suggest that ES can significantly outperform the PSO in terms of both efficiency and success rate. This implies that ES is potentially more powerful in solving NP-hard problems.

However, we have not carried out sensitivity studies of the algorithm-dependent parameters such as the exponent λ in Lévy distribution and the light absorption coefficient γ, which may be fine-tuned to a specific problem. This can form an important research topic for further research. Furthermore, other local search algorithms such as the Newton-Raphson method, sequential quadratic programming and Nelder-Mead algorithms can be used to replace the firefly algorithm, and a comparison study should be carried out to evaluate their performance. It may also show interesting results if the level of uncertainty varies and it can be expected that the higher level of noise will make it more difficult to reach optimal solutions.

As other important further studies, we can also focus on the applications of this hybrid algorithm on the NP-hard traveling salesman problem. In addition, many engineering design problems typically have to deal with intrinsic inhomogeneous materials properties and such uncertainty may often affect the design choice in practice. The application of the proposed hybrid algorithm in engineering design optimization may prove fruitful.

References

[1] Ackley, D.H.: A connectionist machine for genetic hillclimbing. Kluwer Academic Publishers, Dordrecht (1987)
[2] Barthelemy, P., Bertolotti, J., Wiersma, D.S.: A Lévy flight for light. Nature 453, 495–498 (2008)
[3] Bental, A., El Ghaoui, L., Nemirovski, A.: Robust Optimization. Princeton University Press, Princeton (2009)
[4] Bonabeau, E., Dorigo, M., Theraulaz, G.: Swarm Intelligence: From Natural to Artificial Systems. Oxford University Press, Oxford (1999)
[5] Brown, C., Liebovitch, L.S., Glendon, R.: Lévy flights in Dobe Ju/'hoansi foraging patterns. Human Ecol. 35, 129–138 (2007)

[6] Deb, K.: Optimisation for Engineering Design. Prentice-Hall, New Delhi (1995)
[7] Goldberg, D.E.: Genetic Algorithms in Search, Optimisation and Machine Learning. Addison Wesley, Reading (1989)
[8] Kennedy, J., Eberhart, R.C.: Particle swarm optimization. In: Proc. of IEEE International Conference on Neural Networks, Piscataway, NJ, pp. 1942–1948 (1995)
[9] Kennedy, J., Eberhart, R., Shi, Y.: Swarm intelligence. Academic Press, London (2001)
[10] Marti, K.: Stochastic Optimization Methods. Springer, Heidelberg (2005)
[11] Nelder, J.A., Mead, R.: A simplex method for function minimization. Computer Journal 7, 308–313 (1965)
[12] Pavlyukevich, I.: Lévy flights, non-local search and simulated annealing. J. Computational Physics 226, 1830–1844 (2007)
[13] Pavlyukevich, I.: Cooling down Lévy flights. J. Phys. A:Math. Theor. 40, 12299–12313 (2007)
[14] Reynolds, A.M., Frye, M.A.: Free-flight odor tracking in Drosophila is consistent with an optimal intermittent scale-free search. PLoS One 2, e354 (2007)
[15] Shilane, D., Martikainen, J., Dudoit, S., Ovaska, S.J.: A general framework for statistical performance comparison of evolutionary computation algorithms. Information Sciences: an Int. Journal 178, 2870–2879 (2008)
[16] Shlesinger, M.F., Zaslavsky, G.M., Frisch, U. (eds.): Lévy Flights and Related Topics in Phyics. Springer, Heidelberg (1995)
[17] Shlesinger, M.F.: Search research. Nature 443, 281–282 (2006)
[18] Urfalioglu, O., Cetin, A.E., Kuruoglu, E.E.: Levy walk evolution for global optimization. In: Proc. of 10th Genetic and Evolutionary Computation Conference, pp. 537–538 (2008)
[19] Wallace, S.W., Ziemba, W.T.: Applications of Stochastic Programming. SIAM Mathematical Series on Optimization (2005)
[20] Yang, X.S.: Firefly algorithms for multimodal optimization. In: Watanabe, O., Zeugmann, T. (eds.) SAGA 2009. LNCS, vol. 5792, pp. 169–178. Springer, Heidelberg (2009)
[21] Yang, X.S., Deb, S.: Cuckoo search via Lévy flights. In: Proceedings of World Congress on Nature & Biologically Inspired Computing (NaBic 2009), pp. 210–214. IEEE Pulications, India (2009)
[22] Yang, Z.Y., Tang, K., Yao, X.: Large Scale Evolutionary Optimization Using Cooperative Coevolution. Information Sciences 178, 2985–2999 (2008)

CO^{2}RBFN for Short and Medium Term Forecasting of the Extra-Virgin Olive Oil Price

M.D. Pérez-Godoy, P. Pérez-Recuerda, María Pilar Frías, A.J. Rivera,
C.J. Carmona, and Manuel Parras

Abstract. In this paper an adaptation of CO^{2}RBFN, evolutionary COoperative-COmpetitive algorithm for Radial Basis Function Networks design, applied to the prediction of the extra-virgin olive oil price is presented. In this algorithm each individual represents a neuron or Radial Basis Function and the population, the whole network. Individuals compite for survival but must cooperate to built the definite solution. The forecasting of the extra-virgin olive oil price is addressed as a time series forecasting problem. In the experimentation medium-term predictions are obtained for first time with these data. Also short-term predictions with new data are calculated. The results of CO^{2}RBFN have been compared with the traditional statistic forecasting Auto-Regressive Integrated Moving Average method and other data mining methods such as other neural networks models, a support vector machine method or a fuzzy system.

1 Introduction

Radial Basis Function Networks (RBFNs) are an important artificial neural network paradigm [5] with interesting characteristics such as a simple topological structure or universal approximation ability [23]. The overall efficiency of RBFNs has been proved in many areas such as pattern classification [6], function approximation [23] or time series prediction [31].

M.D. Pérez-Godoy · P. Pérez-Recuerda · A.J. Rivera · C.J. Carmona
Department of Informatics, University of Jaén
e-mail: {lperez,pperez,arivera,ccarmona}@ujaen.es

María Pilar Frías
Department of Statistics and Operation Research, University of Jaén
e-mail: mpfrias@ujaen.es

Manuel Parras
Department of Marketing, University of Jaén
e-mail: mparras@ujaen.es

J.R. González et al. (Eds.): NICSO 2010, SCI 284, pp. 113–125, 2010.
springerlink.com   © Springer-Verlag Berlin Heidelberg 2010

An important paradigm for RBFN design is the Evolutionary Computation [3], a general stochastic optimization framework inspired by natural evolution. Typically, in this paradigm each individual represents a whole network (Pittsburgh scheme) that is evolved in order to increase its accuracy.

CO^2RBFN [25] is a evolutionary cooperative-competitive method for the design of RBFNs. In this algorithm each individual of the population represents an RBF and the entire population is responsible for the final solution. The individuals cooperate towards a definitive solution, but they must also compete for survival.

In this paper CO^2RBFN is adapted to solving time series forecasting problems. Concretely a short and medium term forecasting of the extra virgin olive oil price is addressed.

The results obtained using CO^2RBFN are also compared with ARIMA methodology and other hybrid intelligent systems methods such as a Fuzzy System developed with a GA-P algorithm (FuzzyGAP)[28], a MultiLayer Perceptron Network trained using a Conjugate Gradient learning algorithm (MLPConjGrad)[22], a support vector machine (NU-SVR)[9], and a classical design method for Radial Basis Function Network learning (RBFNLMS)[32].

This paper is organized as follows: section 2 discusses generalities about the extra-virgin olive oil price and its prediction, describes the classical ARIMA method and reviews the RBFN design for forecasting problems. In section 3 the extension of CO^2RBFN to time series forecasting is presented. The study and results obtained for the forecast methods are detailed in Section 4. In Section 5, conclusions and future works are outlined.

2 Background

Olive oil has become an important business sector in a continuously expanding market. In 2009, World produced 2,888,000 of tons of olive oil[1], Spain is the first olive oil producing and exporting country and Jaén is the most productive province of Spain, it made 430,000 of tons, the 15% of the total production in the planet.

Agents involved in this sector are interested in the use of forecasting methods for the olive price. This is especially important in the official Market for the negotiation of futures contracts for olive oil (MFAO): a society whose objective is to negotiate an appropriate price for the olive oil at the moment it is to be sold at a fixed time in the future. An accurate prediction of this price in the future could increase the global benefits. In this context, the data provided for the design of the prediction system are the weekly extra-virgin olive oil prices obtained from *Poolred*[2], an initiative of the Foundation for the Promotion and Development of the Olive and Olive Oil located in Jaén, Spain.

The data are a set of regular time-ordered observations of a quantitative characteristic of an individual phenomenon taken at successive periods or points of time, called time series. The problems in which the data are not independent but also have

[1] http://www.mfao.es
[2] http://www.oliva.net/poolred/

a temporal relationship are called time series forecasting problems. Time series forecasting is an active research area and typical paradigm for evaluating it are statistic models [13], such as ARIMA, and data mining methods.

ARIMA [4] stand for Auto-Regressive Integrated Moving Average, a group of techniques for the analysis of time series which generate statistical forecasting models under the assumption of linearity among variables. Data mining is a research area concerned with extracting non-trivial information contained in a database, and has also been applied to time series forecasting. Among data mining techniques, mainly neural networks [8][15][26][30] and fuzzy rule based systems [2][17][18][19][33] have been applied to this kind of problem. In these papers, the presented forecasting problems are mainly addressed as regression problems (see section 4).

Examples of evolutive RBFN design algorithms applied to time series forecasting can be found in [7][12][21][27][29]. However, there are very few algorithms based on cooperative competitive strategies.

The authors have developed a hybrid cooperative-competitive evolutionary proposal for RBFN design, CO^{2}RBFN, applied to the classification problem [25] and have addressed the short-term forecasting of the extra virgin olive oil price [24]. This paper analyzes new data (until December 2008) of this oil price and deals with not only a short-term but also with a new medium-term forecasting of the extra virgin olive oil price, of these new data.

3   CO^{2}RBFN for Time Series Forecasting

CO^{2}RBFN [25], is an hybrid evolutionary cooperative-competitive algorithm for the design of RBFNs. As mentioned, in this algorithm each individual of the population represents, with a real representation, an RBF and the entire population is responsible for the final solution. The individuals cooperate towards a definitive solution, but they must also compete for survival. In this environment, in which the solution depends on the behavior of many components, the fitness of each individual is known as credit assignment. In order to measure the credit assignment of an individual, three factors have been proposed: the RBF contribution to the network output, the error in the basis function radius, and the degree of overlapping among RBFs.

The application of the operators is determined by a Fuzzy Rule-Based System. The inputs of this system are the three parameters used for credit assignment and the outputs are the operators' application probability.

The main steps of CO^{2}RBFN, explained in the following subsections, are shown in the pseudocode in Figure 1.

RBFN initialization. To define the initial network a specified number m of neurons (i.e. the size of population) is randomly allocated among the different patterns of the training set. To do so, each RBF centre, c_i , is randomly established to a pattern of the training set. The RBF widths, d_i, will be set to half the average distance between the centres. Finally, the RBF weights, w_{ij}, are set to zero.

```
1. Initialize RBFN
2. Train RBFN
3. Evaluate RBFs
4. Apply operators to RBFs
5. Substitute the eliminated RBFs
6. Select the best RBFs
7. If the stop condition is not
     verified go to step 2
```

Fig. 1 Main steps of CO^2RBFN

RBFN training. The Least Mean Square algorithm [32] has been used to calculate the RBF weights. This technique exploits the local information that can be obtained from the behaviour of the RBFs.

RBF evaluation. A credit assignment mechanism is required in order to evaluate the role of each RBF ϕ_i in the cooperative-competitive environment. For an RBF, three parameters, a_i, e_i, o_i are defined:

- The contribution, a_i, of the RBF ϕ_i, $i = 1 \ldots m$, is determined by considering the weight, w_i, and the number of patterns of the training set inside its width, pi_i. An RBF with a low weight and few patterns inside its width will have a low contribution:

$$a_i = \begin{cases} |w_i| & if \quad pi_i > q \\ |w_i| * (pi_i/q) & otherwise \end{cases} \qquad (1)$$

where q is the average of the pi_i values minus the standard deviation of the pi_i values.

- The error measure, e_i, for each RBF ϕ_i, is obtained by calculating the Mean Absolute Percentage Error (MAPE) inside its width:

$$e_i = \frac{\sum_{\forall p_i} \left| \frac{f(p_i) - y(p_i)}{f(p_i)} \right|}{npi_i} \qquad (2)$$

where $f(p_i)$ is the output of the model for the point p_i, inside the width of RBF ϕ_i, $y(p_i)$ is the real output at the same point, and npi_i is the number of points inside the width of RBF ϕ_i.

- The overlapping of the RBF ϕ_i and the other RBFs is quantified by using the parameter o_i. This parameter is computed by taking into account the fitness sharing methodology [11], whose aim is to maintain the diversity in the population. This factor is expressed as:

$$o_i = \sum_{j=1}^{m} o_{ij} \qquad (3)$$

where o_{ij} measures the overlapping of the RBF ϕ_i y ϕ_j $j = 1 \ldots m$.

Applying operators to RBFs. In CO^2RBFN four operators have been defined in order to be applied to the RBFs:

- Operator Remove: eliminates an RBF.
- Operator Random Mutation: modifies the centre and width of an RBF in a percentage below 50% of the old width.
- Operator Biased Mutation: modifies the width and all coordinates of the centre using local information of the RBF environment. The technique used follows the recommendations [10] that are similar to those used by the algorithm LMS algorithm. The error for the patterns within the radius of the RBF, ϕ_i, are calculated. For each coordinate of the center and the radius a value Δc_{ij} and Δd_i respectively are calculated. The new coordinates and the new radius are obtained by changing (increasing or decreasing) its old values to a random number (between 5% and 50% of its old width), depending on the sign of the value calculated.

$$\Delta d_i = \sum_k e(\vec{p_k}) \cdot w_i \tag{4}$$

where $e(\vec{p_k})$ is the error for the pattern $\vec{p_k}$.

$$\Delta c_{ij} = sign(c_{ij} - p_{kj}) \cdot e(\vec{p_k}) \cdot w_i \tag{5}$$

- Operator Null: in this case all the parameters of the RBF are maintained.

The operators are applied to the whole population of RBFs. The probability for choosing an operator is determined by means of a Mandani-type fuzzy rule based system [20] which represents expert knowledge about the operator application in order to obtain a simple and accurate RBFN. The inputs of this system are parameters a_i, e_i and o_i used for defining the credit assignment of the RBF ϕ_i. These inputs are considered as linguistic variables va_i, ve_i and vo_i. The outputs, p_{remove}, p_{rm}, p_{bm} and p_{null}, represent the probability of applying Remove, Random Mutation, Biased Mutation and Null operators, respectively.

Table 1 shows the rule base used to relate the described antecedents and consequents. In the table each row represents one rule. For example, the interpretation of the first rule is: If the contribution of an RBF is Low Then the probability of applying the operator Remove is Medium-High, the probability of applying the operator

Table 1 Fuzzy rule base representing expert knowledge in the design of RBFNs

	va	ve	vo	p_{remove}	p_{rm}	p_{bm}	p_{null}		va	ve	vo	p_{remove}	p_{rm}	p_{bm}	p_{null}
	Antecedents			Consequents					Antecedents			Consequents			
R1	L			M-H	M-H	L	L	R6	H			M-H	M-H	L	L
R2	M			M-L	M-H	M-L	M-L	R7		L		L	M-H	M-H	M-H
R3	H			L	M-H	M-H	M-H	R8		M		M-L	M-H	M-L	M-L
R4		L		L	M-H	M-H	M-H	R9		H		M-H	M-H	L	L
R5		M		M-L	M-H	M-L	M-L								

Random Mutation is Medium-High, the probability of applying the operator Biased Mutation is Low and the probability of applying the operator null is Low.

Introduction of new RBFs. In this step, the eliminated RBFs are substituted by new RBFs. The new RBF is located in the centre of the area with maximum error or in a randomly chosen pattern with a probability of 0.5 respectively.

The width of the new RBF will be set to the average of the RBFs in the population plus half of the minimum distance to the nearest RBF. Its weights are set to zero.

Replacement strategy. The replacement scheme determines which new RBFs (obtained before the mutation) will be included in the new population. To do so, the role of the mutated RBF in the net is compared with the original one to determine the RBF with the best behaviour in order to include it in the population.

4   Experimentation and Results

The dataset used in this work have been obtained from *Poolred*[3], an initiative of the Foundation for the Promotion and Development of the Olive and Olive Oil located in Jaén, Spain. The time series dataset contains the weekly extra-virgin olive oil price per kilogram.

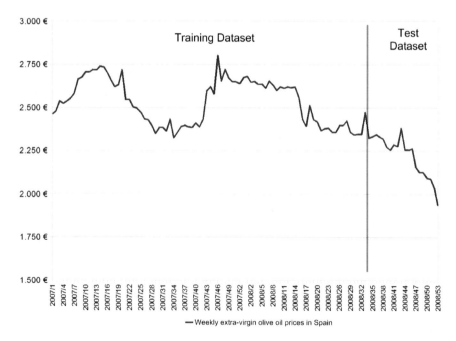

Fig. 2 Weekly extra-virgin olive oil prices in Tons / Euro

---

[3] http://www.oliva.net/poolred/

The task addressed in this work is that of performing two forecast next week and four weeks later of the extra-virgin olive oil price. In this study, the data used are from the 1st week of the year 2007 to the 53nd week of the year 2008 in Spain. The cases in the data set were divided into two subsets: one for training and the other for testing. The data from the 1st week of 2007 to the 33th week of 2008 were used for training. The performance of the different forecastings and methods were tested by estimating the data from the 34nd week to the 53nd week of 2008. Figure 2 shows the time series data and training and test datasets.

As mentioned, experiments carry out predictions with horizons of one week and four weeks. In this way the patterns are heuristically composed of: $(n-3, n-2, n-1, n, n+1)$, when the price to forecast is $n+1$ and must be determined from the past prices $n-3$ to n; $(n-3, n-2, n-1, n, n+4)$, when the price to forecast is $n+4$ and must be determined from the past prices $n-3$ to n.

To estimate prediction capacity, the error considered is the Mean Absolute Percentage Error (MAPE):

$$MAPE = \sum_{i}^{n} (| (f_i - y_i)/f_i |) \tag{6}$$

where f_i is the predicted output of the model and y_i is the desired output.

Other methods used in the experimentation are:

- ARIMA models, also called Box-Jenkins models [4], predict variable's present values from its past values. The development of an ARIMA methodology consists of the search for an ARIMA(p, d, q) model, which is able to generate the time series object of the study. Here p is the value for the auto-regressive parameter, d is the order of differentiation and q is the moving average parameter. ARIMA modeling involves the follow stages: (1) Identification of the model or the initial p, d, and q parameters; (2) Estimation of the p and q parameters; (3) Diagnosis of the residuals in order to investigate model adequacy.

- FuzzyGAP method [28]. A GA-P method [16] uses an evolutionary computation method, a hybrid between genetic algorithms and genetic programming, and optimized to perform symbolic regressions. Each element comprises a chain of parameters and a tree which describes a function, depending on these parameters. The two operators by means of which new members of the population are generated are crossover and mutation. In the GA-P algorithm both operations are performed independently over the tree and the parameter chain.

- MLPConjGrad [22]. MLPConjGrad uses the conjugate-gradient algorithm to adjust weight values of a multilayer perceptron [14]. Compared to gradient descent, the conjugate gradient algorithm takes a more direct path to the optimal set of weight values. Usually, the conjugate gradient is significantly faster and more robust than the gradient descent. The Conjugate gradient also does not require the user to specify learning rate and momentum parameters.

- RBFN-LMS. Builds an RBFN with a pre-specified number of RBFs. By means of the K-Means clustering algorithm it chooses an equal number of points from the training set to be the centres of the neurons. Finally, it establishes a single radius for all the neurons as half the average distance between the set of centres. Once the centres and radio of the network have been fixed, the set of weights is analytically computed using the LMS algorithm [32].

- NU-SVR, the SVM (Support Vector Machine) model uses the sequential minimal optimization training algorithm and treats a given problem in terms of solving a quadratic optimization problem. The NU-SVR, called also v-SVM, for regression problems is an extension of the traditional SVM and it aims to build a loss function [9].

Table 2 ARIMA Model Summary

Parameter	Estimate	Stnd.Error	P-Value
AR(1)	0,906789	0,0461258	0,000000
Mean	2,52871	0,0612143	0,000000

For the ARIMA model has been estimated an ARIMA (1,0,0). In table 1, the Maximum Likelihood Estimation, Standard Errors and P-Values are shown for the parameters of the most appropriate ARIMA model which is fitted to the price time series. When considering the 85 observations the difference equation for the AR(1) model is written as

$$(1 - 0.906789B)(X_t - 2.52871) = \varepsilon_t, \tag{7}$$

with $\varepsilon_t, t = 1,\ldots,n$ the white noise term. The third column in table 1 summarizes the statistical significance of the terms in the forecasting model. Terms with P-Value less than 0.05 are statistically significantly different from zero at 95% confidence level. The P-Value for AR(1) term is less than 0.05, so it is significantly different from 0. In the case of the constant term the P-Value has a similar behavior.

The implementations of the rest data mining methods have been obtained from KEEL [1]. The parameters used in these data mining methods are the values recommended in the literature. For CO_2RBFN the number of executions is 200 and the number of RBFs or individuals in the population is set to 10.

The series have been differentiated to avoid problems related with the stationarity. The predictions have been performed using the differenced data, but errors have been calculated after reconstruct the original series.

The traditional work mode of ARIMA (without updating) is predicting the first value, and then calculate the following values using their own predictions. So it can accumulate a error if the number of test dataset is greater than six or eight samples. That's why for ARIMA work in circumstances similar to the methods of data mining, we will "update" data from test simulating the data mining models. For four weeks forecasting, ARIMA only can use its own predictions with updating.

To obtain the results, algorithms have been executed 10 times and in Table 3 shows the average error MAPE mission and its standard deviation. The figures 3 and 4 show the best prediction achieved by the methods for the test set.

Table 3 Results obtained by different methods forecasting the price of olive oil

Method	MAPE for 1 week forecasting	MAPE for 4 weeks forecasting
Fuzzy-GAP	$0{,}02170 \pm 0{,}00226$	$0{,}03536 \pm 0{,}00461$
MLP ConjGrad	$0{,}02052 \pm 0{,}00041$	$0{,}02970 \pm 0{,}00196$
NU-SVR	$0{,}01936 \pm 0$	$0{,}03003 \pm 0$
RBFN-LMS	$0{,}02111 \pm 0{,}00234$	$0{,}04706 \pm 0{,}00901$
ARIMA (without updating)	$0{,}13036 \pm 0$	-
ARIMA updating	$0{,}02823 \pm 0$	$0{,}06827 \pm 0$
CO^{2}RBFN	$0{,}01914 \pm 0{,}00057$	$0{,}03230 \pm 0{,}00160$

If we analyze the results we can draw the following conclusions:

- The data mining methods have better performance that ARIMA models, which were traditionally used in econometrics to predicting this kind of problem.
- This superiority of data mining methods over ARIMA is even clearer when using ARIMA with traditional methodology (without updating).
- The method proposed by the authors, CO^{2}RBFN, is the best method when the horizon of prediction is one week and is close to the top spot in the forecasting to four weeks.
- CO^{2}RBFN has practically the lowest standard deviation of all non-deterministic methods, which demonstrates the robustness of the method.

Finally, it must be highlighted that the accuracy of the results obtained has been of interest to olive-oil sector experts.

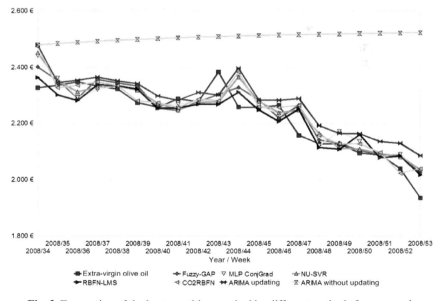

Fig. 3 Forecasting of the best repetition reached by different methods for one week

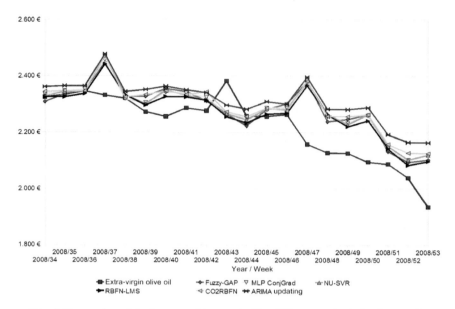

Fig. 4 Forecasting of the best repetition reached by different methods for four weeks

5 Concluding Remarks

This paper presents an application of an evolutionary cooperative-competitive algorithm (CO^2RBFN) to the forecasting of the extra-virgin olive oil price. As important key point of our proposal it is must be highlighted the identification of the role (credit assignment) of each basis function in the whole network. It is defined by three factors are defined and used: the RBF contribution to the network's output, a_i; the error in the basis function radius, e_i; and the degree of overlapping among RBFs, o_i. Another important key is that the application of the evolutive operators is determined by a fuzzy rule-based system which represents expert knowledge of the RBFN design. The inputs of this system are the three parameters used for credit assignment.

A new medium horizon, four weeks, along with a short horizon, one week, have been defined for the forecasting of the extra-virgin olive oil weekly price. The results of CO^2RBFN have been compared with the ones obtained by the well-known classical statistical ARIMA method and a set of reliable data mining methods. The data mining methods applied for the comparison are: MLPConjGrad, a multilayer perceptron network which trains which a conjugate gradient algorithm; FuzzyGAP, a fuzzy system developed with a GA-P algorithm; NU-SVR, a support vector machine method, and RBFNLMS, a radial basis function network trained with the LMS algorithm.

From the results it can be concluded that datamining methods outperforms ARIMA methodology and that CO^2RBFN is the best method in the prediction to one week and is close to the top spot in the forecasting to four weeks. Also lowest standard deviation of CO^{2}RBFN demonstrates the robustness of the method.

As future lines, pre-processing for feature selection and exogenous features like meteorology or econometric data can be taken into account in order to increase the performance of the forecast.

Acknowledgments. Supported by the Spanish Ministry of Science and Technology under the Projects TIN2008-06681-C06-02, the Andalusian Research Plan TIC-3928 and the Project of the University of Jaén UJA-08-16-30.

References

[1] Alcalá-Fdez, J., Sánchez, L., García, S., Del Jesus, M.J., Ventura, S., Garrell, J.M., Otero, J., Romero, C., Bacardit, J., Rivas, V., Fernández, J.C., Herrera, F.: KEEL: A Software Tool to Assess Evolutionary Algorithms for Data Mining Problems. Soft Computing 13(3), 307–318 (2009)

[2] Azadeh, A., Saberi, M., Ghaderi, S.F., Gitiforouz, A., Ebrahimipour, V.: Improved estimation of electricity demand function by integration of fuzzy system and data mining approach. Energy Conversion and Management (2008) doi:10.1016/j.enconman.2008.02.021

[3] Bäck, T., Hammel, U., Schwefel, H.: Evolutionary computation: comments on the history and current state. IEEE Transaction Evolutive Compututation 1(1), 3–17 (1997)

[4] Box, G., Jenkins, G.: Time series analysis: forecasting and control, revised edn. Holden Day, San Francisco (1976)

[5] Broomhead, D., Lowe, D.: Multivariable functional interpolation and adaptive networks. Complex System 2, 321–355 (1998)

[6] Buchtala, O., Klimek, M., Sick, B.: Evolutionary optimization of radial basis function classifiers for data miningapplications. IEEE Transactions on Systems, Man and Cybernetics Part B 35(5), 928–947 (2005)

[7] Chen, C., Wu, Y., Luk, B.L.: Combined genetic algorithm optimization and regularized orthogonal least squares learning for radial basis function networks. IEEE Transaction Neural Networks 10(5), 1239–1243 (1999)

[8] Co, H.C., Boosarawongse, R.: Forecasting Thailand's rice export: Statistical techniques vs. artificial neural networks. Computers and Industrial Engineering 53(4), 610–627 (2007)

[9] Fan, R.E., Chen, P.H., Lin, C.J.: Working set selection using the second order information for training SVM. Journal of Machine Learning Research 6, 1889–1918 (2005)

[10] Ghost, J., Deuser, L., Beck, S.: A neural network based hybrid system for detection, characterization and classification of short-duration oceanic signals. IEEE Jl. Of Ocean Enginering 17(4), 351–363 (1992)

[11] Goldberg, D., Richardson, J.: Genetic algorithms with sharing for multimodal function optimization. In: Grefenstette (ed.) Proc. Second International Conference on Genetic Algorithms, pp. 41–49. Lawrence Erlbaum Associates, Mahwah (1987)

[12] Du, H., Zhang, N.: Time series prediction using evolving radial basis function networks with new encoding scheme. Neurocomputing 71(7-9), 1388–1400 (2008)

[13] Franses, P.H., van Dijk, D.: Non-linear time series models in empirical finance. Cambridge University Press, Cambridge (2000)

[14] Haykin, S.: Neural Networks: A Comprehensive Foundation, 2nd edn. Prentice Hall, Englewood Cliffs (1998)

[15] Hobbs, B.F., Helman, U., Jitprapaikulsarn, S., Konda, S., Maratukulam, D.: Artificial neural networks for short-term energy forecasting: Accuracy and economic value. Neurocomputing 23(1-3), 71–84 (1998)

[16] Howard, L., D'Angelo, D.: The GA-P: A Genetic Algorithm and Genetic Programming Hybrid. IEEE Expert, 11–15 (1995)

[17] Jang, J.R.: ANFIS: Adaptative-Network-based Fuzzy Inference System. IEEE Trans. Systems, Man and Cybernetics 23(3), 665–685 (1993)

[18] Khashei, M., Reza Hejazi, S., Bijari, M.: A new hybrid artificial neural networks and fuzzy regression model for time series forecasting. Fuzzy Sets and Systems 159(7), 769–786 (2008)

[19] Liu, J., McKenna, T.M., Gribok, A., Beidleman, B.A., Tharion, W.J., Reifman, J.: A fuzzy logic algorithm to assign confidence levels to heart and respiratory rate time series. Physiological Measurement 29(1), 81–94 (2008)

[20] Mandani, E., Assilian, S.: An experiment in linguistic synthesis with a fuzzy logic controller. Int. J. Man Mach. Stud. 7(1), 1–13 (1975)

[21] Meng, K., Dong, Z.Y., Wong, K.P.: Self-adaptive radial basis function neural network for short-term electricity price forecasting. IET Generation, Transmission and Distribution 3(4), 325–335

[22] Moller, F.: A scaled conjugate gradient algorithm for fast supervised learning. Neural Networks 6, 525–533 (1990)

[23] Park, J., Sandberg, I.: Universal approximation using radial-basis function networks. Neural Comput. 3, 246–257 (1991)

[24] Pérez, P., Frías, M.P., Pérez-Godoy, M.D., Rivera, A.J., del Jesus, M.J., Parras, M., Torres, F.J.: An study on data mining methods for short-term forecasting of the extra virgin olive oil price in the Spanish market. In: Proceeding of the International Conference On Hybrid Intelligetn Systems, pp. 943–946 (2008)

[25] Pérez-Godoy, M.D., Rivera, A.J., Berlanga, F.J., Jesús, M.J.: CO2RBFN: an evolutionary cooperative-competitive RBFN design algorithm for classification problems. Soft Computing (in press) (2009) doi: 10.1007/s00500-009-0488-z

[26] Pino, R., Parreno, J., Gomez, A., Priore, P.: Forecasting next-day price of electricity in the Spanish energy market using artificial neural networks. Engineering Applications of Artificial Intelligence 21(1), 53–62 (2008)

[27] Rivas, V., Merelo, J.J., Castillo, P., Arenas, M.G., Castellano, J.G.: Evolving RBF neural networks for time-series forecasting with EvRBF. Information Science 165, 207–220 (2004)

[28] Sánchez, L., Couso, I.: Fuzzy Random Variables-Based Modeling with GA-P Algorithms. In: Bouchon, B., Yager, R.R., Zadeh, L. (eds.) Information, Uncertainty and Fusion, pp. 245–256 (2000)

[29] Sheta, A.F., De Jong, K.: Time-series forecasting using GA-tuned radial basis functions. Information Sciencie 133, 221–228 (2001)

[30] Ture, M., Kurt, I.: Comparison of four different time series methods to forecast hepatitis A virus infection. Expert Systems with Applications 31(1), 41–46 (2006)

[31] Whitehead, B., Choate, T.: Cooperative-competitive genetic evolution of Radial Basis Function centers and widths for time series prediction. IEEE Trans. on Neural Networks 7(4), 869–880 (1996)

[32] Widrow, B., Lehr, M.A.: 30 Years of adaptive neural networks: perceptron, madaline and backpropagation. Proceedings of the IEEE 78(9), 1415–1442 (1990)

[33] Yu, T., Wilkinson, D.: A co-evolutionary fuzzy system for reservoir well logs interpretation. Evolutionary computation in practice, 199–218 (2008)

3D Cell Pattern Generation in Artificial Development

Arturo Chavoya, Irma R. Andalon-Garcia, Cuauhtemoc Lopez-Martin, and M.E. Meda-Campaña

Abstract. Cell pattern formation has an important role in both artificial and natural development. This paper presents an artificial development model for 3D cell pattern generation based on the cellular automata paradigm. Cell replication is controlled by a genome consisting of an artificial regulatory network and a series of structural genes. The genome was evolved by a genetic algorithm in order to generate 3D cell patterns through the selective activation and inhibition of genes. Morphogenetic gradients were used to provide cells with positional information that constrained cellular replication in space. The model was applied to the problem of growing a solid French flag pattern in a 3D virtual space.

1 Introduction

In biological systems, development is a fascinating and very complex process that involves following a sequence of genetically programmed events that ultimately produce the developed organism. One of the crucial stages in the development of an organism is that of pattern formation, where the fundamental body plans of the individual are outlined. Recent evidence has shown that gene regulatory networks play a central role in the development and metabolism of living organisms [13]. Researchers in biological sciences have confirmed that the diverse cell patterns created during the developmental stages are mainly due to the selective activation and inhibition of very specific regulatory genes.

On the other hand, artificial models of cellular development have been proposed over the years with the objective of understanding how complex structures and patterns can emerge from one or a small group of initial undifferentiated cells

Arturo Chavoya · Irma R. Andalon-Garcia · Cuauhtemoc Lopez-Martin ·
M.E. Meda-Campaña
Universidad de Guadalajara, Periférico Norte 799 - L308
Zapopan, Jal., Mexico CP 45000
e-mail: {achavoya,agi10073,cuauhtemoc,emeda}@cucea.udg.mx

J.R. González et al. (Eds.): NICSO 2010, SCI 284, pp. 127–139, 2010.
springerlink.com   © Springer-Verlag Berlin Heidelberg 2010

[7, 21, 22, 24]. In this paper we propose an artificial cellular growth model that generates 3D patterns by means of the selective activation and inhibition of development genes under the constraints of morphogenetic gradients. Cellular growth is achieved through the expression of structural genes, which are in turn controlled by an Artificial Regulatory Network (ARN) evolved by a Genetic Algorithm (GA). The ARN determines at which time steps cells are allowed to grow and which gene to use for reproduction, whereas morphogenetic gradients constrain the position at which cells can replicate. Both the ARN and the structural genes make up the artificial cell's genome. In order to test the functionality of the development program found by the GA, the evolved genomes were applied to a cellular growth testbed based on the Cellular Automata (CA) paradigm that has been successfully used in the past to develop simple 2D and 3D geometrical shapes [8]. The model presented in this work was applied to a 3D version of what is known as the *French flag problem*. The 2D version of this problem has traditionally been used in biology —and more recently in computer science— to model the determination of cell patterns in tissues, usually through the use of morphogenetic gradients to help determine cell position.

The paper starts with a section describing the French flag problem with a brief description of models that have used it as a test case. The next section describes the cellular growth testbed used to evaluate the evolved genomes in their ability to form the desired patterns, followed by a section presenting the morphogenetic gradients that constrain cell replication. The artificial cell's genome is presented next, followed by a section describing the GA and how it was applied to evolve the genomes. Results are presented next, followed by a section of conclusions.

2 The French Flag Problem

The problem of generating a French flag pattern was first introduced by Wolpert in the late 1960s when trying to formulate the problem of cell pattern development and regulation in living organisms [30]. This formulation has been used since then by some authors to study the problem of artificial pattern development. More specifically, the problem deals with the creation of a pattern with three sharp bands of cells with the colors and order of the French flag stripes.

Lindenmayer and Rozenberg used the French flag problem to illustrate how a grammar-based L-System could be used to solve the generation of this particular pattern when enunciated as the production of a string of the type $a^n b^n c^n$ over the alphabet $\{a, b, c\}$ and with $n > 0$ [23]. On the other hand, Herman and Liu [18] developed an extension of a simulator called CELIA [1] and applied it to generate a French flag pattern in order to study synchronization and symmetry breaking in cellular development.

Miller and Banzhaf used what they called Cartesian genetic programming to evolve a cell program that would construct a French flag pattern [25]. They tested the robustness of their programs by manually removing parts of the developing pattern. They found that several of their evolved programs could repair to some extent the damaged patterns. Bowers also used this problem to study the phenotypic robust-ness of his embryogeny model, which was based on cellular growth with diffusing chemicals as signaling molecules [4].

Gordon and Bentley proposed a development model based on a set of rules evolved by a GA that described how development should proceed to generate a French flag pattern [16]. The morphogenic model based on a multiagent system de-veloped by Beurier et al. also used an evolved set of agent rules to grow French and Japanese flag patterns [3]. On the other hand, Dever et al. proposed a neural network model for multicellular development that grew French flag patterns [14] . Even models for developing evolvable hardware have benefited from the French flag problem as a test case [17, 28].

More recently, Knabe et al. [20] developed a model based on the CompuCell3D package [12] combined with a genetic regulatory network that controlled cell pa-rameters such as size, shape, adhesion, morphogen secretion and orientation. They were able to obtain final 2D patterns with matches of over 75% with respect to a 60×40 pixel target French flag pattern.

3   Cellular Growth Testbed

Cellular automata were chosen as models of cellular growth, as they provide a sim-ple mathematical model that can be used to study self-organizing features of com-plex systems [29]. CA are characterized by a regular lattice of N identical cells, an interaction neighborhood template η, a finite set of cell states Σ, and a space- and time-independent transition rule ϕ which is applied to every cell in the lattice at each time step.

In the cellular growth testbed used in this work, a $13 \times 13 \times 13$ regular lattice with non-periodic boundaries was used. The set of cell states was defined as $\Sigma = \{0, 1\}$, where 0 can be interpreted as an empty cell and 1 as an occupied or active cell. The interaction neighborhood η considered was a 3D Margolus template (Fig. 1), which has previously been used with success to model 3D shapes [31]. In this template there is an alternation of the block of cells considered at each step of the CA al-gorithm. At odd steps, the seven cells shown to the left and the back in the figure constitute the interaction neighborhood, whereas at even steps the neighborhood is formed by the mirror cells of the previous block.

The CA rule ϕ was defined as a lookup table that determined, for each local neighborhood, the state (empty or occupied) of the objective cell at the next time step. For a binary-state CA, these update states are termed the rule table's "output bits". The lookup table input was defined by the binary state value of cells in the local interaction neighborhood, where 0 meant an empty cell and 1 meant an occu-pied cell and the parity bit p determined which of the two blocks of cells was being considered for evaluation [8]. The output bit values shown in Fig. 1 are only for illustration purposes; the actual values for a predefined shape, such as a cube, are found by a GA.

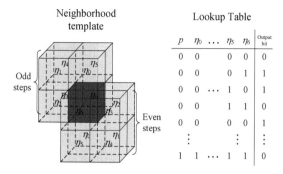

Fig. 1 Cellular automaton's 3D Margolus neighborhood template and the associated lookup table. The parity bit p in the lookup table determines which block of the neighborhood template is being considered for evaluation. The objective cell is depicted as a darker cube in the middle of the template

4 Morphogenetic Gradients

Ever since Turing's seminal article on the theoretical influence of diffusing chemical substances on an organism's pattern development [27], the role of these molecules has been confirmed in a number of biological systems. These organizing substances were termed *morphogens*, given their involvement in driving morphogenetic processes. In the present model, morphogenetic gradients were generated similar to those found in the eggs of the fruit fly *Drosophila*, where orthogonal gradients offer a sort of Cartesian coordinate system [5]. These gradients provide reproducing cells with positional information in order to facilitate the spatial generation of patterns. The artificial morphogenetic gradients were set up as suggested in [24], where morphogens diffuse from a source towards a sink, with uniform morphogen degradation throughout the gradient.

Before cells were allowed to reproduce in the cellular growth testbed, morphogenetic gradients were generated by diffusing the morphogens from one of the CA boundaries for 1000 time steps. Initial morphogen concentration level was set at 255 arbitrary units, and the source was replenished to the same level at the beginning of each cycle. The diffusion factor was 0.20, i.e. at each time step every grid position diffused 20% of its morphogen content and all neighboring positions received an equal amount of this percentage. This factor was introduced to avoid rapid morphogen depletion at cell positions and its value was experimentally determined to render a smooth descending gradient. The sink was set up at the opposite boundary of the lattice, where the morphogen level was always set to zero. At the end of each time step, morphogens were degraded at a rate of 0.005 throughout the CA lattice. Three orthogonal gradients were defined in the CA lattice, one for each of the main Cartesian axes (Fig. 2). In the figures presented in this work the following conventions are used: in the 3D insets the positive x axis extends to right, the positive y axis is towards the back of the page, the positive z axis points to the top, and the axes are rotated 45 degrees to the left to show a better perspective.

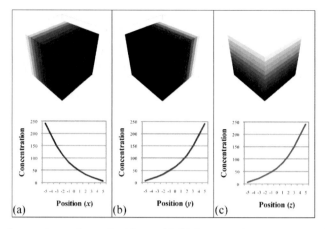

Fig. 2 Morphogenetic gradients. Positions with highest morphogen concentration are depicted in white; darker tones mean lower concentrations. (**a**) Left to right (*x axis*); (**b**) back to front (*y axis*); (**c**) top to bottom (*z axis*)

5   Genome

Genomes are the repository of genetic information in living organisms. They are encoded as one or more chains of DNA, and they regularly interact with other macromolecules, such as RNA and proteins. Artificial genomes are typically coded as strings of discrete data types. The genome used in this model was defined as a binary string starting with a series of ten regulatory genes, followed by a number of structural genes (Fig. 3).

5.1   Regulatory Genes

The series of regulatory genes at the beginning of the genome constitutes an Artificial Regulatory Network. ARNs are computer models whose objective is to emulate the gene regulatory networks found in nature. ARNs have previously been used to study differential gene expression either as a computational paradigm or to solve particular problems [2, 7, 15, 19, 26]. The gene regulatory network implemented in this work is an extension of the ARN presented in [9], which in turn is based on the model proposed by Banzhaf [2].

In the present model, each regulatory gene consists of a series of eight inhibitor/enhancer sites, a series of five regulatory protein coding regions, and three morphogen threshold activation sites that determine the allowed positions for cell reproduction (Fig. 3). Inhibitor/enhancer sites are composed of a 12-bit function defining region and a regulatory site. Regulatory sites can behave either as an enhancer or an inhibitor, depending on the configuration of the function defining bits associated with them. If there are more 1's than 0's in the defining bits region, then the regulatory site functions as an enhancer, but if there are more 0's than 1's, then

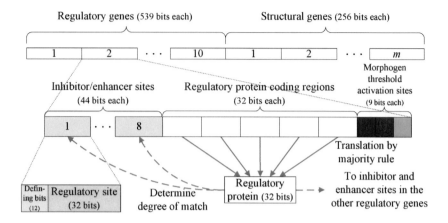

Fig. 3 Genome structure and regulatory gene detail. Regulatory genes make up an artificial regulatory network, whereas structural genes contain the lookup tables that control cell reproduction. The number of structural genes m depends on the pattern to be generated and whether or not structural genes are duplicated, as explained in Sect. 7. For the final simulations, $m = 6$

the site behaves as an inhibitor. Finally, if there is an equal number of 1's and 0's, then the regulatory site is turned off [10].

Regulatory protein coding regions "translate" a protein using the majority rule, i.e. for each bit position in these regions, the number of 1's and 0's is counted and the bit that is in majority is translated into the regulatory protein. The regulatory sites and the individual protein coding regions all have the same size of 32 bits. Thus the protein translated from the coding regions can be compared on a bit by bit basis with the regulatory site of the inhibitor and enhancer sites, and the degree of matching can be measured. As in [2], the comparison was implemented by an XOR operation, which results in a "1" if the corresponding bits are complementary. Each translated protein is compared with the inhibitor and enhancer sites of all the regulatory genes in order to determine the degree of interaction in the regulatory network. The influence of a protein on an enhancer or inhibitor site is exponential with the number of matching bits. The strength of enhancement en or inhibition in for gene i with $i = 1,...,n$ is defined as

$$en_i = \frac{1}{v} \sum_{j=1}^{v} c_j e^{\beta \left(u_{ij}^+ - u_{max}^+ \right)} \text{ and} \tag{1}$$

$$in_i = \frac{1}{w} \sum_{j=1}^{w} c_j e^{\beta \left(u_{ij}^- - u_{max}^- \right)}, \tag{2}$$

where n is the total number of regulatory genes, v and w are the total number of active enhancer and inhibitor sites, respectively, c_j is the concentration of protein j, β is a constant that fine-tunes the strength of matching, u_{ij}^+ and u_{ij}^- are the number of matches between protein j and the enhancer and inhibitor sites of gene i, respectively, and u_{max}^+ and u_{max}^- are the maximum matches achievable (32 bits) between a protein and an enhancer or inhibitor site, respectively [2].

Once the *en* and *in* values are obtained for all regulatory genes, the corresponding change in concentration c for protein i in one time step is calculated using

$$\frac{dc_i}{dt} = \delta \left(en_i - in_i \right) c_i , \qquad (3)$$

where δ is a constant that regulates the degree of protein concentration change.

Protein concentrations are updated and if a new protein concentration results in a negative value, the protein concentration is set to zero. Protein concentrations are then normalized so that total protein concentration is always the unity. Parameters β and δ were set to 1.0 and 1.0×10^6, respectively, as previously reported [11].

The morphogen threshold activation sites provide reproducing cells with positional information as to where they are allowed to grow in the CA lattice. There is one site for each of the three orthogonal morphogenetic gradients described in Sect. 4. These sites are 9 bits in length, where the first bit defines the allowed direction (above or below the threshold) of cellular growth, and the next 8 bits code for the morphogen threshold activation level, which ranges from 0 to $2^8 - 1 = 255$. If the site's high order bit is 0, then cells are allowed to replicate below the morphogen threshold level coded in the lower order eight bits; if the value is 1, then cells are allowed to reproduce above the threshold level. Since in a regulatory gene there is one site for each of the orthogonal morphogenetic gradients, for each set of three morphogen threshold activation levels, the three high order bits define in which of the eight relative octants cells expressing the associated structural gene can reproduce.

5.2 Structural Genes

Structural genes code for the particular shape grown by the reproducing cells and were obtained using the methodology presented in [8]. Briefly, the CA rule table's output bits from the cellular growth model described in Sect. 3 were evolved by a GA in order to produce predefined 3D shapes. A structural gene is interpreted as a CA rule table by reading its bits as output bits of the CA rule. As mentioned in Sect. 3, at each time step of the CA run, an empty objective cell position can be occupied by an active cell (output bit $= 1$) depending on the configuration of the cells in the Margolus neighborhood block $(\eta_0, ..., \eta_6)$ and on the value of the parity bit p.

A structural gene is always associated with a corresponding regulatory gene, i.e. structural gene number 1 is associated with regulatory gene number 1 and its related translated protein, and so on. However, in a particular genome there can be less structural genes than regulatory genes; as a result, some regulatory genes are not associated with a structural gene and their role is only to participate in the activation or inhibition of other regulatory genes without directly activating a structural gene.

A structural gene was defined as being active if and only if the regulatory protein translated by the associated regulatory gene was above a certain concentration threshold. The value chosen for the threshold was 0.5, since the sum of all protein concentrations is always 1.0, and there can only be a protein at a time with a concentration above 0.5. As a result, at most one structural gene can be expressed at a particular time step in a cell. If a structural gene is active, then the CA lookup table coded in it is used to control cell reproduction. Structural gene expression is visualized in the cellular growth model as a distinct external color for the cell.

6 Genetic Algorithm

Genetic algorithms are search and optimization methods based on ideas borrowed from natural genetics and evolution. A GA starts with a population of chromosomes representing vectors in search space. Each chromosome is evaluated according to a fitness function and the best individuals are selected. A new generation of chromosomes is then created by applying genetic operators on selected individuals from the previous generation. The process is repeated until the desired number of generations is reached or until the desired individual is found.

For the present work, chromosomes represent either the output bits from a CA rule table to be evolved to generate a simple form such a cube, or an ARN whose objective is to activate structural genes in a particular order to produce a multicolored shape such as a French flag pattern.

The GA in this paper uses tournament selection with single-point crossover and mutation as genetic operators. As in a previous report, we used the following parameter values [11]. The initial population consisted of 1000 binary chromosomes whose bit values were chosen at random. Tournaments were run with sets of 3 individuals randomly selected from the population. Crossover and mutation rates were 0.60 and 0.15, respectively. Finally, the number of generations was set at 50, as there was no significant improvement after this number of generations.

The fitness function used by the GA was defined as

$$Fitness = \frac{1}{k} \sum_{i=1}^{k} \frac{ins_i - \frac{1}{2} outs_i}{des_i} \, , \qquad (4)$$

where k is the number of different colored shapes, each corresponding to an expressed structural gene, ins_i is the number of active cells inside the desired shape i with the correct color, $outs_i$ is the number of active cells outside the desired shape i, but with the correct color, and des_i is the total number of cells inside the desired shape i. The range of values for this function is $[0, 1]$ with a fitness value of 1 representing a perfect match.

7 Results

The GA described in Sect. 6 was used in all cases to obtain the CA's rule tables that made up the structural genes for specific simple patterns and to evolve the ARNs for the desired multicolored pattern. After an evolved genome was obtained, an initial active cell containing it was placed in the center of the CA lattice and was allowed to reproduce for 60 time steps in the cellular growth testbed described in Sect. 3, controlled by the gene activation sequence found by the GA. In order to grow the desired structure with a predefined color and position for each cell, the regulatory genes in the ARN had to evolve to be activated in a precise sequence and for a specific number of iterations. Not all GA experiments produced a genome capable of generating the desired pattern.

In order to grow a solid 3D French flag pattern, three different structural genes were used. Expression of the first gene creates the white central cube, while the other two genes drive cells to extend the lateral walls to the left and to the right simultaneously, expressing the blue and the red color, respectively. These two last genes do not necessarily code for a cube, since they only extend a wall of cells to the left and to the right for as many time steps as they are activated and when unconstrained, they produce a symmetrical pattern along the x axis. The independent expression of these three genes is shown in Fig. 4. The two genes that extended the lateral walls were activated after a central white cube was first produced. In order to generate the desired French flag pattern, cells expressing one of these two genes should only be allowed to reproduce on each side of the white central cube (left for the blue cube and right for the red cube). This behavior was to be achieved through the use of genomes where the morphogen threshold activation sites evolved to allow growth only in the desired portions of the 3D CA lattice.

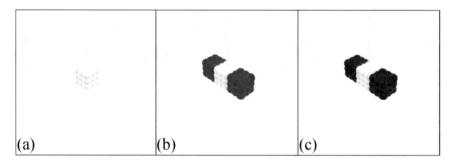

(a)   (b)   (c)

Fig. 4 Expression of the three genes used to create a 3D French flag pattern. (**a**) Create central white cube; (**b**) extend blue lateral walls; (**c**) extend red lateral walls. The last two genes were activated after the creation of a white central cube

However, when trying to evolve a genome to produce the 3D French flag pattern, it was found that the GA could not easily evolve an activation sequence that produced the desired pattern. Using the same approach as in [6], in order to increase the likelihood for the GA to find an appropriate genome, instead of using one series of three structural genes, a tandem of two identical series of three structural genes was used, for a total of six structural genes. In that manner, for creating the central white cube, the genome could express either structural gene number 1 or gene number 4, and for the left blue and right red cubes, it could use genes 2 or 5, or genes 3 or 6, respectively. Thus, the probability of finding an ARN that could express a 3D French flag pattern was significantly increased.

Figure 5 shows a $9 \times 3 \times 3$ solid French flag pattern grown from the expression of the three different structural genes mentioned above. The graph of the corresponding ARN protein concentration change is shown in Fig. 5(e). Starting with an initial white cell (a), a white central cube is formed from the expression of gene number 4 (b), the left blue cube is then grown (c), followed by the right red cube (d). The evolved morphogenetic fields where cells are allowed to grow are depicted in the figure as a translucent volume for each of the three structural genes.

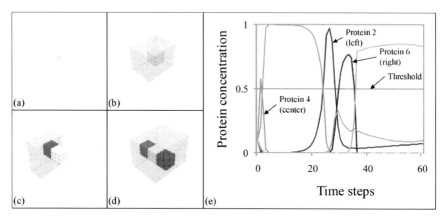

Fig. 5 Growth of a 3D French flag pattern. (**a**) Initial cell; (**b**) central white cube with morphogenetic field for gene 4 (cube); (**c**) central white cube and left blue cube with morphogenetic field for gene 2 (extend blue lateral walls); (**d**) finished flag pattern with morphogenetic field for gene 6 (extend red lateral walls); (**e**) graph of protein concentration change from the genome expressing the French flag pattern; the unlabeled lines correspond to proteins from regulatory genes that are not associated with structural genes

It is clear from the figure that for the genes that extend the wall of cells to the sides, the corresponding morphogenetic fields limited growth to the desired direction (left for blue cells and right for red cells). It should also be noted that the left blue cube is formed from the activation of the second gene from the first series of structural genes, while the other two genes are expressed from the second series of the tandem.

8 Conclusions

The results presented in this paper show that a GA can give reproducible results in evolving an ARN to grow predefined simple 3D cellular patterns starting with a single cell. In particular, simulations showed that the combination of a GA and CA with a 3D Margolus interaction neighborhood was a feasible choice for modeling 3D pattern generation.

In general, the framework developed proved to be suitable for generating simple patterns, but more work is needed to explore generation of more complex structures. It is also desirable to study cellular structure formation allowing cell death and cell displacement, as in actual cellular growth. Furthermore, in order to build a more accurate model of the growth process, the use of a more realistic physical environment may be necessary. The long-term goal of this work is to study the emergent properties of the artificial development process. It is conceivable that highly complex structures will one day be built from the interaction of myriads of simpler entities controlled by a development program.

References

[1] Baker, R.W., Herman, G.T.: Celia - a cellular linear iterative array simulator. In: Proceedings of the fourth annual conference on Applications of simulation. Winter Simulation Conference, pp. 64–73 (1970)

[2] Banzhaf, W.: Artificial regulatory networks and genetic programming. In: Riolo, R.L., Worzel, B. (eds.) Genetic Programming Theory and Practice, ch. 4, pp. 43–62. Kluwer, Dordrecht (2003)

[3] Beurier, G., Michel, F., Ferber, J.: A morphogenesis model for multiagent embryogeny. In: Rocha, L.M., Yaeger, L.S., Bedau, M.A., Floreano, D., Goldstone, R.L., Vespignani, A. (eds.) Proceedings of the Tenth International Conference on the Simulation and Synthesis of Living Systems (ALife X), pp. 84–90 (2006)

[4] Bowers, C.: Simulating evolution with a computational model of embryogeny: Obtaining robustness from evolved individuals. In: Capcarrère, M.S., Freitas, A.A., Bentley, P.J., Johnson, C.G., Timmis, J. (eds.) ECAL 2005. LNCS (LNAI), vol. 3630, pp. 149–158. Springer, Heidelberg (2005)

[5] Carroll, S.B., Grenier, J.K., Weatherbee, S.D.: From DNA to Diversity: Molecular Genetics and the Evolution of Animal Design, 2nd edn. Blackwell Science, Malden (2004)

[6] Chavoya, A.: Cell pattern generation in artificial development. In: Rossi, C. (ed.) Brain, Vision and AI, In-Teh, Croatia, ch. 4, pp. 73–94 (2008)

[7] Chavoya, A.: Artificial development. In: Foundations of Computational Intelligence. Volume 1: Learning and Approximation (Studies in Computational Intelligence), vol. 8, pp. 185–215. Springer, Heidelberg (2009)

[8] Chavoya, A., Duthen, Y.: Using a genetic algorithm to evolve cellular automata for 2D/3D computational development. In: GECCO 2006: Proceedings of the 8th annual conference on Genetic and evolutionary computation, pp. 231–232. ACM Press, New York (2006)

[9] Chavoya, A., Duthen, Y.: An artificial development model for cell pattern generation. In: Randall, M., Abbass, H.A., Wiles, J. (eds.) ACAL 2007. LNCS (LNAI), vol. 4828, pp. 61–71. Springer, Heidelberg (2007)

[10] Chavoya, A., Duthen, Y.: Use of a genetic algorithm to evolve an extended artificial regulatory network for cell pattern generation. In: GECCO 2007: Proceedings of the 9th annual conference on Genetic and evolutionary computation, p. 1062. ACM Press, New York (2007)

[11] Chavoya, A., Duthen, Y.: A cell pattern generation model based on an extended artificial regulatory network. BioSystems 94(1), 95–101 (2008)

[12] Cickovski, T., Aras, K., Swat, M., Merks, R.M.H., Glimm, T., Hentschel, H.G.E., Alber, M.S., Glazier, J.A., Newman, S.A., Izaguirre, J.A.: From genes to organisms via the cell: A problem-solving environment for multicellular development. Computing in Science and Eng. 9(4), 50–60 (2007)

[13] Davidson, E.H.: The Regulatory Genome: Gene Regulatory Networks in Development And Evolution, 1st edn. Academic Press, London (2006)

[14] Devert, A., Bredeche, N., Schoenauer, M.: Robust multi-cellular developmental design. In: GECCO 2007: Proceedings of the 9th annual conference on Genetic and evolutionary computation, pp. 982–989. ACM, New York (2007)

[15] Eggenberger, P.: Evolving morphologies of simulated 3D organisms based on differential gene expression. In: Harvey, I., Husbands, P. (eds.) Proceedings of the 4th European Conference on Artificial Life, pp. 205–213. Springer, Heidelberg (1997)

[16] Gordon, T.G.W., Bentley, P.J.: Bias and scalability in evolutionary development. In: GECCO 2005: Proceedings of the 2005 conference on Genetic and evolutionary computation, pp. 83–90. ACM, New York (2005)

[17] Harding, S.L., Miller, J.F., Banzhaf, W.: Self-modifying cartesian genetic programming. In: GECCO 2007: Proceedings of the 9th annual conference on Genetic and evolutionary computation, pp. 1021–1028. ACM, New York (2007)

[18] Herman, G.T., Liu, W.H.: The daughter of Celia, the French flag and the firing squad. In: WSC 1973: Proceedings of the 6th conference on Winter simulation, p. 870. ACM, New York (1973)

[19] Joachimczak, M., Wróbel, B.: Evo-devo in silico: a model of a gene network regulating multicellular development in 3D space with artificial physics. In: Bullock, S., Noble, J., Watson, R., Bedau, M.A. (eds.) Artificial Life XI: Proceedings of the Eleventh International Conference on the Simulation and Synthesis of Living Systems, pp. 297–304. MIT Press, Cambridge (2008)

[20] Knabe, J.F., Nehaniv, C.L., Schilstra, M.J.: Evolution and morphogenesis of differentiated multicellular organisms: autonomously generated diffusion gradients for positional information. In: Artificial Life XI: Proceedings of the Eleventh International Conference on the Simulation and Synthesis of Living Systems, pp. 321–328. MIT Press, Cambridge (2008)

[21] Kumar, S., Bentley, P.J.: On Growth, Form and Computers. Academic Press, London (2003)

[22] Lindenmayer, A.: Mathematical models for cellular interaction in development, Parts I and II. Journal of Theoretical Biology 18, 280–315 (1968)

[23] Lindenmayer, A., Rozenberg, G.: Developmental systems and languages. In: STOC 1972: Proceedings of the fourth annual ACM symposium on Theory of computing, pp. 214–221. ACM, New York (1972)

[24] Meinhardt, H.: Models of Biological Pattern Formation. Academic Press, London (1982)

[25] Miller, J.F., Banzhaf, W.: Evolving the program for a cell: from French flags to Boolean circuits. In: Kumar, S., Bentley, P.J. (eds.) On Growth, Form and Computers, pp. 278–301. Academic Press, London (2003)

[26] Reil, T.: Dynamics of gene expression in an artificial genome - implications for biological and artificial ontogeny. In: Floreano, D., Mondada, F. (eds.) ECAL 1999. LNCS, vol. 1674, pp. 457–466. Springer, Heidelberg (1999)

[27] Turing, A.M.: The chemical basis of morphogenesis. Philosophical Transactions of the Royal Society of London. Series B, Biological Sciences 237(641), 37–72 (1952)

[28] Tyrrell, A.M., Greensted, A.J.: Evolving dependability. J. Emerg. Technol. Comput. Syst. 3(2), 7 (2007)

[29] Wolfram, S.: Statistical mechanics of cellular automata. Reviews of Modern Physics 55, 601–644 (1983)

[30] Wolpert, L.: The French flag problem: a contribution to the discussion on pattern development and regulation. In: Waddington, C. (ed.) Towards a Theoretical Biology, pp. 125–133. Edinburgh University Press, New York (1968)

[31] Wu, P., Wu, X., Wainer, G.A.: Applying cell-devs in 3D free-form shape modeling. In: Sloot, P.M.A., Chopard, B., Hoekstra, A.G. (eds.) ACRI 2004. LNCS, vol. 3305, pp. 81–90. Springer, Heidelberg (2004)

Partial Imitation Rule in Iterated Prisoner Dilemma Game on a Square Lattice

Degang Wu, Mathis Antony, and K.Y. Szeto*

Abstract. A realistic replacement of the general imitation rule in the Iterated Prisoner Dilemma (IPD) is investigated with simulation on square lattice, whereby the player, with finite memory, can only imitate those behaviors of the opponents observed in past games. In contrast to standard practice where all the possible behaviors of the opponents are accessible, the new partial imitation rule assumes that the player can at most access those behaviors of his opponent observed in the past few moves. This partial imitation of the behavior in IPD shows very different out-comes in the long time behavior of the games, such as the ranking of various strategies. In particular, the well known tit-for-tat (TFT) strategy loses its importance.

1 Introduction

Game theory [1] has attracted the attention of many scientists working in complex systems as an experimental playground in computer simulation of multi-agent systems is now easily available [2]. Since the introduction of evolutionary game theory by Maynard Smith and Price [3, 4], one of the important issues of this theory is to understand the spontaneous cooperation towards a more efficient outcome with agent interactions in the absence of a central planner [5, 6]. Among the many games, the most studied example by political scientists and sociologists is the Prisoner's Dilemma, as it provides a simple example of the difficulties of cooperation [7]. Prisoner Dilemma (PD) is described by the following set of rules. When two players play a PD game, each of them can choose to cooperate (C) or defect (D). Each player will gain a payoff depending jointly on his choice and the opponent's choice.

Degang Wu · Mathis Antony · K.Y. Szeto
Department of Physics,
Hong Kong University of Science and Technology,
Clear Water Bay, Hong Kong, HKSAR
e-mail: phszeto@ust.hk

* Corresponding author.

J.R. González et al. (Eds.): NICSO 2010, SCI 284, pp. 141–150, 2010.
springerlink.com

Cooperation yields a payoff $R(S)$ if the opponent cooperates (defects) and defection yields $T(P)$ if the opponent cooperates (defects). R is the Reward for cooperation, S is the Sucker's payoff, T is the Temptation to defect and P is the Punishment. Typically, $T > R > P > S$ and $2R > T + P$. PD game is a non zero sum game because one player's loss does not equal the opponent's gain. In order to reduce the amount of parameters, one can follow Nowak et al. [8] and use the following simplified payoff table A,

$$A = \begin{pmatrix} R & S \\ T & P \end{pmatrix} = \begin{pmatrix} 1 & 0 \\ b & 0 \end{pmatrix}. \tag{1}$$

In this setup, there remains a free parameter $b(=T)$ which should be in the range $(1,2)$. The tragedy behind this simple PD game is that the best strategy for a selfish individual, who chooses to defect, will result in mutual defection. This entails the worst collective effect for the society. In this game, the expectation of defection (D) is greater than the expectation of cooperation (C), independent of the opponent's strategy, even though mutual cooperation yields a higher total payoff for the society. The only state where no player can gain more by unilaterally changing its own strategy, the state called the Nash Equilibrium, occurs when all players defect. Hence, if the players use the simple imitation rule so that the players will adapt the strategy of a more successful player, the dominant strategy is defection. In order to further investigate the emergence of cooperation, a variant of the PD game is to consider a set of players located on a lattice and play the so-called spatial PD game (SPDG). In this case, cooperators can support each other in more than one dimension [9]. There are other approaches which will favor the survival of cooperation, as can be found in the recent work of Hebling et al [10] and Nowak [6].

The total income of player i in the spatial PD game can be described by a two-state Potts model Hamiltonian [2, 11]:

$$H_i = \sum_{j(i)} \underline{S}_i^T A \underline{S}_j \quad \text{with} \underline{S}_i^T, \underline{S}_j \in \left\{ \vec{C}, \vec{D} \right\} \text{ and } \vec{C} = \begin{pmatrix} 1 \\ 0 \end{pmatrix}, \vec{D} = \begin{pmatrix} 0 \\ 1 \end{pmatrix} \tag{2}$$

Here is the state vector for player j who is a neighbor of player i and the state vector can be either one of the two unit vectors . The summation runs over all the neighbors of the player i sitting at node i, while the neighborhood is defined by the topology of the given network. We will also give the players the ability to remember a fixed number of the most recent events and supply each player with a rule to decide what move they should take to respond to a history. We call this rule a strategy. A complete strategy covers all the possible situations but in a real game only a subset of the strategy will be used. Players will adapt their strategies, imitating other more successful players following a certain imitation rule. The usual imitation rule assumes that the player will copy all the strategies of his idol, who is a more successful opponent in his encounter. However, if only a subset of all the strategies of the idol has been used, it is unrealistic for the player to copy all the strategies, including those that have never been observed. A realistic modification on the imitation rule is to copy only those strategies that have been observed. The modification of the traditional imitation rule is necessitated by the fact that all players can only have finite memory. This simple observation, together with the existence of a generally large set of possible strategies, motivates us to consider a new imitation rule. We call it a "partial imitation rule", as it permits the player to imitate at most those strategies his idol has used. In real life, a player cannot even remember all the observed moves of his idol. We will formulate our representation of the strategy and the numerical methods used in setting up the spatial iterated PD game in Section 2. We then present a detailed discussion on the various imitation rules in Section 3. The results of our simulation are summarized and discussed in Section 4. We conclude with some discussion on the implication of partial imitation rule and discuss future works in the final section.

2   Methods

2.1   Memory Encoding

A two-player PD game yields one of the four possible outcomes because each of the two independent players has two possible moves, cooperate (C) or defect (D). To an agent i, the "outcome" of playing a PD game with his opponent, agent j, can be represented by an ordered pair of strategies $s_i s_j$. Here s_i can be either C for "cooperate" or D for "defect". Thus, there are four possible scenarios for any one game between them: $\{s_i s_j\}$ takes on one of these four outcomes $\{CC, CD, DC, DD\}$. In general, for n games, there will be a total of $4n$ possible scenarios. A particular pattern of these n games will be one of these $4n$ scenarios, and can be described by an ordered sequence of the form $S_{i1} S_{j1} \ldots S_{in} S_{jn}$. This particular ordered sequence of outcomes for these n games is called a history of games between these two players, which consists of n pairs of outcome $\{S_i S_j\}$, with the leftmost one being the first game played, while the rightmost one being the outcome of the last gamed played, or the most recent outcome. For example, an ordered sequence of strategy pairs $DDDDDDCC$ represents that the two players cooperate right after the past three defection $\{DD\}, \{DD\}, \{DD\}$. Note the convention for the outcome is that the in the pair $\{s_i s_j\}$, s_i is the move made by agent i, who is the player we address, while s_j is the move made by agent j, the opponent of our player.

We say that a player has a memory of fixed-length m, when this player can remember only m-pairs of outcomes. Obviously, a "Memory" is a sub-sequence of a history. In a PD-game with a fixed memory-length m, the players can get access to the outcomes of the past m games and decide the response to the specific outcomes in the present game. For example, for an agent with two-game memory ($m = 2$), given a history represented by $DDDDDDCC$, the memory of the player consists of only the substring $DDCC$. Because a given memory can be represented by a unique sequence of strategies, a memory can be conveniently designate by a unique number. In this paper, cooperation is represented by 1 and defection 0. Thus, the memory $DDCC$ can be represented by the binary number 0011 or the decimal number 3. The number of all the possible memory, given that the agent can memorize the outcomes of the last m games, is $4m$. Next, we must address the beginning of the game between our players.

Let's consider a non-trivial example when $m = 3$. In this case there are $64 = 4m = 4 \cdot 3$ possible histories of the strategies used by the two players. Following a method proposed by Bukhari and Haider[12], we reserve one bit for the first move of our player: $\{D, C\}$, and use two more bits for the second move of our player when confronted with the two possibilities of the first move of the opponent $\{D, C\}$. (Our player can choose C or D when the opponent's first move is D, and our player also can choose C or D when the opponent's first move is C. Thus we need two more bits for our player). To account for the four possible scenarios of the last two moves of the opponents: $\{DD, DC, CD, CC\}$, we need to reserve 4 more bits to record the third move of our player. Thus, for a PD game played by prisoners who can remember 3 games, a player will need $1 + 2 + 4 = 7$ bits to record his first three moves. After this initial stage, the strategy sequence for our player will need to respond to the game history with a finite memory. Since there are a total of $64 = 4m = 4 \cdot 3$ possible Memory, i.e., 64 possible outcomes of the last three games, our player will need 64 more bits. In conclusion, the length of the strategy sequence is $7 + 64 = 71$ and there are a total of , possible strategies. Thus the space of strategies for a $m = 3$ game is already very large. Let's now denote the ensemble of m-step memory as M_m, then the total number of bits required to encode the possible strategy sequence is $b(m) = 2m - 1 + 4m$ and the total number of possible strategy sequences is $|Mm| = 2b(m)$. Table 1 summarizes the enumeration of the encoding of the possible strategies for $m = 1$. The representation of the strategy sequence in $M1$ is denoted as $S_0|S_1 S_2 S_3 S_4$. Here $b(1) = 5$ and there are a total of 32 possible strategies, since each S_i can have two possible choices (C or D) for $i = 0, ..., 4$. For $m = 2$, we have $b(2) = 19$ and $|M2| = 524288$, allowing for an exhaustive enumeration of all possible strategies [13]. For $m = 3$, we see that the $|M3|$ is $2^{71} = 2.4 \cdot 10^{21}$, which is already very large.

Table 1 Representation of Strategy Sequence in M_1

Memorized History	First Move	DD	DC	CD	CC
Players' Strategy	S_0	S_1	S_2	S_3	S_4

2.2 Monte Carlo Simulation

In this paper, agents will be placed on a square lattice of size LxL, with periodic boundary condition. Each agent only interacts with its four nearest neighbors. For one confrontation we randomly choose an agent i and a neighbor j of i and let them play F games with each other. We can compute the payoff $U(i)$ and $U(j)$ of agent i and j over these games in this confrontation. The payoff parameters used are $T = 5.0, R = 3.0, P = 1.0, S = 0.0$, which are widely used and allow meaningful comparison with the existing results. Agent i will then imitate agent j with probability

$$P(\text{i imitates j}) = \frac{1}{1 + \exp\left(\frac{U_i - U_j}{K}\right)} \tag{3}$$

K is similar to the temperature and represents the thermal noise level. The larger corresponds to smaller noise. We use $K = 100$. The reason that we decide that in one confrontation between agent i and j, they have to play $F(> 1)$ games is that memory effect will not be evident unless there is some repeated encounter between the two players to let them learn about the selected strategies used. However, a fixed number for F is rather artificial. Different pairs of players may play different number of games. Furthermore, we find that fixing F does affect the results in a complex manner. In order to test the strategies for different F, we introduce a probability parameter p for a player to stop playing games with his chosen opponent. We further define one generation of the PD game on the square lattice when all LxL confrontations are completed. With this stopping probability p, one effectively control the average number of games played between pair of players, thereby determining F. The choice of F and the rest of the procedure in one independent simulation can be described by the pseudo code in algorithm 3 for a given p.

Algorithm 3. Iterated SPDG algorithm.

$P := 0.05$
$F := 1$
for $i = 0$ to 100 **do**
    while $rand() > p$ % where rand() generates a random number in $[0, 1)$ drawn from a
    uniform distribution **do**
        $F := F + 1$
    end while
end for
for $j = 0$ to $L * L$ **do**
    randomly pick one site A
    A plays with its neighbors, each confrontation lasts for F games
    randomly pick one site B from A's neighborhood
    B plays with its neighbors, each confrontation lasts for F games
    if $rand() < \left\{1.0 + exp\left(\frac{A.payoff - B.payoff}{K}\right)\right\}^{-1}$ **then**
        A imitates B using different imitation rules
    end if
end for

---

3   Imitation Rule

The standard imitation rule for the spatial PD game without memory is that the focal agent i will adopt the pure strategy of a chosen neighbor depending on payoff. The generalized imitation rule for PD game with memory is adopting the entire set of strategy sequences. We call such imitation rule the traditional imitation rule (tIR). In this way, tIR impose that condition that every agent has complete information about the entire set of the strategy sequence of all its neighbors. Such assumption of complete information is unrealistic since the focal agent only plays a few games with its neighbors while the space of strategies used by the neighbor is generally astronomically larger than F. A more realistic situation is that the focal agent i only has partial information about the strategies of his neighbors. In this paper, every agent only knows a subset of the strategy sequence used by a chosen neighbor. For a pair of players (i, j), playing approximately F games, the focal player i will only observed a set $(S_j(i, j))$ of strategy sequences actually used by agent j. This set $S_j(i, j)$ is much smaller than the entire set of strategies available to agent j. With this partial knowledge of the strategies of the neighbors, the new imitation rule for agent i is called the partial imitation rule. We will give an example to illustrate the difference between partial imitation rule and the traditional one. Let's consider an agent i with $C|DDDD$ strategy confronts another agent j with the Tit-for-Tat strategy $(S_0|S_1S_2S_3S_4 = C|DCDC)$ and agent i decides to imitate the agent j's strategy. In tIR, we assume that agent i somehow knows all the five bits of Tit-for-Tat though in the confrontation with agent j only four bits of Tit-for-Tat have been used. On the other hand, with partial imitation rule (pIR), when a $C|DDDD$ agent confronts a Tit-for-Tat agent, the $C|DDDD$ will know only four bits of Tit-for-Tat $(S_0|S_1S_2S_3S_4 = C|DCDC)$, i.e., $S_0 = C$, $S_1 = D$, $S_2 = C$, $S)3 = D$ (c.f. table 1). Thus, when agent i imitates agent j using pIR, agent i will become $(C|DDDC)$, which corresponds to a Grim Trigger instead of Tit-for-Tat $(C|DCDC)$. We call this new imitation rule the type 1 partial imitation rule, denoted by pIR1. In a more relaxed scenario, we can slightly loosen the restriction on the access of our focal agent i to the information of neighbors' strategy sequences. If we denote the subset of agent j's strategy sequence used during the confrontation between agent i and agent j as $S_j(i, j)$, then we can assume that agent i knows the larger subset of strategy sequences of agent j described by

$$G_j(i, j) = \bigcup_{k \in \Omega(j)} S_j(k, j) \tag{4}$$

where $\Omega(j)$ denotes the nearest neighbors of agent j. Note that this set of strategy sequences of agent j is substantially larger than $S_j(i, j)$, but still should generally be much smaller than the entire set of strategies of player j. In pIR1, we provide agent i information on agent j defined by the set $S_j(i, j)$. We now introduce a second type of partial imitation rule, denoted by pIR2, if we replace $S_j(i, j)$ by the much larger set $G_j(i, j)$.

We now illustrate pIR2 with an example using the notation of table 1. Consider an always-cooperating agent i $(C|CCCC)$ confronting a Grim Trigger $(C|DDDC)$ agent j, who has four neighbors. One of them of course is the always cooperating agent i. Let's assume that the remaining three neighbors of agent j are always-defecting $(D|DDDD)$. Let's call these three neighbors agent a, b, and c. In the confrontation between agent i (who is $C|CCCC$) and agent j (Grim Trigger), S_0 and S_4 of Grim Trigger are used. However, in the confrontation between agent j (Grim Trigger) and its three neighbors (agent a, b and c), who are $D|DDDD$, agent j will use S_0, S_1

and S_3 of Grim Trigger. With pIR1, agent i imitates agent j, but the result will be unchanged as they will use C for S_0 and S_4 of Grim Trigger based on the set $S_j(i, j)$. However, for pIR2, agent i imitates agent j and changes from $C|CCCC$ to the Grim Trigger agent, which results in a change of its S_0, S_1, S_3 and S_4 to the corresponding bits of Grim Trigger, giving the new strategy of agent i as $C|DCDC$. This is not a Grim Trigger. Finally, if we use tIR, the traditional imitation rule, we of course will replace agent i with Grim Trigger ($C|DDDC$). We see from this example, the result of tIR, pIR1 and pIR2 are all different.

4 Results

We first test our algorithm of SPDG with the published results [13]. We initialize our strategy sequence with each element assigned cooperation or defection at equal probability and reproduce results similar to figure 3a in [13] in figure 1 using the traditional imitation rule. Here, Tit-For-Tat (TFT) and Grim- Trigger (GT) dominate at long time. These two strategies together with Pavlov and $C|CCDC$ are the only four surviving strategies in the long run. In figure 2(a) we use partial imitation rule 1 (pIR1) and in 2(b), we use pIR2. In both cases, only GT dominates and the concentration of TFT is reduced greatly to the level of Pavlov and $C|CCDC$. Results are independent of the lattice size, provided that it is sufficiently large so that every strategy in M_1 can be visited several times. We next discuss the importance of game

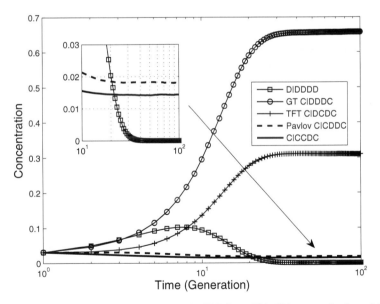

Fig. 1 Concentration of important strategies in SPDG on 100x100 square lattice with M_1. Result is averaged over 1000 independent simulations, with $K = 0.01$, using traditional Imitation Rule (tIR)

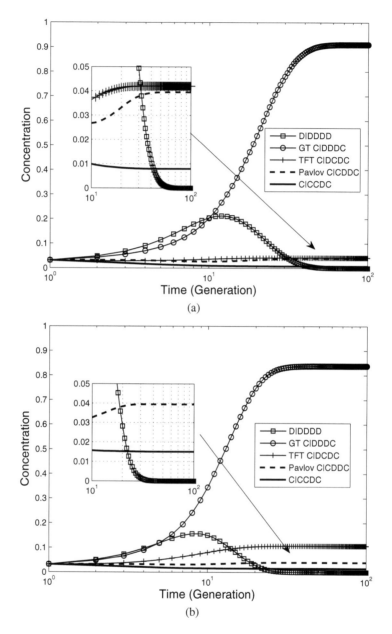

Fig. 2 Concentration of important strategies in SPDG on 100x100 square lattice with M_1. Result is averaged over 1000 independent simulations, with $K = 0.01$, using partial Imitation Rule 1 (pIR1) in figure 2(a) and partial Imitation Rule 2 (pIR2) in figure 2(b)

sampling in terms of the value of p, i.e. the probability to end one confrontation at time t. Our numerical experiments show that p affects the concentrations of all the strategies regardless of the imitation rule used. When $p = 1$, agents will always co-operate or defect without making use of the memory mechanism. When p is smaller than 1, agents can use their memory to access the benefits of different strategies. Recall that we have in general 32 strategies for M_1. For sufficiently small p, our numerical results indicate that the concentrations of these 32 strategies reach a constant value at long time. In this paper, we choose p to be 0.05, but in fact when p is smaller than 0.2, the results will not differ much.

5 Conclusion

Our introduction of memory effects on the players in spatial PD game indicates the importance of the imitation rule used in the learning process of the players. We started our spatial PD game with traditional imitation rule, which makes the unrealistic assumption that the players have a complete access of his opponent's strategies. When this assumption is relaxed and implemented with the partial imitation rules that assume the players only have the information on a selected subset of his opponent's strategies, the long time behavior of the concentration of various strategies are very different. Indeed, for the traditional imitation rule, where TFT and GT dominate at long time, only GT remains dominant when partial imitation rule is used. With the traditional imitation rule, TFT and GT dominate at the long run, while with either partial imitation rule 1 or 2, only GT dominate the population and the concentration of TFT is remarkably smaller than that with the traditional imitation rule. This work shows that with memory, more realistic imitation rules may have an impact on the concentration of the surviving strategies such as TFT and GT. In the scenario we set up in the paper, GT is favored by the partial imitation rules. This result has important implication of previous studies on PD game as partial knowledge of the opponents' strategies should be the norm rather than the exception in real life. In future work, we will investigate more on the generalization of our observation to longer memory cases.

Acknowledgements. K.Y. Szeto acknowledges the support of CERG grant 602506 and 602507.

References

[1] von Neumann, J., Morgenstern, O.: Theory of Games and Economic Behaviour. Princeton University Press, Princeton (1944)
[2] Szabo, G., Fath, G.: Evolutionary games on graphs. Physics Reports 446(4-6), 97–216 (2007)
[3] Smith, J.M., Price, G.M.: The logic of animal conflict. Nature 246, 15–18 (1973)
[4] Smith, J.M.: Evolution and the Theory of Games. Cambridge University Press, Cambridge (1982)

[5] Ohtsuki, H., Hauert, C., Lieberman, E., Nowak, M.A.: A simple rule for the evolution of cooperation on graphs and social networks. Nature 441, 502–505 (2006)

[6] Nowak, M.A.: Five Rules for the Evolution of Cooperation. Science 314(5805), 1560–1563 (December 8, 2006)

[7] Axelrod, R.: The Evolution of Cooperation. Basic Books, New York (1984)

[8] Nowak, M.A., May, R.M.: The spatial dilemmas of evolution. Int. J. of Bifurcation and Chaos 3(1), 35–78 (1993)

[9] Szabo, G., Vukov, J., Szolnoki, A.: Phase diagrams for an evolutionary prisoner's dilemma game on two-dimensional lattices. Phys. Rev. E 72(4), 047107 (2005)

[10] Helbing, D., Lozano, S.: Routes to cooperation and herding effects in the prisoner's dilemma game (May 2009)

[11] Ariosa, D., Fort, H.: Extended estimator approach for 2x2 games and its mapping to the Ising Hamiltonian. Phys. Rev. E 71, 016132 (2005)

[12] Bukhari, S., Adnan, H.A.S.: Using genetic algorithms to develop strategies for the prisoners dilemma. Asian Journal of Information Technology 8(5), 866–871 (2006)

[13] Baek, S.K., Kim, B.J.: Intelligent tit-for-tat in the iterated prisoner's dilemma game. Physical Review E (Statistical, Nonlinear, and Soft Matter Physics) 78(1), 011125 (2008)

A Dynamical Game Model for Sustainable Development

D. Dumitrescu and Andrei Sîrghi

Abstract. The paper addresses the possibility to combine nature inspired optimization techniques, Dynamical Systems and Game Theory in order to solve a complex real-world problem. A computational model for *Sustainable Development* (SD) problem, called *Dynamical Game for Sustainable Development* (DGSD) is proposed. This model combines ideas from *Dynamical Systems* and *Game Theory* in a new paradigm for adaptive behavior of systems. The actors of SD: *Economy, Environment* and *Society*, are viewed as evolvable systems. The main aim is to ensure a balanced coevolution of SD actors. A chain of control points are used to guide the evolution toward system equilibrium (*sustainability*). Each control point is represented as a three player game. In order to guide system to *sustainability*, the local equilibrium at each control point is used to determine further development strategies for SD actors. The local equilibrium is conceived as a game equilibrium. Several kinds of equilibria are possible. For detecting these equilibria an evolutionary approach may be used.

Introduction

The concept and methodology of sustainable development [3] appeared over the past few decades as result to a set of interdependent issues like: *climate change, pollution control, preservation of biodiversity*, etc. Crises, degradation and risks affecting human health, social and economic stability have fostered public suspicions on the evolution of technology and economic growth. These suspicions gave rise to this new concept, and further, a new branch of science. SD concept lead us to principles of organizing and controlling the development and complex interactions

D. Dumitrescu · Andrei Sîrghi
Center for the Complexity Studies, Babeş-Bolyai University,
Department of Computer Science, Babeş-Bolyai University,
Cluj-Napoca, Romania
e-mail: ddumitr@cs.ubbcluj.ro, andreisirghi@yahoo.com

J.R. González et al. (Eds.): NICSO 2010, SCI 284, pp. 151–162, 2010.
springerlink.com

between society, production activities and natural resources in such a way that results in a constructive coexistence between these three big areas.

The paper proposes a new mathematical model for SD problem. This model, called *Dynamical Game for Sustainable Development* (DGSD), combines ideas from Game Theory and Dynamical Systems in a new paradigm for self-organizing coevolutive systems. We use *Dynamical Systems* to describe the evolution of each area included in SD. *Game Theory* is used to model the decision process of our model in each control point. Each control point represents a three player game, corresponding to SD actors. DGSD model is intended to show how much, the decisions taken in one area influence the other areas of the region.

The conflict of interests between players and limited amount of resources in the region, enforce players to be very "careful" in the process of decision making. If a player does not consider this principle of "carefulness", and as result, exploits resources from the other areas, his further decisions are constrained by big limits in resource usage imposed by the other players. Therefore, the area represented by this player collapses.

The principle of *carefulness* represents an important factor of self-organization which ensures a balanced systems coevolution for achieving SD goals.

The DGSD model has an adaptive behavior. This behavior is ensured by the game, which finds compromise solutions between the actors in any situation of region development described by a control point. The goal of these compromise solutions is to converge region development to SD. Each compromise represents the game equilibrium in a specific point of region development. To obtain game equilibrium, we consider an evolutionary approach based on generative relations [5].

1   Sustainable Development Problem

At the base of SD concept stays the principle that objectives of *Society*, *Economy* and *Environment* should be complementary and interdependent in the development process of a region.

The problem related to SD which we propose to solve in this paper can be defined as:

Create a mathematical model having the next characteristics:

1. Represents the real development process of a region from the three aspects corresponding to the major areas: *Economy*, *Environment* and *Society*;
2. Suitable for the robust prediction of the future state and behavior of the real process;
3. Valuable in the real decision making process;
4. Every area has a particular set of objectives and decision functions;
5. Proposes strategies and control elements that leads region to Sustainable Development.

Our aim is to develop such a model by considering a specific region and supposing to have complete information about economy, environment and society.

2 Related Work

SD problem received high attention from its first apparition, but until now does not exist a powerful mathematical model that can be used to represent this problem. The result of researches of almost all communities that analyze this problem is a set of indicators which can be used to measure the quality of sustainable development process for specific regions.

There are two widely accepted methods to measure the sustainable development of a country:

1. Sustainable Development Gauging Matrix (SDGM) [11]. The measure technique of SDGM consists in the aggregation of three dimension indices: economic (I_{ec}), ecological (I_e) and social (I_s) in the index of sustainable development (I_{sd}). Further, each of these indices is calculated by using other six global indices widely used in Statistics communities.

2. IPAT equation [1]. Expresses the relationship between technological innovation and environmental impact. *IPAT* states that human impact (*I*) on the environment equals the product of population (*P*), affluence (*A* - consumption per capita) and technology (*T* - environmental impact per unit of consumption).

3 Dynamical Systems of Areas Evolution

To describe the evolution of SD areas we use a system of dynamical models. Each area included in SD has an own model of evolution, which contains one or more dynamical functions, and interacts with the models of the other areas. Every particular model has two types of parameters:

- internal parameters - that are indices of the area represented by the model and which describe the evolution of the area, and
- external parameters - that are important in decision making process and indicates the dependences between the current area and the other areas. These parameters represent a key element in the process of SD self-organization because they are used by individual areas to influence the other areas.

The DGSD model is extensive, and it can be used with different evolution models of the areas, depending on different circumstances. In this paper we work with an abstract model which reflects the basic relations between SD areas and their structures. This model can be simple replaced by a more descriptive one in specific cases.

The abstract model is built as a system of individual functions of evolution for sustainable development areas: *Economy*, *Environment* and *Society*. The correlations between particular functions are very important for our approach. They represent the base criteria to analyze and control the sustainability of region development. Further we describe the functions of areas evolution and their correlations.

Economy plans the optimal amount of products outcome by choosing corresponding quantities of capital ($K(t)$), nonrenewable natural resources ($h(t)$), renewable natural resources ($r(t)$) and social capital or labor ($l(t)$). The function of economic development (denoted *EC*) may be represented as a dynamical system given by a particular *production function* [9] *ec*:

$$EC(t+1) = ec(K(t),h(t),r(t),l(t)). \tag{1}$$

where *t* stands for time period ($t = [t_0, T]$).

The production function *ec* in the *Cobb-Douglas* [7] form is:

$$\begin{cases} ec(K,h,r,l) = AK(t)^\alpha h(t)^\beta r(t)^\gamma l(t)^\delta \\ \alpha+\beta+\gamma+\delta = 1 | \alpha,\beta,\gamma,\delta \in (0,1] \end{cases}, \tag{2}$$

where *A* represents *total factor productivity*, and the exponents α,β, γ, and δ represent the elasticities of production related to capital, nonrenewable resources, renewable resources and labor respectively.

Accumulated capital stock evolution depends on rate of capital deprecation (σ), and economic products consumption ($c(t)$):

$$K(t+1) = (1-\sigma)K(t) + ec(K(t),h(t),r(t),l(t)) - c(t). \tag{3}$$

The goal of the *Economy* is to maximize production in condition of sustainability which implies an activity constrained by actual and future benefits of all areas. The optimization problem of this area may be represented as:

$$\begin{cases} ec(K,h,r,l) \to max \\ \text{subject to sustainability (SD) constraints.} \end{cases} \tag{4}$$

Environment tries to achieve a sustainable trajectory in the development of nonrenewable (R_n) and renewable (R_r) resources stocks by restricting as much as possible the natural resources consumption. *Environment* development may be represented as a dynamical system which represents the evolution of natural resources stocks:

$$\begin{cases} R_n(t+1) = R_n(t) - h(t) \\ R_r(t+1) = R_r(t) - r(t) + g(R_r(t) - r(t)) \end{cases}, \tag{5}$$

where *g* represents *Environment*'s regenerative capacity and can have multiple forms [2].

The goal of *Environment* is to maximize stock of renewable resources and to preserve actual stock of nonrenewable resources by imposing limits in resources consumption:

$$\begin{cases} R_n \sim const \\ h(t) \to h_n(t) \\ R_r \to max \\ r(t) \to r_n(t) \end{cases}, \tag{6}$$

where $h_n(t)$ is the limit of nonrenewable resources consumption imposed by the *Environment* to *Economy*, and $r_n(t)$ is the limit of renewable resources consumption.

The *Environment* and the *Economy* influence the living conditions in the region, which can be suitable or not for people life. Analyzing these conditions, the *Society* has to choose, to stay in this system or not.

Society's goal is to achieve normal values for indicators of social development such as: birthrate, mortality rate, migration rate and unemployment rate. *Society* development may be represented as the dynamical system:

$$S(t+1) = S(t) + s(R_n(t), R_r(t), \omega_m(t), \omega_b(t), l(t), c(t)), \tag{7}$$

where $S(t)$ represents society size in period t, and s represents *society growth*. Society growth depends on natural resources availability, mortality rate ($\omega_m(t)$), birthrate ($\omega_b(t)$), used labor and economic products consumption. A simple form of society growth function can be represented as:

$$s = \omega_b(t)S(t) - \omega_m(t)S(t) - m(u_r(t), R_n(t), R_r(t), c(t))S(t), \tag{8}$$

where $m(u_r(t), R_n(t), R_r(t), c(t))$ represents migration rate, and is influenced by unemployment rate (u_r), per capita consumption ($c(t)/S(t)$), and per capita natural resources availability($R_n(t)/S(t)$ and $R_r(t)/S(t)$).

The goal of the *Society* is to maximize the living conditions in the region. This goal may be expressed as the system:

$$\begin{cases} \omega_m \rightarrow \omega_{nm} \\ \omega_b \rightarrow \omega_{nb} \\ m \rightarrow nm \\ u_r \rightarrow u_{nr} \end{cases}, \tag{9}$$

where ω_{nm} is the normal rate of mortality, ω_{nb} is the normal birthrate, nm is the normal migration rate and u_{nr} is the normal unemployment rate.

The areas goals are mutually contradictory. Thus a sustainable strategy must coordinate and manage the development process of these areas in a balanced manner, which must result in the long-term viability of the system. In following Section we analyze how these goals can be balanced using Game Theory.

4 Sustainable Development Game

At the beginning of each development iteration t, the SD actors should propose a development strategy for their specific areas. Within DGSD model, Game Theory is used to represent the decision process of SD actors in each control point of region evolution. The SD game involves three players, or agents: *Economy (EC)*, *Environment (EV)* and *Society (S)*. Using this game, each agent chooses best available development strategy by combining the information about the state of their area with their "belief" about the behavior of the other agents.

4.1 Extended Form of the SD Game

SD decision process may be represented as an extended form game by using a tree which levels corresponds to the player information sets (see Fig. 1).

Player EC has two pure strategies:

1. to choose a quantity of natural resources that follow environmental standards. Let us denote this pure strategy E;
2. to exploit environment resources - strategy denoted NE.

EC is represented by the root of the game tree, as depicted in Fig. 1.

The second information set in the game tree is denoted by EV and represents the *Environment*. EV has two nodes:

1. n_1, which represents the behavior, or potential actions of *Environment* when *Economy* plays E, and
2. n_2, that represent the behavior of EV when *Economy* plays NE.

EV assigns the probability x to the node n_1 and the probability $(1-x)$ to the node n_2. This means that EV "*believes*" that EC will choose a good environmental policy with probability x.

Environment has two pure strategies:

1. to be suitable for human life and for economy (to restrict as much as possible natural resource consumption). Let us call this strategy ST;
2. to be not suitable - strategy NST.

The third player in the game is the *Society*. This player has two information sets:

1. l_1, which corresponds to the choice E of economic agent, and
2. l_2, which corresponds to the choice NE.

Each set has two nodes: l_{11} and l_{12} for l_1, and l_{21} and l_{22} for l_2. l_{11} and l_{21} follow the decision ST of the *Environment* agent. The nodes l_{12} and l_{22} follow the pure strategy NST of the *Environment*.

Society has two pure strategies:

1. to stay in this region, to live and to work here - strategy denoted by L;
2. to escape from the region - strategy denoted by NL.

In SD game, *Economy* plans the optimal production quantity for the next period of region development, by choosing either an economic strategy that follows environmental standards (E), or one strategy that destroys the environment (NE). Then *Environment* must move. *Environment* move can be suitable for people life and for economy (strategy ST) or not (strategy NST). But *Environment* is not informed about EC choice, it has just a belief about behavior of EC. Selecting an appropriate strategy, the *Environment* imposes the admissible values of resource usage for EC. Eventually, the *Society* must choose to live in this system (strategy L), or not (strategy NL), without any information about the other players moves, excluding its belief.

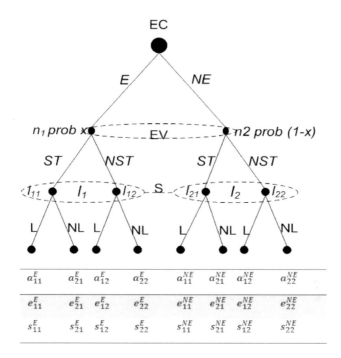

Fig. 1 SD Game Tree: tree representation of the sustainable development game

By choosing a strategy, each agent controls the development of their area, and influences the development of the other areas through external parameters of dynamical models presented in Section 3. Economy plans the development strategy by choosing corresponding quantities of natural resources, labor and capital, thus influencing Environment (through natural resources consumption) and Society (through used labor). Environment controls the quantities of natural resources that can be used by Economy, and can impose a penalty for natural resources overconsumption to Economy. Therefore, Environment influences the Economy through the limits in natural resources usage, and the Society through the availability of natural resources per capita. Finally, Society can influence the Economy and Environment through migration.

4.2   Payoffs in the SD Game

Vector (a,e,s) from SD Game Tree (see Fig. 1) represents player payoffs, where real numbers a^i_{hk}, $i \in \{E, NE\}$, $h,k \in \{1,2\}$ represents the payoff of EC corresponding to different strategy profiles, analogously, e^i_{hk}, s^i_{hk} represents the payoffs corresponding to the *Environment* (EV) and *Society* (S) respectively. *Environment* assigns probability x to play the game given by strategy E of *Economy* and $(1-x)$ to play the game given by strategy *NE*. Similarly *Society* assigns its belief parameters $(q, 1-q)$,

$0 \le q \le 1$ for information set l_1 and $(q', 1 - q')$, $0 \le q' \le 1$ for information set l_2. The probabilities q and q', represents the belief of *Society* that region will be suitable for people life, when *Economy* plays E, or NE respectively. The belief of economic agent is represented by probabilities $P_x(ST)$ and $P_x(NST)$. $P_x(ST)$ is the belief of *Economy* that the region will be suitable for economic activity. $P_x(NST)$ represents *Economy*'s belief that region will not be suitable for economic activity in next period, and is computed as: $P_x(NST) = 1 - P_x(ST)$.

Expected values for different strategies of each agent may now be computed. The expected value that *Environment* assign to the strategy ST is:

$$E_{EV}(ST) = x[qe_{11}^E + (1 - q)e_{21}^E] + (1 - x)[q'e_{11}^{NE} + (1 - q')e_{21}^{NE}]. \tag{10}$$

Expected value of the *Environment* for the strategy NST is:

$$E_{EV}(NST) = x[qe_{12}^E + (1 - q)e_{22}^E] + (1 - x)[q'e_{12}^{NE} + (1 - q')e_{22}^{NE}]. \tag{11}$$

The actual and future benefit of the *Economy* depends on the states of the *Environment* and *Society*. Analyzing these states and short term and long term goals, the economic agent chooses the most profitable strategy. The expected value that EC assigns to the strategy E is:

$$E_{EC}(E) = P_x(ST)[qa_{11}^E + (1 - q)a_{21}^E] + P_x(NST)[qa_{12}^E + (1 - q)a_{22}^E]. \tag{12}$$

Expected value for the strategy NE is:

$$\begin{aligned} E_{EC}(NE) = &P_x(ST)[q'a_{11}^{NE} + (1 - q')a_{21}^{NE}] \\ &+ P_x(NST)[q'a_{12}^{NE} + (1 - q')a_{22}^{NE}] - p(h(t), r(t)), \end{aligned} \tag{13}$$

where $p(h(t), r(t))$ represents the penalty paid by *Economy* if the consumption of natural resources is larger than admissible quantity of resource consumption imposed by the *Environment*.

The expected value assigned by *Society* to the strategy L is:

$$E_S(L) = q(s_{11}^E + s_{12}^E) + q'(s_{11}^{NE} + s_{12}^{NE}). \tag{14}$$

Expected value that *Society* assigns to the strategy NL is:

$$E_S(NL) = (1 - q)(s_{21}^E + s_{22}^E) + (1 - q')(s_{21}^{NE} + s_{22}^{NE}). \tag{15}$$

In order to simplify the representation of SD decision process, we considered each player has to choose a pure strategy. In real life each player usually prefer to play a *mixed strategy game* [10]. Using mixed strategies, players control the intensity of their SD policy which is situated between two limits: sustainable development policy, and unsustainable development policy. An important remark about DGSD model is that the concept of *game strategy* is abstract. In presented game, each player has two strategies, but the intensity of their strategies can be very different, and each strategy can be instantiated from a large set of values.

5 Sustainable Development Game Equilibrium

The development strategies of SD areas are given by the game which is played in each control point. In SD game, we assume that each player chooses the best available strategy. A strategy containing the best choice of each player represents the *Game Equilibrium* (or *Nash Equilibrium*). Game equilibrium depends, in general, on the potential choices of the other players. Each player has to form a hypothesis about the behavior of the concurrent players. For each game, Nash Equilibrium always exists for mixed strategies. Therefore SD game always has an equilibrium, which means, the players always find a compromise solution according to sustainable development criteria.

For each development iteration, every player in SD game chooses a combination between the two available pure strategies. In other words, each player plays the first pure strategy with a probability p and the other strategy with the probability $(1-p)$.

Hence *Economy* plays E with the probability p_{EC}, and *NE* with the probability $(1 - p_{EC})$. *Environment* plays ST with the probability p_{EV}, and *NST* with the probability $(1 - p_{EV})$. Finally, *Society* chooses L with probability p_S, and *NL* with the probability $(1 - p_S)$.

To compute game equilibrium, an evolutionary technique is considered [4] [5] [6]. The main advantage of this technique is the possibility to model the game having different types of rationality. Depending on these rationalities, multiple types of equilibrium exist: Nash, Pareto, Mixed. Therefore the most representative type of equilibrium may be chosen. To keep the model simple, we consider only the Nash Equilibrium.

Applying this evolutionary technique the Nash equilibrium of the SD game is detected. Detected equilibrium is represented by the set $\{p_{EC}, p_{EV}, p_S\}$. At each iteration, or control point, this set actually redirects the development of the region to a direction formed by the combination of the new goals of players. This combination represents a compromise solution between the players goals. The equilibrium corresponding to the control points actually enforces the players to follow sustainable criteria of development.

6 System Equilibrium

An important concept to study DGSD model is *balanced system evolution* based on SD criteria. In essence, DGSD system evolution is balanced if weak perturbations cause just small variations in the trajectories with respect to desired SD trajectory . The most commonly SD trajectory is that of equilibrium.

Basically, equilibrium of a dynamical system corresponds to a situation where the evolution is stopped or, as in our case, has a stable behavior in the sense that the system states become steady. In this situation we can say that the SD system acquired a sustainable development behavior. The direction of development for DGSD system is guided by its decision process for each iteration. The only way for the system to achieve equilibrium is to take optimal and viable decisions at each development iteration.

Mathematically, this steady state can be described through a system which includes the goals of areas included in SD. This system is described by equations: 4, 6 and 9.

7   Numerical Experiments

We analyze the behavior of our model in two situations:

Situation 1 is described by: *Environment* possesses sufficient stocks of renewable and nonrenewable resources for over 50 iterations, *Economy* has a capital stock for about 10 iterations, *Society* is described by normal values for almost all indicators.

Situation 2 is characterized by: the stock of nonrenewable resources is sufficient for about 30 iterations, the initial stock of renewable resources is sufficient for 2 iterations, and capital stock will be consumed in 2 iterations.

Each situation represents a start state which describes a region in terms of DGSD model. Applying DGSD model on each situation we obtain the evolution of region which is guided by DGSD model toward sustainable development. The evolution of the region for this two situations is presented in Fig. 2 and Fig. 3.

Four indices are used to represent the general sustainability of the region, and the sustainabilities of individual areas: Economy, Environment and Society. Each sustainability index of individual areas shows the state of the area for a development period, relative to the best possible state and the worst possible state. General sustainability of the region represents the arithmetic average of the indices for individual areas. Each sustainability index takes values from the interval [0, 1].

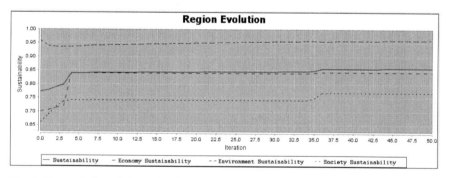

Fig. 2 The evolution of the region in Situation 1. *Economy* starts with an increase in evolution which causes society development and general sustainability to grow. But *Economy* increase collapses *Environment* evolution. Hence, the amount of natural resources which can be consumed by *Economy* is restricted. Further the region development tends to have a stable behavior with small variations

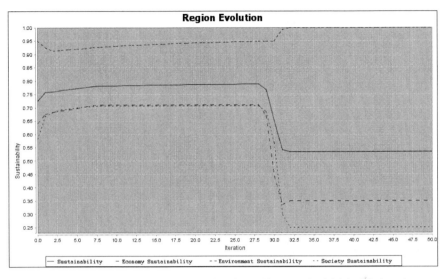

Fig. 3 The evolution of the region in Situation 2. In this case, the initial variations in areas evolution are weaker, but region achieves stability in a longer interval. After iteration 28 the development indicators of *Economy* and *Society* decrease significantly, because of nonrenewable resources deficiency. In this case, model cannot propose viable solutions to continue region evolution in a sustainable manner

The evolution diagrams show how *Environment*, *Society* and *Economy* contribute to the general sustainability of region. Here is easy to observe control principles which were described in SD game and dynamical systems:

- when *Economy* tend to have an explosive evolution, the *Environment* restricts the natural resources consumption;
- an increase in economic or environment sustainability results in a smaller increase in society sustainability;
- all areas involved in SD model participate with the same rate in general sustainability of analyzed region.

8 Conclusion and Further Work

A new model for Sustainable Development problem is proposed. This model, called Dynamical Game for Sustainable Development (DGSD), combines nature inspired optimization techniques, ideas from Dynamical Systems and Game Theory. Dynamical Systems were used to describe the evolution of each area included in SD. Using Game Theory we modeled the control points of DGSD model, as a game between these areas. Game equilibrium is computed using an evolutionary technique based on generative relations. Game equilibrium controls the evolution of each area in conformance with region sustainability principles.

Game Theory captures very well strategic situations of each player involved in the game and the conflicts existing between the areas. We proved that using game equilibrium, each area involved in region development is enforced to implement just "sustainable" decisions, otherwise they are constrained by the other areas.

However, game equilibrium alone, does not guarantee to drive the region to maximum sustainability in all the cases. It finds just the local optimum, and as it is, does not use the idea or concept of long term optimum. Nash equilibrium imposes players to be selfish, but in our case the players must put the interest of the whole region development over their individual interest. In other words, the player must consider an *altruistic behavior* related to the development of the region where they act.

To avoid the drawback of local optimum we intend to combine our model with approaches that guarantees the convergence of the model to the global optimum, which in our case is maximum sustainability. Further, we intend to integrate DGSD model with different decision making approaches. We may consider an alternative approach for decision process which is based on Public Good Games(PGG) [8]. PGGs describe social behavior through public goods, and models players diversity depending on their percentage of collaboration.

We hope that DGSD model is a good start point to study Sustainable Development. Even if with present form, this model does not provide best strategies to drive any region to SD development, there are a lot of possibilities to extend and to improve it, preserving its fundamental ideas.

References

[1] Chertow, M.R.: The IPAT Equation and Its Variants. Journal of Industrial Technology 4.4, 13–29 (2001)
[2] De Lara, M., Doyen, L.: Sustainable Management of Natural Resources. Mathematical Models and Methods. Springer, Berlin (2008)
[3] Dalal-Clayton, B., Bass, S.: Sustainable Development Strategies. Earthscan Publications Ltd., London (2002)
[4] Dumitrescu, D., Lung, R.I., Mihoc, T.D.: Evolutionary Equilibria Detection in Non-cooperative Games. In: Giacobini, M., Brabazon, A., Cagnoni, S., Di Caro, G.A., Ekárt, A., Esparcia-Alcázar, A.I., Farooq, M., Fink, A., Machado, P. (eds.) EvoCOMNET. LNCS, vol. 5484, pp. 253–262. Springer, Heidelberg (2009)
[5] Dumitrescu, D., Lung, R.I., Mihoc, T.D.: Generative Relations for Evolutionary Equilibria Detection. In: GECCO 2009, pp. 1507–1512 (2009)
[6] Dumitrescu, D., Lung, R.I.: ESCA: A new Evolutionary-Swarm Cooperative Algorithm. In: NICSO 2007, pp. 105–114 (2007)
[7] Gujarati, D.: Basic Econometrics. McGraw-Hill, Columbus (2003)
[8] Janssen, M.A., Ahn, T.: Adaptation vs. Anticipation in Public-Good Games (2003), http://www.allacademic.com/meta/p64827_index.html
[9] Mishra, S.K.: A Brief History of Production Functions. Social Science Research Network (2007)
[10] Osborne, M.J., Rubinstein, A.: A Course in Game Theory. MIT Press, Cambridge (1994)
[11] Zgurovsky, M.Z.: Sustainable Development Global Simulation: Opportunities and Treats to the Planet, Kyiv (2007)

Studying the Influence of the Objective Balancing Parameter in the Performance of a Multi-Objective Ant Colony Optimization Algorithm

A.M. Mora, J.J. Merelo, P.A. Castillo, J.L.J. Laredo,
P. García-Sánchez, and M.G. Arenas

Abstract. Several multi-objective ant colony optimization (MOACO) algorithms use a parameter λ to balance the importance of each one of the objectives in the search. In this paper we have studied two different schemes of application for that parameter: keeping it constant, or changing its value during the algorithm running, in order to decide the configuration which yields the best set of solutions. We have done it considering our MOACO algorithm, named hCHAC, and two other algorithms from the literature, which have been adapted to solve the same problem. The experiments show that the use of a variable value for λ yields a wider Pareto set, but keeping a constant value for this parameter let to find better results for any objective.

1 Introduction

The *military unit path-finding problem* consists in getting the best path for a military unit, from an origin to a destination point in a battlefield, keeping a balance between route speed and safety, considering the presence of enemies (which can fire against the unit) and taking into account some properties and restrictions which make the problem more realistic. Being speed (important if the unit mission requires arriving as soon as possible to the target) and safety (important when the enemy forces are not known or when the unit effectives are very valuable), the two main criteria that the commander of a unit takes into account inside a battlefield in order to accomplish the mission with success.

To solve this problem we designed an Ant Colony Optimization algorithm [4] adapted to deal with two objectives (see [3] for a survey on multi-objective optimization), named hCHAC [10] so it is a Multi-Objective Ant Colony Optimization Algorithm (MOACO [5]).

A.M. Mora · J.J. Merelo · P.A. Castillo · J.L.J. Laredo · P. García-Sánchez · M.G. Arenas
Dpto. Arquitectura y Tecnología de Computadores. University of Granada, Spain
e-mail: {amorag, jmerelo, pedro, juanlu,
          pgarcia, mgarenas}@geneura.ugr.es

J.R. González et al. (Eds.): NICSO 2010, SCI 284, pp. 163–176, 2010.
springerlink.com   © Springer-Verlag Berlin Heidelberg 2010

As most of the metaheuristics, it considers a set of parameters which determines the behaviour in the search, usually having an effect in the exploration and exploitation balance. But in hCHAC case, there is also an additional (and key) parameter, named λ. It was introduced in the algorithm rule of decision, to set the relative importance of each one of the objectives in the search. But actually, it sets the area in the space of solutions, that each of the ants explores, yielding somewhat of 'specialised' ants in each one of the objectives (or both of them).

Some of the parameters of hCHAC and their influence in the search, were analysed in a previous work [8] using statistical methods, reaching some conclusions in addition to the best set of values for them.

In the present work the λ parameter has been studied, but statistics have not been applied in the analysis, because λ just weights the relative importance of one objective with respect to the other (there are just two objectives in this problem) and does not take concrete (and sometimes hard-coded) values as the rest of parameters. Moreover, two different parameter schemes can be applied in the algorithm, and the aim of the analysis is to study their influence in the search and decide which one is the best scheme.

The rest of the paper is structured as follows. Firstly, the problem to solve and its modelling in a simulator environment is briefly described in Section 2. Then the hCHAC and the literature algorithms (adapted to solve the problem) are introduced respectively in Sections 3 and 4. The parameter to study (λ)) is commented in Section 5. Section 6 shows the performed analysis, by presenting some problem instances and the results of the experiments. Finally, in Section 7 the conclusions and the future work in this line are exposed.

2   The Military Unit Bi-criteria Pathfinding Problem

The MUPFP-2C is modelled considering that the unit has got a *level of energy (health)* and a *level of resources*, which are consumed when it moves along the path, so the problem objectives are adapted to minimize the resources and energy consumption. The problem is located inside a battlefield which is modelled as a grid of hexagonal cells with a *cost in resources*, which represents the difficulty of going through it, and a *cost in energy/health*, which means the unit depletes its human resources or vehicles suffer damage when crossing over the cell (*no combat casualties*). Both costs depend on the cell type. Besides, moving between cells with different heights also costs resources (more if it goes up), and falling in a weapons impact zone depletes energy. All these features are represented using different colors in the maps.

Figure 1 shows an example of real world battlefield and the information layer associated to it, which has been created using a custom-made application [1].

We consider *fast* paths (if speed is constant) when the total cost in resources is low (it is not very difficult to travel through the cells, so it takes little time). On the other hand *safe* paths, have associated a low cost in energy/health.

See [10] for further details about the problem definition and modelling.

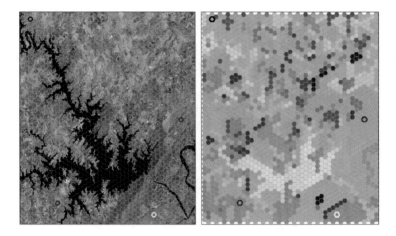

Fig. 1 Example Map (45x45 cells). The image on the left-hand side is a real world picture showing a lake surrounded by some hills and lots of vegetation. On the right-hand side it is shown its associated information layer. The different shades in the same color models height (light color) and depth (dark color). There are two enemies with red border, an origin point (in the top-left corner of the images) with black border and a destination point (in the bottom-right) with yellow border

3   hCHAC Algorithms

There were designed some algorithms to solve the commented problem, all of them were included in the so-called $hCHAC^1$ family. In this work two of them will be considered.

The main approach, also known as **hCHAC** [9] is an Ant Colony System (ACS) [4] adapted to deal with two objectives (MOACO) [3, 5]. Since it is an ant algorithm, the problem is transformed into a graph where each node corresponds to a cell in the map, and each edge between two nodes is the connection between neighbor cells in the map. Every edge has associated two weights, which are the costs in resources and health that going through that edge causes to the unit.

In every iteration, the ants separately build a complete path (solution), between the origin and destination points (if possible), by travelling through the graph. To guide this movement they use a State Transition Rule (STR) which combines two kinds of information: pheromone trails (learnt information) and heuristic knowledge.

The MUPFP-2C has two independent objectives to minimize. These objectives are named f, minimization of the resources consumed in the path (fast path) and s, minimization of the energy/health consumed in the path (safe path).

hCHAC uses two pheromone matrices (τ_f, τ_s) and two heuristic functions (η_f, η_s) (one per objective), a single colony, and two STRs: (*Combined State Transition*

---

[1] Which means *Compañía de Hormigas ACorazadas* (Armored Ant Company) with the prefix 'hexa' due to the grid topology where it works.

Rule, CSTR), similar to the one proposed in [6] and (*Dominance State Transition Rule, DSTR*), which ranks neighbouring cells according to how many (of the neighbours) they dominate.

The local and global pheromone updating formulae are based in the MACS-VRPTW algorithm proposed in [2], with some changes due to the use of two pheromone matrices. Finally, there are two evaluation functions (used to assign a global cost value to every solution found) named F_f (minimization of resources consumption) and F_s (minimization of energy consumption).

The definition and formulae of all these features can be found in [9].

The CSTR uses the λ parameter, so it is the rule which we are going to consider in the present study. It is defined as:

If $(q \le q_0)$

$$j = \arg\max_{j \in N_i} \left\{ \tau_f(i,j)^{\alpha \cdot \lambda} \cdot \tau_s(i,j)^{\alpha \cdot (1-\lambda)} \cdot \eta_f(i,j)^{\beta \cdot \lambda} \cdot \eta_s(i,j)^{\beta \cdot (1-\lambda)} \right\} \tag{1}$$

Else

$$P(i,j) = \begin{cases} \dfrac{\tau_f(i,j)^{\alpha \cdot \lambda} \cdot \tau_s(i,j)^{\alpha \cdot (1-\lambda)} \cdot \eta_f(i,j)^{\beta \cdot \lambda} \cdot \eta_s(i,j)^{\beta \cdot (1-\lambda)}}{\sum_{u \in N_i} \tau_f(i,u)^{\alpha \cdot \lambda} \cdot \tau_s(i,u)^{\alpha \cdot (1-\lambda)} \cdot \eta_f(i,u)^{\beta \cdot \lambda} \cdot \eta_s(i,u)^{\beta \cdot (1-\lambda)}} & if \ j \in N_i \\ \\ 0 & otherwise \end{cases} \tag{2}$$

In that rule, $q_0 \in [0,1]$ is the standard ACS parameter and q is an uniformly random selected value in [0,1]. τ_f, τ_s and η_f, η_s are the previously commented matrices and functions. α and β are the usual (in ACO algorithms) weighting parameters for pheromone and heuristic information respectively, and N_i is the current feasible neighbourhood for the node i. As can be seen, the λ parameter is used to weight the terms (pheromone and heuristic values) related to each one of the objectives, using its value 'λ' for the first term and the complementary '$(1 - \lambda)$' for the second.

This state transition rule works as follows: when an ant is building a solution path and is placed at one node i, a random number q in [0,1] is generated, if $q \le q_0$ the best neighbor j is selected as the next node in the path (Equation 1). Otherwise, the algorithm decides which node is the next by using a roulette wheel considering $P(i,j)$ as probability for every feasible neighbour j (Equation 2).

The DSTR has not been taken into account in this work, because it does not use the λ parameter, since the objectives are considered as completely independent in the rule, and they are not weighted and combined (it is not an aggregative function).

The other proposed algorithm which has been studied in this work is **hCHAC-4** [7], a redefinition of the bi-criteria hCHAC focused to deal with four objectives, since each one of the main considered criteria can be subdivided into two sub-objectives, this way: *speed* can be defined as distance to target point minimization (straight paths) and cost in resources minimization; while *safety* can be understood as visibility[2] to enemies and cost in energy/health minimizations.

So, the four objectives to minimize, have been considered separately: *resources consumption* (**r**), *distance to target point* (**d**), *energy consumption* (**e**) and *visibility to enemies* (**v**); the two first are related to speed, and the others to safety.

hCHAC-4 is also a MOACO algorithm that works in a graph (which models the battlefield), but considering four weights in each edge. It is again an ACS, so it uses the q_0 parameter in the STR. All the elements of the algorithm have been adapted to deal with four objectives, so there are four heuristic functions, four pheromone matrices, and four evaluation functions. In addition, and as in hCHAC, there are two different state transition rules, which work considering four objectives this time.

The CSTR-4 is similar to the CSTR of hCHAC, but involving four terms (one per objective). Each one of these terms is defined as follows:

$$T_x(i,j) = \tau_x(i,j)^\alpha \cdot \eta_x(i,j)^\beta \tag{3}$$

τ is the correspondent pheromone trails matrix, η is the heuristic function, and $x = r, d, e, v$. α and β are the usual (in ACO algorithms) weighting parameters for pheromone and heuristic information respectively. So, the STR for four objectives (CSTR-4) is:

If $(q \leq q_0)$

$$j = \arg\max_{j \in N_i} \left\{ T_r(i,j)^\lambda \cdot T_d(i,j)^\lambda \cdot T_e(i,j)^{(1-\lambda)} \cdot T_v(i,j)^{(1-\lambda)} \right\} \tag{4}$$

Else

$$P(i,j) = \begin{cases} \dfrac{T_r(i,j)^\lambda \cdot T_d(i,j)^\lambda \cdot T_e(i,j)^{(1-\lambda)} \cdot T_v(i,j)^{(1-\lambda)}}{\displaystyle\sum_{u \in N_i} T_r(i,u)^\lambda \cdot T_d(i,u)^\lambda \cdot T_e(i,u)^{(1-\lambda)} \cdot T_v(i,u)^{(1-\lambda)}} & if\ j \in N_i \\[4ex] 0 & otherwise \end{cases} \tag{5}$$

Where all the parameters and terms are the same as in the CSTR of hCHAC. As can be seen, the λ parameter is also used in this equation. It sets the importance of all objectives at the same time, since they are related with speed (resources consumption and distance to target point) or with safety (energy consumption or visibility to enemies). The rule works as the previously commented STR.

Again, the DSTR (for four objectives this time) will not be considered in this study since it does not use the λ parameter.

---

[2] It is considered a *cost in visibility* with regard to the enemies, which is minimum if the unit is hidden (at a point) to all the enemies, and it increases exponentially when it is visible to any (or some) of them. With no enemy present, it is computed taking into account whether it is visible to the surrounding cells in a radius, calculating a score, so the higher number of cells can see the unit, the higher the score is.

4 Algorithms to Compare

In previous works we considered two MOACO algorithms to make results comparisons which those yielded by hCHAC family methods. They had been presented by other authors in literature and were adapted to solve the MUPFP-2C. Both of them use the λ parameter in the STR to weight the objectives and to address the search performed by the ants. We will consider them in this study in order to get more general conclusions.

The first one is ***MOACS*** (Multi-Objective Ant Colony System), which was proposed by Baran et al. [2], to solve the Vehicle Routing Problem with Time Windows (VRPTW). It uses a single pheromone matrix for both objectives (instead of one per objective).

It has been adapted to solve the MUPFP-2C [9], so it considers the same heuristic and evaluation functions (see them in [10]), but different STR and pheromone updating formulae. The STR is similar to the CSTR in hCHAC, but using only one pheromone matrix (as we previously said). It is defined as follows:

If $(q \le q_0)$

$$j = \arg\max_{j \in N_i} \left\{ \tau(i,j) \cdot \eta_f(i,j)^{\beta \cdot \lambda} \cdot \eta_s(i,j)^{\beta \cdot (1-\lambda)} \right\} \tag{6}$$

Else

$$P(i,j) = \begin{cases} \dfrac{\tau(i,j) \cdot \eta_f(i,j)^{\beta \cdot \lambda} \cdot \eta_s(i,j)^{\beta \cdot (1-\lambda)}}{\displaystyle\sum_{u \in N_i} \tau(i,u) \cdot \eta_f(i,u)^{\beta \cdot \lambda} \cdot \eta_s(i,u)^{\beta \cdot (1-\lambda)}} & if \ j \in N_i \\[4ex] 0 & otherwise \end{cases} \tag{7}$$

Where all the terms and parameters are the same as in Equation 2. This rule also uses λ to balance the importance of the objectives in the search. The rule works as we previously explain for hCHAC and hCHAC-4.

Since *MOACS* is an ACS, there are two levels of pheromone updating, local and global. There has been defined new equations (in respect to the author's original algorithm definition) for both tasks, in addition to a new reinitialization mechanism, which can be also consulted in [9].

The evaluation functions for each objective F_f and F_s, are the same as in the previous approaches.

The second algorithm is ***BiAnt*** (BiCriterion Ant), which was proposed by Iredi et al. [6] as a solution for a multi-objective problem with two criteria (the Single Machine Total Tardiness Problem, SMTTP). It is an Ant System (AS) which uses just one colony, and two pheromone matrices and heuristic functions (one per objective). So, the STR is similar to the CSTR of hCHAC, but without consider the q_0 parameter, it is:

$$P(i,j) = \begin{cases} \dfrac{\tau_f(i,j)^{\alpha \cdot \lambda} \cdot \tau_s(i,j)^{\alpha \cdot (1-\lambda)} \cdot \eta_f(i,j)^{\beta \cdot \lambda} \cdot \eta_s(i,j)^{\beta \cdot (1-\lambda)}}{\displaystyle\sum_{u \in N_i} \tau_f(i,u)^{\alpha \cdot \lambda} \cdot \tau_s(i,u)^{\alpha \cdot (1-\lambda)} \cdot \eta_f(i,u)^{\beta \cdot \lambda} \cdot \eta_s(i,u)^{\beta \cdot (1-\lambda)}} & if\ j \in N_i \\[4mm] 0 & otherwise \end{cases} \qquad (8)$$

Where all the terms and parameters are again the same as in Equation 2. In addition, the rule uses the λ parameter to weight the objectives in the search. The rule works in the same way that hCHAC CSTR but without consider the random number q, just directly calculating the probability for the feasible nodes using this formula and using a roulette wheel to choose the next node in the path.

The definition of the heuristic and evaluation functions are the same as in hCHAC. But, the pheromone updating scheme is different since BiAnt is an AS. So it is just performed a global pheromone updating, including evaporation in all nodes and contribution just in the edges of the best paths to the moment (those included in the Pareto Set (PS)).

5   The λ Parameter

As shown in the equations applied in the algorithms, there are some parameters considered in their expressions. Most of them were previously analysed [8] in order to determine their influence on the behaviour of the main algorithm (hCHAC) and the best set of values that they should take to yield the best solutions.

However, there is a very important parameter which has not been studied yet. It is present in all the STRs (of the commented algorithms) and, since these rules are the most important factor in every ACO algorithm, this parameter is key in the algorithm performance. It is λ, the parameter which determines the importance of each one of the objectives in the STR.

$\lambda \in [0,1]$, has been to the moment user-defined[3], determining which objective has higher priority and how much. If the user decides to search for a fast path, λ will take a value close to 1, on the other hand, if he wants a safe path, close to 0.

This value has been considered as constant during the algorithm for all ants, so the algorithms always search in the same zone of the space of solutions (the zone related to the chosen value for λ). That is, all the ants search in the same area of the Pareto Front (PF) [3], yielding solutions (in average) with similar costs in both objectives. These cost would maintain the relationship determined by the weight set through the λ value.

This was the initial idea applied in hCHAC and in hCHAC-4 [7, 10], and this scheme has been also implemented in the adapted (to the MUPFP-2C) versions of MOACS and BiAnt, commented in previous section.

These algorithms were initially defined [2, 6] with a different policy for λ, which consists in assign a different value for the parameter to each ant h, following the expression:

---

[3] This algorithms have been applied inside a simulator where an user can be determine in advance the importance of each objective in the search.

$$\lambda_h = \frac{h-1}{m-1} \qquad \forall h \in [1,m] \qquad (9)$$

Considering that there are m ants, the parameter takes an increasing value that goes from 0 for the first ant to 1 for the last one. This way, the algorithms search in all the possible areas of the space of solutions (each ant is devoted to a zone of the PF).

This is the recommended scheme for solving classical MO problems, in which the biggest (and fittest) Pareto Set (PS) is wanted to be obtained, but the MUPFP-2C is addressed to get a set of solutions according to the user decision, that is, a set of solutions with the desired relationship of importance between the objectives.

So, the idea could be to use this search scheme and to restrict the yielded solutions using λ once the final PS has been obtained.

This way, the aim of the study is to decide the best scheme for applicating λ: the *constant scheme*, where the value for the parameter is set at the beginning of the algorithm for all the ants; and the *variable scheme*, where every ant considers its own value for the parameter during the search and the user-criteria is applied at the end of the run for restrict the solutions in the PS.

6   Experiments and Results

In order to study the different configurations for λ, some problems have been solved using each one of the commented algorithms, considering in addition each one of the schemes: constant and variable λ application.

So, the experiments have been performed in three different (and realistic) maps, modelled from some screens of the PC Game Panzer General™. These maps are *PG-Forest Map* (Figure 2), *PG-River Map* (Figure 3) and *PG-Mountain Map* (Figure 4).

All the algorithms have been run in these maps using the same parameter values: $\alpha=1$, $\beta=2$, $\rho=0.1$ and $q_0=0.4$ (tending to an exploitative search more as usual in ACO algorithms). The λ parameter has taken values 0.9 and 0.1 in the constant scheme to consider one objective with higher priority than the other.

All these MOACOs yield a set of non-dominated solutions, but less than usual in this kind of algorithms, since it only searches in the region of the ideal Pareto front determined by the λ parameter. In addition, we usually only consider one (which is chosen by the military staff following their own criteria and the features of each problem).

The considered evaluation functions are: F_f (minimization of the resources consumed in the path, or fast path), and F_s (minimization of the energy consumed in the path, or safe path). As a reminder, even in the hCHAC-4 algorithm, the final solutions are evaluated using these functions to compare with the yielded results by the other algorithms. 30 independent runs (1500 iterations and 30 ants) have been performed with each one of the algorithms, using both schemes for λ, and searching for the fastest and safest paths (in two different runs) in the case of the constant scheme, and searching for both types of paths in the variable scheme.

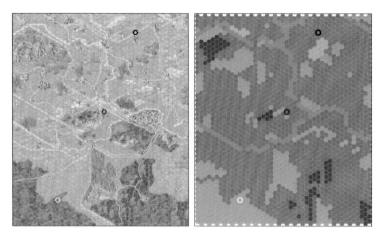

Fig. 2 PG-Forest Map. 45x45 cells map where some patches of forest are shown, there are also some small villages and hills. The unit is located at the north (black border cell), the target at the south (with yellow border), and there is one enemy placed in the centre (red border cell). On the right figure it is shown the underlying information layer which models the map on the left figure

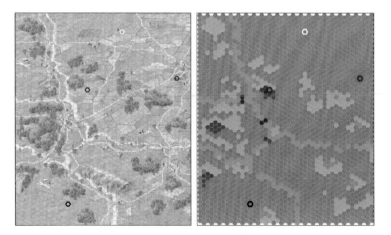

Fig. 3 PG-River Map. 45x45 cells map where it is modelled an scenery with some villages and cities, there are also some rivers and bridges, a patch of forest and some hills. The unit is placed at the south (black border cell) and the target point at the north (with yellow border). There are two enemies (cells with red border), one of them firing at the zone surrounding him and also at some bridges (cells in red color), and the other one watching over on the top of a hill. On the right figure it is shown the underlying information layer which models the map on the left

At the end of every run, and depending on the scheme, some of the solutions in the PS have been chosen, so in the constant approach the best solution in the correspondent PS (fast paths PS or safe paths PS), looking at the cost related to

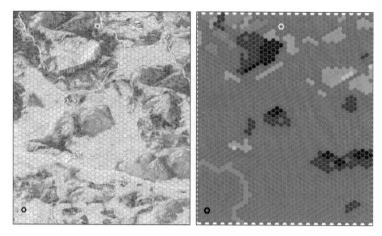

Fig. 4 PG-Mountain Map. 45x45 cells map modelling a mountainous zone, with many mountains, hills, hollows and valleys. The problem unit is placed at the south-west (with black border) and the target point at the north (with yellow border). There is no known enemy. The right figure shows the underlying information layer which models the map on the left

the preferred objective is selected: the one with the smallest F_f cost in the case of fast paths (speed objective with higher priority), and the one with the smallest F_s cost in the case of safe paths (safety objective with higher priority). In the variable approach, the best solutions depending on each one of these costs are selected, but just in a single PS (containing fast and safe paths). Once the best solutions in all the runs have been chosen, the mean and standard deviation of all of them have been calculated and presented in the Tables 1, 2 and 3.

Table 1 λ study results for the four algorithms in PG-Forest Map. 1500 iterations, 50 ants

			Fastest ($\lambda=0.9$)		Safest ($\lambda=0.1$)	
			F_f	F_s	F_f	F_s
Constant λ	hCHAC	Best	**68.50**	295.40	80.50	**7.30**
		Mean	**77.88** ±7.84	166.20 ±131.08	84.67 ±3.64	**8.02** ±0.55
	hCHAC-4	Best	70.00	305.50	89.00	8.30
		Mean	79.52 ±6.67	322.28 ±37.59	110.33 ±14.18	**73.94** ±**86.66**
	MOACS	Best	74.00	286.00	89.50	8.20
		Mean	83.03 ±**5.08**	227.03 ±111.94	101.95 ±6.71	9.29 ±0.53
	BiAnt	Best	84.50	297.00	146.50	13.90
		Mean	123.82 ±32.86	320.49 ±75.44	158.75 ±32.25	284.47 ±152.01
Variable λ	hCHAC	Best	**68.50**	295.40	80.50	**7.30**
		Mean	80.60 ±6.36	85.84 ±118.90	84.98 ±3.34	8.11 ±0.51
	hCHAC-4	Best	74.50	285.90	96.00	9.30
		Mean	93.62 ±10.27	290.07 ±101.75	110.98 ±15.78	195.59 ±89.03
	MOACS	Best	77.50	286.20	92.50	8.20
		Mean	86.53 ±5.51	214.02 ±110.22	97.28 ±5.66	**9.16** ±**0.68**
	BiAnt	Best	101.00	238.80	129.00	12.30
		Mean	138.37 ±32.13	314.30 ±111.90	145.47 ±29.88	288.32 ±111.31

Table 2 λ study results for the four algorithms in PG-River Map. 1500 iterations, 50 ants

			Fastest (λ=0.9)		Safest (λ=0.1)	
			F_f	F_s	F_f	F_s
Constant λ	hCHAC	Best	61.00	244.90	74.00	27.30
		Mean	66.42 ±3.29	225.19 ±90.26	84.68 ±4.89	28.36 ±0.48
	hCHAC-4	Best	66.00	285.20	81.00	28.00
		Mean	71.70 ±3.70	316.66 ±58.73	98.13 ±15.99	108.46 ±63.79
	MOACS	Best	64.00	304.90	77.00	27.60
		Mean	70.77 ±2.43	294.66 ±79.44	93.60 ±6.92	29.23 ±0.67
	BiAnt	Best	74.00	256.00	116.50	41.20
		Mean	100.27 ±16.71	279.70 ±153.73	135.90 ±31.96	287.33 ±135.75
Variable λ	hCHAC	Best	64.50	235.30	72.00	27.10
		Mean	68.23 ±3.41	178.12 ±47.92	82.37 ±5.48	28.14 ±0.54
	hCHAC-4	Best	68.50	295.40	111.00	50.90
		Mean	81.40 ±10.01	302.24 ±46.05	101.22 ±19.55	212.00 ±56.22
	MOACS	Best	64.50	295.00	76.00	27.50
		Mean	71.67 ±2.90	244.90 ±61.14	91.00 ±6.67	28.97 ±0.66
	BiAnt	Best	75.50	316.00	139.50	43.50
		Mean	119.63 ±33.12	325.58 ±143.87	128.55 ±33.90	272.38 ±150.70

Table 3 λ study results for the four algorithms in PG-Mountains Map. 1500 iterations, 50 ants

			Fastest (λ=0.9)		Safest (λ=0.1)	
			F_f	F_s	F_f	F_s
Constant λ	hCHAC	Best	74.36	352.66	80.53	336.18
		Mean	76.43 ±0.99	352.39 ±8.98	81.66 ±2.49	354.61 ±11.86
	hCHAC-4	Best	75.99	365.25	82.75	360.59
		Mean	84.33 ±5.81	398.48 ±32.09	88.88 ±6.45	395.40 ±29.07
	MOACS	Best	79.15	378.63	85.31	351.86
		Mean	84.45 ±2.73	388.24 ±15.40	87.09 ±2.27	382.93 ±17.44
	BiAnt	Best	87.57	397.36	96.47	415.56
		Mean	116.65 ±20.98	528.56 ±96.95	138.06 ±26.57	620.65 ±117.80
Variable λ	hCHAC	Best	75.84	362.32	86.12	308.06
		Mean	78.45 ±1.53	340.12 ±15.50	83.43 ±2.12	322.75 ±14.60
	hCHAC-4	Best	82.93	404.62	87.66	359.07
		Mean	89.57 ±4.37	415.88 ±28.63	91.52 ±5.93	408.38 ±26.80
	MOACS	Best	80.23	370.26	89.42	354.34
		Mean	83.99 ±2.18	386.22 ±16.12	85.80 ±3.68	376.42 ±13.45
	BiAnt	Best	86.58	417.52	86.58	417.52
		Mean	124.29 ±28.31	570.14 ±135.73	124.39 ±28.36	570.13 ±135.73

In these tables, data is grouped mainly into two big columns, depending on the criteria with the higher priority (fastest or safest), so the cost function corresponding to this criteria is the most interesting (F_f in the fastest case, and F_s in the safest case) and the other one takes always worse values, since it corresponds to the secondary objective in the search.

Table 1 shows that results corresponding to the constant scheme are better in general; the best cost only slightly, but clearly on average. In general, the costs associated to the preferred objective are better in the constant scheme, sometimes the best solutions are the same (as in Table 1 for hCHAC), but the mean and standard deviation demonstrate that they are worse.

There is a fact to point out, the mean results in the variable scheme for the objective which is not being minimised (the less important), are better than in the constant case. It is reasonable since in this case, the ants search in the whole space of solutions, yielding good solutions not only in one of the objectives, but in both of them. This should be interesting in most MO problems, where the aim is finding a good solution in all the objectives, or yielding as much solutions in the PF as possible, but in the MUPFP-2C, the user just want one solution in each case (or a set of solutions) which minimises the most important objective. So the best option would be the constant scheme.

In the hCHAC-4 case, the differences are more remarkable looking at the best solutions and much more remarkable looking at the means. The reason is that considering four objectives to minimise, makes the search space bigger, so it is quite difficult to get a good set of solutions in the same time. In fact, it is more complicated to yield good solutions considering just one objective (the preferred one), if the algorithm explores the whole space of solutions (variable scheme), than if the algorithm search just in a concrete area (constant scheme). But again, better solutions for the secondary objective are obtained.

Relating to MOACS, there are small differences between the results of both schemes, being a bit better the constant approach, except in one case in which the mean for the variable configuration is better, but the standard deviation is worse, so in an averaged sense, results favours again to the constant scheme.

BiAnt shows stronger differences favouring to the constant configuration too, due to the higher exploration factor associated to the algorithm (since it is an AS), as can be seen in the high standard deviation of its results. So this approach takes advantage of the extra exploitation factor that adds the constant λ scheme.

Looking at the Table 2 results, they are quite similar to those commented (on the previous table), being better in general for the constant configuration. But this time there are some exceptions in which the variable scheme yields better solutions, even considering the mean and standard deviation (MOACS case). This happens always in the F_s cost for the searching of the safest paths, since there are just a few number of safe paths, all of them moving in a concrete area of the map, since both enemies are watching over the greater part of the scenery. So just the left-side zone, which is far from them, and behind some forest patches and hills is a safe zone. This way, the most of the safest solutions move through that area (they are quite similar), and the variable scheme explores some more possibilities inside the good ones.

The third map (which results are showed in Table 3) is an special case, since there are no known enemies on it. So both, the fastest and the safest paths, are quite straight from the origin to the target point (moving through the most hidden cells, to their environment, in the safest case). This means that in the constant scheme just a small area in the search space is explored (the one surrounding the most straight solution), but in the variable approach, there is a higher exploration around this straight zone, which yields better solutions on average when the algorithms search for safe solutions. This map has been included in the study to show that the application of a variable scheme could be better in maps with a 'very restricted' set of solutions, where there are a small zone of the space of solutions to explore due to the problem definition.

7 Conclusions and Future Work

In this paper, two different schemes of application for the parameter which sets the relevance of two objectives, in four MOACOs designed to solve the military unit path-finding problem, considering speed and safety, have been analysed. This parameter is known as λ, and it is applied in the State Transition Rule of the algorithms.

The *variable scheme* consists in assign a different value to each of the ants in the algorithm in order to search in the whole space of solutions (every ant searches in an area). The other configuration, named *constant scheme*, determines in advance one value which is used by all the ants, so they search in the same area of the space of solutions. Some experiments have been performed over three realistic maps and the general conclusion reached is the constant approach yields better solutions most of times, following the user (of the application which applies the algorithms) criteria.

In general the use of the constant λ scheme implies a higher exploitation of solutions (since all the ants search in a concrete area). On the other hand, the variable λ scheme adds an exploration factor to the algorithms, since each ant searches in a different area of the space of solutions, yielding different solutions (better sometimes) and a bigger Pareto Set (in the ideal case). The second approach would be better for most of the multi-objective algorithms, to solve ordinary or common multi-objective problems, since the aim is to find as much solutions as possible (the biggest PS). But in the problem addressed in this work, the aim is to get the best solution considering a set priority for both objectives, so the constant scheme has demonstrated to be better. This way, the two algorithms taken from the literature and adapted to solve this problem (MOACS and BiAnt), show that the constant configuration yields better results in this case, but they were defined considering a variable approach, since their aim were to solve general MO problems.

There are some ideas as future work in this line. The first one is to test both schemes in some other problems, considering the best set of values for the other parameters in each one of the algorithms, since some of them should show some differences between the results yielded by both schemes when they have a correct exploitation factor, as BiAnt case. Another approach consists in the implementation of the proposed algorithms to solve common MO problems. In this case, we will perform a new experimental set to determine the best λ application scheme.

Finally, we would like to implement the auto-evaluation of the scenario to fix the best set of values before the running of the algorithm. Also including the decision of the most appropriate scheme for applicating λ.

Acknowledgements

This paper has been funded in part by the Spanish MICYT projects NoHNES (Spanish Ministerio de Educación y Ciencia - TIN2007-68083) and TIN2008-06491-C04-01 and the Junta de Andalucía P06-TIC-02025 and P07-TIC-03044.

References

[1] Mini-Simulator hCHAC (2008),
    http://forja.rediris.es/frs/download.php/1355/mss_chac.zip
[2] Barán, B., Schaerer, M.: A multiobjective ant colony system for vehicle routing problem
    with time windows. In: IASTED International Multi-Conference on Applied Informat-
    ics. IASTED IMCAI, vol. 21, pp. 97–102 (2003)
[3] Coello, C.A.C., Veldhuizen, D.A.V., Lamont, G.B.: Evolutionary Algorithms for Solv-
    ing Multi-Objective Problems. Kluwer Academic Publishers, Dordrecht (2002)
[4] Dorigo, M., Stützle, T.: The ant colony optimization metaheuristic: Algorithms, appli-
    cations, and advances. In: Glover, F., Kochenberger, G. (eds.) Handbook of Metaheuris-
    tics, pp. 251–285. Kluwer, Dordrecht (2002)
[5] García-Martínez, C., Cordón, O., Herrera, F.: An empirical analysis of multiple objec-
    tive ant colony optimization algorithms for the bi-criteria TSP. In: Dorigo, M., Birattari,
    M., Blum, C., Gambardella, L.M., Mondada, F., Stützle, T. (eds.) ANTS 2004. LNCS,
    vol. 3172, pp. 61–72. Springer, Heidelberg (2004)
[6] Iredi, S., Merkle, D., Middendorf, M.: Bi-criterion optimization with multi colony ant
    algorithms. In: Zitzler, E., Deb, K., Thiele, L., Coello Coello, C.A., Corne, D.W. (eds.)
    EMO 2001. LNCS, vol. 1993, pp. 359–372. Springer, Heidelberg (2001)
[7] Mora, A., Merelo, J., Laredo, J., Castillo, P., Sánchez, P., Sevilla, J., Millán, C., Torrecil-
    las, J.: hCHAC-4, an ACO algorithm for solving the four-criteria military path-finding
    problem. In: Krasnogor, N., Nicosia, G., Pavone, M., Pelta, D. (eds.) Proceedings of the
    International Workshop on Nature Inspired Cooperative Strategies for Optimization.
    NICSO 2007, pp. 73–84 (2007)
[8] Mora, A., Merelo, J., Castillo, P., Laredo, J., Cotta, C.: Influence of parameters on the
    performance of a moaco algorithm for solving the bi-criteria military path-finding prob-
    lem. In: WCCI 2008 Proceedings, pp. 3506–3512. IEEE Press, Los Alamitos (2008)
[9] Mora, A.M., Merelo, J.J., Millán, C., Torrecillas, J., Laredo, J.L.J., Castillo, P.A.:
    Comparing aco algorithms for solving the bi-criteria military pathfinding problem. In:
    Almeida e Costa, F., Rocha, L.M., Costa, E., Harvey, I., Coutinho, A. (eds.) ECAL
    2007. LNCS (LNAI), vol. 4648, pp. 665–674. Springer, Heidelberg (2007)
[10] Mora, A.M., Merelo, J.J., Millán, C., Torrecillas, J., Laredo, J.L.J., Castillo, P.A.: En-
    hancing a MOACO for solving the bi-criteria pathfinding problem for a military unit
    in a realistic battlefield. In: Giacobini, M. (ed.) EvoWorkshops 2007. LNCS, vol. 4448,
    pp. 712–721. Springer, Heidelberg (2007)

HC12: Highly Scalable Optimisation Algorithm

Radomil Matousek

Abstract. In engineering as well as in non-engineering areas, numerous optimisation problems have to be solved using a wide range of optimisation methods. Soft-computing optimisation procedures are often applied to problems for which the classic mathematical optimisation approaches do not yield satisfactory results. In this paper we present a relatively new optimisation algorithm denoted as HC12 and demonstrate its possible parallel implementation. The paper aims to show that HC12 is highly scalable and can be implemented in a cluster of computers. As a practical consequence, the high scalability substantially reduces the computing time of optimisation problems.

1 Introduction

This paper describes the possibility of parallelizing the HC12 optimisation algorithm. Designed in 1995 [3], HC12 algorithm was well described e.g. in [4, 5]. This original algorithm uses a hill-climbing approach [6], more precisely: for a given optimisation problem, in each iteration step i, a solution $(\mathbf{A}_{kernel,i})$ exists to which a neighbourhood of further possible solutions is generated using a fixed pattern. From this neighbourhood, a best solution is chosen for iteration step $i + 1$, which will again be used to generate a new solution $(\mathbf{A}_{kernel,i+1})$. The algorithm stops if no best solution can be found, that is, if (for a minimisation problem)

$$\min\left(f(\mathbf{A}_{kernel,i})\right) \leq \min\left(f(\mathbf{A}_{kernel,i+1})\right), \tag{1}$$

where i is the iteration number and f is the objective function. As Chapter 2 and Chapter 4 shows, the HC12 algorithm is, among others, designed to lend itself to

Radomil Matousek
Brno University of Technology, Faculty of Mechanical Engineering,
Department of Applied Computer Science,
Technická 2, Brno 616 69, Czech Republic
e-mail: matousek@fme.vutbr.cz

J.R. González et al. (Eds.): NICSO 2010, SCI 284, pp. 177–183, 2010.
springerlink.com

good parallelization. Thus the paper aims to verify this property by implementing the algorithm in a cluster of computers. The test has been designed to verify and demonstrate a computation-time reduction in problems for which "this is already significant" in terms of time. As a testing problem, the F6 optimisation problem has been chosen for which, given a complexity and the parameters of the SW and HW environment, an average single-processor computing time of 45 minutes has been achieved.

2 HC12 Algorithm

A mathematical description of the algorithm can be found in [5]. Here the basic ideas are summarized:

- The solution of a given optimisation problem is represented by a binary vector A.
- This binary vector $\mathbf{A}$ codes k real parameters of the optimisation problem, that is, the real input parameters x_i (where $i \in \{1...k\}$) of the objective function. This provides a basis for discretizing the Domain of Definition (DoD) of the problem parameters to be found. The degree of discretization depends on the size of the binary string being proportional to 2^s where s is the number of bits per parameter.
- The procedure used to decode the binary string is given by the way the optimisation problem is formulated. For F6, the binary vector $\mathbf{A}$ is decoded using the Gray coding on a vector of integer parameters $(int_1,...,int_k)$ where $k = 50$ in our case. Next, using the DoD, this vector is further transformed into a vector of real parameters x_i.
- In the first iteration, a binary vector $\mathbf{A}_{kernel,1}$ and a neighbourhood to fit a fixed pattern are randomly generated. With HC12, this is a neighbourhood with distances 1 and 2 from vector $\mathbf{A}_{kernel}$ in the sense of the Hamming metric. In each iteration, the best solution is chosen as the new basis. The procedure is repeated until condition (1) is satisfied.

The Hamming distance ρ_H between two binary vectors of equal length is the number of positions for which the corresponding symbols are different. Let $\mathbf{a}, \mathbf{b}$ are binary vectors of length N and a, b its elements, then the Hamming distance can be calculated as follows:

$$\rho_H(\mathbf{a}, \mathbf{b}) = \sum_{i=1}^{N} |a_i - b_i| \qquad (2)$$

The principle of decoding a binary string to a vector of real parameters is shown in Fig. 1, the generating of a neighbourhood using a four-bit vector $\mathbf{A}_{kernel}$, is shown in Fig. 2. It follows from the principle that the cardinality (size) of the neighbourhood for a Hamming distance of 1 corresponds to the length N of the binary vector thus growing linearly with the length of the binary vector. On the other hand, the cardinality (size) of a neighbourhood for a Hamming distance of 2 corresponds to the combination number $(N, 2)$. It is exactly this type of neighbourhood that causes an unwelcome combinatorial expansion of the neighbourhood generated whose size grows exponentially with the length of the binary vector.

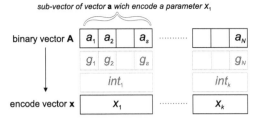

Fig. 1 Parameters' encoding scheme (binary and Gray code, integer and real parameters)

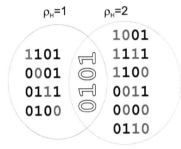

Fig. 2 An example of 4-bits neighbourhood generating for $\rho_H = 1$ and $\rho_H = 2$

3   The F6 Test Function

To demonstrate the performance of the HC12 algorithm and to verify its scalability in a cluster of computers, a multimodal function has been used which is usually used to test the power of both standard and soft-computing optimisation algorithms referred to as the Rastrigino F6 function.

$$F_6(\mathbf{x}) = 10n + \sum_{i=1}^{n} \left(x_i^2 - 10\cos(2\pi x_i)\right)$$

$$-5.12 \leqslant x_i \leqslant -5.12, \ \min F_6(\mathbf{x}) = F_6(0,...,0) = 0$$

(3)

As follows from definition (3), this function is continuous and smooth, but has a large number of maxima and minima. This number grows as a power given by the dimension of the function. In the present example, the optimisation problem is solved for 50 variables with an identical domain of [-5.12, 5.12]. The discretization chosen yields a calculation precision of $eps = 0.01$ for each parameter x_i.

It should be stressed that, for a minimisation problem, the F6 function represents 1.1739e+052 possible extremes on a given domain. Moreover, each of the variables to be found is encoded using 10 bits, which, given the number of parameters, results in a 500-bit binary string. If a rude-force algorithm were used to find an optimum solution, such a length would require 3.2734e+150 possible variants! This clearly demonstrates the enormous time needed to solve such an optimisation problem. As the first iteration of the HC12 algorithm is of stochastic nature, all the tests have been made for 100 algorithm runs with the average taken as a result since the median and average are very close.

To be able to analyse the scalability, for each configuration (the number of computers clustered), we used the same "random" vector configurations of the first iteration $A_{kernel,1}$ The below table shows an example of the results of a given optimisation problem.

Table 1 Record of the optimisation process of looking for a minimum of the F6 function

Iteration	Objective Function Value
1	884.3852097349522
2	805.3840740133274
3	730.8040415793919
...	...
30	121.2632026930635
...	...
60	21.1934433716936
...	...
100	0.6542459906284
...	...
122	0.0000000000000

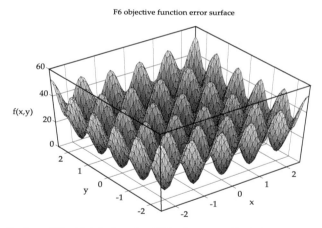

Fig. 3 Visualization of Rastrigin's function F6 in the range (DoD) from -5 to 5. The global optimum is positioned at point [0, 0]

4 Parallel Implementation

The optimization method was implemented as parallel application in Java and deployed on a cluster of computers.

The computers had AMD Opteron CPUs running at 2.6GHz and Linux operating system, the Java platform used was SUN 64-bit JDK version 1.6.0_14 with the Server VM.

The application used a master-slave architecture, where slaves registered with the master and the master then assigned them chunks of work. The master and slave parts communicated using the RMI (Remote Method Invocation) mechanism provided by the Java platform.

The master part coordinated repeated iterations, where each iteration involved:

- taking the so far best known N-bit vector as a starting vector;
- finding all N-bit vectors with Hamming distance of 1 from the starting vector - there are just N of them, so parallelisation is not needed;
- converting all of them into vectors of floating point numbers with double precision, computing the value of the objective function and finding the minimum value, eventually replacing the starting vector with the new minimum;
- dividing the $N(N-1)/2$ N-bit vectors with Hamming distance of 2 from the starting vector into chunks of equal size and assigning them to slaves to compute the value of the objective function for each of them, returning the best one found;
- collecting the best found bit vectors from slaves and selecting the best of them;
- if the best found bit vector is better than the starting point, it uses it as a starting point for the next iteration, otherwise it ends by equation (1).

This list shows that the master part makes synchronization points at the end of each iteration, so the Amdahl's law predicts that the speedup cannot be fully linear with growing number of CPUs. But for growing number of bits in vectors, the number of vectors processed by the parallel part of the algorithm is growing faster compared to the number of vectors processed by the serial part, and the maximum possible speedup is improving.

5 Results and Conclusions

The results of a parallel implementation of the optimization process for 500 bits, 50 dimensions by 10 bits each, and objective function F6 are presented by Table 2 and Fig. 3. An average count of iterations to reach the best HC12 solution was 122 (an example of iteration run is in the Table 1). A code was found to scale nearly linearly on a cluster of computing machines if each machine runs just one instance of the slave part. However, when more than one slave was located on the same machine, the scalability was much worse. For example, a machine with 8 dual-core AMD Opteron CPUs was tried with 1, 2, 4, 8, and 16 instances of the slave part. When using 8 slaves, the computation was only about 4 times faster than when using 1 slave. When using 16 slaves (the machine had 16 CPU cores so that improvement was possible) the computation was even slower than when using only 8 slaves. This

Table 2 Scalability vs. computing time

CPU[a]	time	speed up
1	2722 [s]	1.00 x
2	1400 [s]	1.94 x
3	948 [s]	2.87 x
4	697 [s]	3.91 x
5	571 [s]	4.77 x
6	479 [s]	5.69 x
7	437 [s]	6.23 x
8	382 [s]	7.13 x
9	336 [s]	8.10 x

[a] The number of CPUs in cluster (AMD Opteron, two cores).

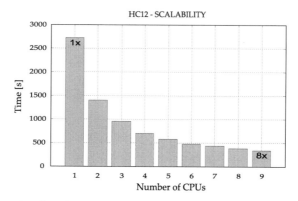

Fig. 4 Computing time depending on the number of CPUs in the cluster of computers

result is likely to be caused by scalability issues of multiprocessor machines, and does not provide any information about the scalability of the optimization method itself.

The testing of a parallel implementation if the HC12 algorithm has proved un-equivocally that the algorithm is highly scalable within a cluster of computers. For a given optimisation problem, the difference between the computing time on a single computer (about 45 minutes) and 9 computers (about 5 minutes) is about 40 minutes as given by the measurements (see Table 2). This acceleration may be considered significant. As the favourable properties of the HC12 optimisation algorithm or its implementations within other soft-computing methods (such as GAHC algorithm [5]) have already been demonstrated several times for a single CPU, the authors will further focus on the research and applications of its parallel implementations. In this light, this paper demonstrating a high scalability of the HC12 algorithm may be thought of as one of the pilot papers concerning parallel implementations of the HC12 algorithm.

Acknowledgements. The access to the MetaCentrum supercomputing facilities provided under the research intent MSM6383917201 is highly appreciated. This work was supported by the Czech Ministry of Education in the frame of MSM 0021630529 "Intelligent Systems in Automation" and by the Grant Agency of the Czech Republic No.: 102/091668 "Control Algorithm Design by means of Evolutionary Approach".

References

[1] Amdahl, G.: Validity of the Single Processor Approach to Achieving Large-Scale Computing Capabilities. In: AFIPS Conference Proceedings, vol. (30), pp. 483–485 (1967)

[2] Battiti, R., Tecchiolli, G.: Local search with memory: Benchmarking RTS. journal of Operations Research Spectrum 17(2/3), 67–86 (1995)

[3] Matousek, R.: GA (GA with HC mutation) – implementation and application, Master thesis (in Czech), Brno University of Technology, Brno, Czech Republic (1995)

[4] Matousek, R.: GAHC: A Hybrid Genetic Algorithm. In: Proceedings of the 10th Fuzzy Colloquium in Zittau, Zittau, pp. 239–244 (2002) ISBN: 3-9808089-2-0

[5] Matousek, R.: GAHC: Improved Genetic Algorithm. In: Krasnogor, et al. (eds.) Nature Inspired Cooperative Strategies for Optimization (NICSO 2007). Springer book series, vol. 129, XIV, p. 114 (12p). Springer, Berlin (2008)

[6] Mitchell, M., Holland, J.H.: When Will a Genetic Algorithm Outperform Hill Climbing? In: Cowan, J.D., Tesauro, G., Alspector, J. (eds.) Advances in Neural Information Processing Systems, vol. 6. Morgan Kaufmann, San Mateo (1994)

[7] Zhou, R., Hansen, E.A.: Breadth-First Heuristic Search. In: 14th International Conference on Automated Planning and Scheduling (ICAPS 2004), Whister, British Columbia, Canada (2004)

Adaptive Evolutionary Testing: An Adaptive Approach to Search-Based Test Case Generation for Object-Oriented Software

José Carlos Bregieiro Ribeiro, Mário Alberto Zenha-Rela,
and Francisco Fernández de Vega

Abstract. Adaptive Evolutionary Algorithms are distinguished by their dynamic manipulation of selected parameters during the course of evolving a problem solution; they have an advantage over their static counterparts in that they are more reactive to the unanticipated particulars of the problem. This paper proposes an adaptive strategy for enhancing Genetic Programming-based approaches to automatic test case generation. The main contribution of this study is that of proposing an Adaptive Evolutionary Testing methodology for promoting the introduction of relevant instructions into the generated test cases by means of mutation; the instructions from which the algorithm can choose are ranked, with their rankings being updated every generation in accordance to the feedback obtained from the individuals evaluated in the preceding generation. The experimental studies developed show that the adaptive strategy proposed improves the test case generation algorithm's efficiency considerably, while introducing a negligible computational overhead.

1 Introduction

The application of Evolutionary Algorithms (EAs) to test data generation is often referred to as Evolutionary Testing (ET) [15] or Search-Based Testing (SBT) [10]. ET is an emerging methodology for automatically generating high quality test data; it is, however, a difficult subject – especially if the aim is to implement an automated

José Carlos Bregieiro Ribeiro
Polytechnic Institute of Leiria – Leiria, Portugal
e-mail: jose.ribeiro@estg.ipleiria.pt

Mário Alberto Zenha-Rela
University of Coimbra – Coimbra, Portugal
e-mail: mzrela@dei.uc.pt

Francisco Fernández de Vega
University of Extremadura – Mérida, Spain
e-mail: fcofdez@unex.es

J.R. González et al. (Eds.): NICSO 2010, SCI 284, pp. 185–197, 2010.
springerlink.com   © Springer-Verlag Berlin Heidelberg 2010

solution, viable with a reasonable amount of computational effort, which is adaptable to a wide range of test objects. This paper proposes an adaptive methodology for the ET of Object-Oriented (OO) software.

EAs are powerful, yet general, methods for search and optimization. Their generality comes from the unbiased nature of the standard operators used, which perform well for problems where little or no domain knowledge is available [1]. However, if knowledge about a problem is available, a bias can be introduced directly into the problem so as to remove (or penalize) undesirable candidate solutions and improve the efficiency of the search. Unfortunately, *a priori* knowledge about the intricacies of the problem is frequently unavailable. Having little information about a problem does not, however, necessarily prevent introducing an appropriate specific bias into an evolutionary problem; for many tasks, it is possible to dynamically adapt aspects to anticipate the regularities of the environment and improve solution optimization or acquisition speed. Adaptive EAs are distinguished by their dynamic manipulation of selected parameters or operators during the course of evolving a problem solution [5]. They have an advantage over their standard counterparts in that they are more reactive to the unanticipated particulars of the problem and, in some formulations, can dynamically acquire information about regularities in the problem and exploit them.

Typically, EAs maintain a population of candidate solutions rather than just one current solution; in consequence, the search is afforded many starting points, and the chance to sample more of the search space than local searches. Mutation is the main process through which new genetic material is introduced during an evolutionary run with the intent of diversifying the search and escaping local maxima. The main contribution of this study is that of proposing an adaptive strategy for promoting the introduction of relevant instructions into the existing test cases by means of mutation; the set of instructions from which the algorithm can choose is ranked, with their rankings being updated every generation in accordance to the feedback obtained from the individuals evaluated in the preceding generation.

This article is organized as follows. The next Section starts by providing background on ET and Adaptive EAs. Section 3 details the Adaptive ET strategy proposed. The experiments conducted in order to validate and observe the impact of the Adaptive ET technique are discussed in Section 4. The concluding Section presents some final considerations and summarizes the most relevant contributions.

2 Background and Terminology

Software Testing is the process of exercising an application to detect errors and to verify that it satisfies the specified requirements. When performing unit-testing, the goal is to warrant the robustness of the smallest units – the *test objects* – by testing them in an isolated environment. Unit-testing is performed by executing the test objects in different scenarios using relevant *test cases*. Classes and objects are typically considered to be the smallest units that can be tested in isolation in OO programs [17]. An object stores its state in fields and exposes its behaviour through methods.

A unit-test case for OO software consists of a *Method Call Sequence (MCS)*, which defines the test scenario. During test case execution, all participating objects are created and put into particular states through a series of method calls. Each test case focuses on the execution of one particular public method – the *Method Under Test (MUT)*. In most situations, an OO class is not standalone; testing a single class involves other classes – i.e., classes that appear as parameter types in the method signatures of the *Class Under Test (CUT)*. The set of classes which are relevant for testing a particular class is called the *Test Cluster* [16]. The Test Cluster for a given class can be obtained by performing a transitive static analysis of the signatures of the public methods of this class; each data type (class, interface, or primitive type) encountered during this analysis is added to the Test Cluster. After all methods of the CUT have been included in the Test Cluster, the analysis continues by evaluating all the public methods of the Test Cluster types which have not yet been considered; once all method signatures have been analysed in this manner, the Test Cluster contains all relevant types.

Genetic Programming (GP) is usually associated with the evolution of tree structures; it focuses on automatically creating computer programs by means of evolution [7], and is thus especially suited for representing and evolving Test Programs. GP algorithms maintain a population of candidate solutions, which is iteratively recombined and mutated in order to evolve successive generations of individuals. An individual's probability of being selected for reproduction is associated to its *fitness*, which quantifies the optimality of a solution; the idea is to favour the fitter individuals in the hope of breeding better offspring. Within the tree genomes, the leaf nodes are called *terminals* (and can be inputs to the program, constants or functions with no arguments), whereas the non-leaf nodes are called *non-terminals* (functions taking at least one argument). The *Function Set* is the set of functions from which the GP system can choose when constructing trees. Non-typed GP approaches are, however, unsuitable for representing test programs for OO software, because any element can be a child node in a parse tree for any other element without having conflicting data types; in contrast, *Strongly-Typed Genetic Programming (STGP)* [11] allows the definition of types in the Function Set, which causes the initialization process and the various genetic operations to only construct syntactically correct trees, which can be translated to compilable programs.

Still, syntactically correct and compilable MCSs may still abort prematurely, if a runtime exception is thrown during execution [17]. Test cases can thus be separated in two classes: *feasible* test cases are effectively executed, and terminate with a call to the MUT; *unfeasible* test cases terminate prematurely because a runtime exception is thrown by an instruction of the MCS.

Several methodologies to the ET of OO software have been proposed, focusing on the usage of distinct EAs – e.g., Genetic Algorithms [4, 6, 15], GP [14], STGP [12, 13, 17, 18]. However, to the best of our knowledge, there are no studies on the possibility of applying Adaptive EAs to ET problems.

The action of determining the variables and parameters of an EA to suit the problem has been termed *adapting* the algorithm to the problem; in EAs this can be performed dynamically, while the algorithm is searching for a solution. *Adaptive*

EAs provide the opportunity to customize the EA to the problem and to modify the configuration and the strategy parameters used while the problem solution is sought. This enables incorporating domain information into the EA more easily, and allows the algorithm itself to select those parametrization which yield better results. Also, these values can be modified during the run of the Adaptive EA to suit the situation during that part of the run. Adaptive EAs have already been applied to solve several problems; interesting review articles include [1, 5]. In [5], the authors proposed a classification based on the adaptation type and adaptation level of the Adaptive EA. The type of adaptation consists of two main categories: static and dynamic. *Static adaptation* is where the strategy parameters have a constant value throughout the run of the EA; consequently, an external agent or mechanism (e.g., the user) is needed to tune the desired strategy parameters and choose the most appropriate values. *Dynamic adaptation* happens if there is some mechanism which modifies a strategy parameter without external control (e.g., by means of some deterministic rule or some form of feedback from the EA).

3 Adaptive Evolutionary Testing

With STGP approaches, MCSs are encoded (and evolved) as STGP trees; each tree subscribes to a Function Set, which must be specified beforehand and defines the STGP nodes and establishes the constraints involved in the trees' construction. In other words, the Function Set contains the set of instructions from which the algorithm can choose when building the MCSs that compose test cases.

The Function Set can be defined completely automatically based solely on the Test Cluster information [16]. The definition of the Test Cluster is, therefore, of paramount importance to the algorithm's performance and accuracy. If the Test Cluster consists of many classes (or if it is composed of few classes which possess a high number of public methods), the Function Set can be extremely large. With an increasing size of the Function Set (and hence an increasing size of the search space) the probability that the "right" methods appear in a candidate test sequence decreases – and so does the efficiency of the evolutionary search. Conversely, if a more conservative strategy is employed, the Test Cluster may not include all the classes needed to attain full coverage, thus compromising effectiveness. As such, the selection of the classes and methods to be included in the Test Cluster – and, consequently, in the Function Set – must be carefully pondered, and adequate strategies must be employed for defining the Test Cluster and sampling the search domain.

Still, there are good reasons to suppose that there is no one strategy, however clever, recursive, or self-organizing that will be optimal for all problem domains. The Test Cluster parametrization process is heavily problem-specific and, as such, it usually falls on the users' hands. Leaving this task to the user has, however, several drawbacks. Namely: the users' mistakes in setting the parameters could be sources of errors and/or suboptimal performance; parameter tuning costs a lot of time; and the optimal parameter value may vary during the evolution [5].

What's more, the users' choices are inevitably biased, and performance is (arguably) often compromised for the sake of accuracy; in the particular case of ET problems, not doing so could result in the impossibility of obtaining suitable test sets, in conformity to the criteria defined. In [16], Wappler suggested strategies for addressing the problem of large Function Sets, that result from large Test Clusters with classes that possess many methods:

- Performing a static analysis so as to eliminate all the functions in the Function Set that correspond to methods which are neither object-creating nor state-changing. An Input Domain Reduction strategy, based on the concept of Purity Analysis, that meets this suggestion has already been proposed in [13].
- Defining a distance-based heuristic, that prevents the methods from those Test Cluster classes that are associated to the CUT via several other classes from being transformed to functions of the Function Set. Such an heuristic would have to be heavily problem-specific, and decisions would have to be made statically and *a priori* – potentially compromising the success of the search. It seems difficult to implement an automated solution for this idea without compromising generality.
- Naming classes whose methods shall not be transformed to functions of the Function Set. This idea exploits the user's knowledge of the CUT, and suffers from the drawbacks mentioned above.

We propose a different strategy, based on the concept of dynamically adapting the Function Set's constraints selection probabilities. During an evolutionary run, it is possible to perceive that the introduction of certain instructions should be favoured. By allowing the constraints' selection probabilities to fluctuate throughout the search, with basis on the feedback obtained by the behaviour of the individuals produced and evaluated previously, the introduction of interesting genetic material will be promoted. This strategy allows mitigating the negative effects of including a large number of entries into the Test Cluster; also, it allows a higher degree of freedom when defining the Test Cluster, by minimizing the impact of redundant, irrelevant or erroneous choices.

Mutation plays a central role on the diversification of the search and on the exploration of the search space; it basically consists of selecting a mutation point in a tree, and substituting the sub-tree rooted at the point selected with a newly generated sub-tree [7]. Previous studies indicate that better results can be attained if the mutation operator is assigned a relatively high probability of selection [13].

Mutation is, in fact, the main process by which new genetic material is introduced during the evolutionary search. In the particular case of ET of OO software problems, it allows the introduction of new sequences of method calls into the generated test cases, so as to allow trying out different objects and states in the search for full structural coverage. Also, it is clear that during an evolutionary run, it is possible to perceive that some method calls are more relevant than others, e.g. because they had been less prone to throw runtime exceptions and their introduction will likely contribute to test case feasibility, or simply because they have been used less frequently and their introduction will promote diversity (precisely the main task of the Mutation operator).

Table 1 Example Function Set and Type Set

Function Name	Return Type	Child Types
void print(Object)	TREE	OBJECT
String intToStr(Integer)	STRING (rank:0.2)	INT
"Foo"	STRING (rank:0.8)	
"Bar"	STRING (rank:0.2)	
Integer add(Integer,Integer)	INT (rank:0.8)	INT, INT
0	INT (rank:0.4)	
1	INT (rank:0.6)	
Set Types: OBJECT = [STRING, INT]		

Whenever mutation occurs, a new (sub-)tree must be created; usually, one of the standard tree builders (e.g., Grow, Full, Half-Builder or Uniform) is used to generate these trees [8]. We propose employing Luke's Strongly-Typed Probabilistic Tree Creation 2 (PTC2) algorithm [9] to perform this task, so as to take advantage of the built-in feature that allows assigning probabilities to the selection of constraints. What's more, we have modified this algorithm in order to be able to dynamically update the constraints' probabilities during the evolutionary run.

The Strongly-Typed Probabilistic Tree Creation 2 algorithm works as follows: it picks a random position in the horizon of the tree (i.e., unfilled child node positions), fills it with a non-terminal (thus extending the horizon), and repeats this process until the number of nodes (non-terminals) in the tree, plus the number of unfilled node positions, is greater or equal to the requested tree size. Finally, the remaining horizon is filled with terminals. The tree size is provided by the user.

PTC2 provides uniform distribution of functions and has very low computational complexity [8]. Also – and most interestingly – PTC2 has provisions for picking non-terminals with a certain probability over other non-terminals of the same return type, and terminals over other terminals likewise. In order to illustrate the methodology followed by this algorithm, let us consider a simple problem which includes a Function Set (Table 1) composed of seven entries (or constraints), defining three non-terminal nodes – void print(String), String intToStr(Integer), Integer add (Integer, Integer) – and four terminal nodes – "Foo", "Bar", 0 and 1. Also, it defines three atomic types – TREE, STRING and INT – and one set type – OBJECT, which includes both INT and STRING. The TREE type is used as return type of the STGP tree.

The constraint selection rankings are also defined. "Foo" is given a rank of 0.8, and "Bar" a rank of 0.2, for example; this means that, if the PTC2 algorithm is required to select a terminal node with a STRING return type, it will select constraint "Foo" with a probability of 80% and "Bar" with a probability of 20%. If, however, it is required to select a terminal node with an OBJECT return type, PTC2 uniformly distributes the rankings of the STRING and INT atomic types, with the constraints probabilities being defined as follows: "Foo"–40%; "Bar"–10%; 0–20%; 1–30%. Continuing with this example, if required to grow a tree of size 3,

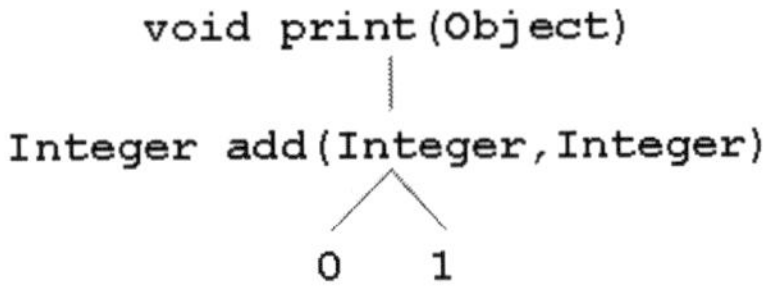

```
        void print(Object)                    Integer i1 = 0;
                |                              Integer i2 = 1;
   Integer add(Integer,Integer)               Integer i3 = add(i1, 12);
                /\                             print(i3);
               0  1
```

Fig. 1 Example STGP tree *(left)* and corresponding Method Call Sequence *(right)*

the PTC2 algorithm would build the tree depicted in Figure 1 with a 19.2% chance: 100% probability of selecting the root node, times 80% probability of selecting the non-terminal constraint `Integer add(Integer, Integer)` as an OBJECT type provider for the root node, times 40% chance of choosing 0 as the first terminal of type INT, times 60% chance of selecting 1 as the second terminal.

The dynamic adaptive strategy described in the following Subsection aims at dynamically tuning the Function Set's constraints selection rankings, so as to promote the creation of sub-trees, for insertion in the population via mutation, that favour both feasibility and diversity.

3.1 Dynamic Adaptation Strategy

Let the *constraint selection ranking* of constraint c in generation g be identified as ρ_c^g. Also, let λ be the *runtime exceptions caused factor*, σ be the *runtime exceptions caused by ancestors factor*, and γ be the *constraint diversity factor*. Then, ρ_c^g is updated, at the beginning of each generation, in accordance to the following Equation.

$$\rho_c^g = \rho_c^{g-1} - \lambda_c^{g-1} - \sigma_c^{g-1} - \gamma_c^{g-1} \qquad (1)$$

That is, the constraint selection ranking ρ_c^g of a given constraint c in generation g is calculated as being the constraint selection ranking ρ of the previous generation, minus the λ factor of the previous generation (with $\lambda \in [0,1]$), minus the σ factor of the previous generation (with $\sigma \in [0,1]$), minus the γ factor of the previous generation (with $\gamma \in [-1,1]$).

In order to calculate the normalized constraint selection ranking ρ'^g_c, if the minimum ρ_c^g in generation g is negative, the data is firstly shifted by adding all numbers with the absolute of the minimum ρ_c^g; then, ρ'^g_c is normalized into the range of $[0,1]$ as follows.

$$n'^g_c = \frac{n_c^g}{n_{MAX}^g - n_{MIN}^g} \qquad (2)$$

The following subsections detail the procedure used for calculating the λ, σ, and γ factors.

Runtime Exceptions Caused Factor. Let E_c^g be the set of runtime exceptions caused by constraint c in generation g, and T^g be the set of trees produced in generation g, with $\left|E_c^g\right|$ and $\left|T^g\right|$ being their cardinalities. Then, λ is calculated as follows.

$$\lambda_c^g = \frac{|E_c^g|}{|T^g|} \qquad (3)$$

That is, the λ factor is equal to the number of runtime exceptions thrown by instructions corresponding to constraint c, dividing by the total number of trees. It should be noted that only a single runtime exception may be thrown by MCS (i.e., by tree). This factor's main purpose is that of penalizing the ranking of constraints corresponding to instructions that have caused runtime exceptions to be thrown in the preceding generation. This factor is normalized into the range of $[0, 1]$ using Equation 2.

Runtime Exceptions Caused by Ancestors Factor. Let X_c^g be the set of runtime exceptions thrown by ancestors of constraint c in generation g, and $x_{ca}^g \in X_c^g$ be a runtime exception thrown by an ancestor of level a, with $a \in \{2 = parent, 3 = grandparent, \ldots\}$ being the *ancestry level* of the constraint that threw the exception. Also, let A_c^g be the multiset containing the ancestry levels of $x_{ca}^g \in X_c^g$. Then, σ is calculated as follows.

$$\sigma_c^g = \sum_{a \in A_c^g} a^{-1} \qquad (4)$$

That is, the σ factor is equal to the sum of the inverses of the ancestry levels of the ancestors of constraint c that threw runtime exceptions. This factor's main purpose is that of penalizing the ranking of constraints corresponding to instructions which have participated in the composition of sub-trees (i.e., sub-MCSs) that have caused runtime exceptions to be thrown in the preceding generation; the higher the ancestry level, the lower the penalty. This factor is normalized into the range of $[0, 1]$ using Equation 2.

Constraint Diversity Factor. Let C^g be a multiset containing the number of times each constraint appeared in generation g, and c^g be the number of times constraint c appeared in generation g. Also, let m_{C^g} be the mean of the values contained in multiset C^g, and $d_c^g = c^g - m_{C^g}$ be the deviation of constraint c in generation g, and $r_d^g = d_{MAX}^g - d_{MIN}^g$ be the range of deviation for generation g. Then, γ_c^g is calculated as follows.

$$\gamma_c^g = \frac{d_c^g}{r_d^g} \qquad (5)$$

The γ factor's main purposes are those of allowing constraints to recover their ranking if they have been being used infrequently, and penalizing the ranking of constraints which have been selected too often.

4   Experimental Studies

The adaptive strategy described in the preceeding Section was embedded into eCrash [13], an automated ET tool, with the objective of observing the impact of this technique on both the efficiency and effectiveness of the test case generation process.

eCrash's approach to ET involves representing and evolving test cases for OO software using the STGP paradigm. The methodology for evaluating the quality of test cases includes instrumenting the MUT, and executing it using the generated test cases – with the intention of collecting trace information with which to derive coverage metrics. The MUTs are represented internally by weighted Control-Flow Graphs (CFGs); the strategy for favouring test cases that exercise problematic structures involves re-evaluating the weight of CFGs' nodes every generation. The aim is that of efficiently guiding the search process towards achieving full structural coverage – i.e., generating a set of test cases that traverse all CFG nodes of the MUT. A thorough description of *eCrash* can be found in [13].

The Java `Vector` and `BitSet` classes (JDK 1.4.2) were used as test objects. The rationale for employing these classes is related with the fact that they represent "real-world" problems and, being container classes, possess the interesting property of containing explicit state, which is only controlled through a series of method calls [2]. Additionally, they have been used in several other case studies described in literature (e.g., [2, 15, 18]), providing an adequate test object set in the lack of common benchmark cluster that can be used to test and compare different techniques [3].

4.1   Setup

The experiments were executed using an Intel Core2 Quad 2.60GHz processor with 4.0 GB RAM desktop, with four test case generation processes running in parallel. 20 runs were executed for each of the 67 MUTs – in a total of 820 runs for the `Vector` class and 520 runs for the `BitSet` Class. Half of these runs were executed employing the adaptive strategy proposed, and half using a "static" approach for comparison purposes. The only difference between the adaptive and the static runs was that, in the latter, the constraints' rankings remained unaltered throughout the evolutionary search. Since the same seeds were used in both the adaptive and non-adaptive runs, and because *eCrash* is deterministic, the discrepancies in the results will solely mirror the impact of the adaptive technique employed.

A single population of 10 individuals was used; the rationale for selecting a relatively small population size had to do with the adaptive algorithm's need of obtaining frequent feedback. The search stopped if an ideal individual was found or after 200 generations. For the generation of individuals, 3 child sources were defined: strongly-typed versions of Mutation (selection probability: 40%) and Crossover (selection probability: 30%), and a simple Reproduction operator (selection probability: 30%). The selection method was Tournament Selection with size 2. The tree builder algorithm was PTC2 (for the reasons explained in Section 3), with the maximum and minimum tree depths being defined as 1 and 4. The constraints' ranking were initialized with the value 1.0, and were updated at the beginning of every generation (before individuals were produced), in accordance to Equation 1.

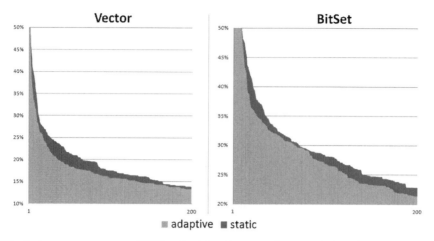

Fig. 2 Average percentage of CFG nodes left to be covered per generation, for Vector and BitSet classes, with and without adaptation

4.2 Results

Table 2 depicts the percentage of successful runs (i.e., runs in which a test set attaining full structural coverage was found) for the MUTs of the Vector and BitSet classes, with and without adaptation. The graphs shown in Figure 2 contain the percentage of CFG nodes remaining per generation using the adaptive and the static techniques for the classes tested; their inclusion enables the analysis of the strategy's impact during the course of the search.

The results depicted in Table 2 clearly indicate that the tests case generation process's performance is improved by the inclusion of the Adaptive ET methodology proposed. The adaptive strategy outperformed the static approach for 28.4% of the MUTs tested, whereas the latter only surpassed the former in 5.9% of the situations. In terms of the average success rate, the adaptive strategy enhances results by 3% for the Vector class; the improvement is even more significant for the BitSet class, with the results meliorating 11%. What's more, the adaptive strategy allowed attaining full structural coverage in some situation in which the success rate had been of 0% using the non-adaptive strategy – namely, for the Object remove(int) and List subList(int,int) MUTs of the Vector class, and for the int length() and boolean intersects(BitSet) MUTs of the BitSet class; these observations indicate that this strategy is specially suited for overcoming some difficult state problems.

The graph shown in Figure 2 also provides clear indication that the evolutionary search benefits from the inclusion of the adaptive approach described. For the Vector's MUTs, the average number of nodes remaining when the Adaptive ET approach is used decreases as much as 6% during the initial generations; for the BitSet class, the contrast is the results is less perceptible, but the adaptive approach still manages to attain a 3% improvement at certain stages.

Table 2 Percentage of runs to attain full structural coverage, for the MUTs of the `Vector` and `BitSet` classes, with and without adaptation

Vector				BitSet		
MUT	*adaptive*	*static*		*MUT*	*adaptive*	*static*
void add(int,Object)	80%	**90%**		boolean get(int)	**90%**	60%
boolean add(Object)	100%	100%		BitSet get(int,int)	0%	0%
Object get(int)	100%	100%		int hashCode()	100%	100%
int hashCode()	100%	100%		Object clone()	0%	0%
Object clone()	0%	0%		void clear(int, int)	0%	0%
int indexOf(Object)	100%	100%		void clear()	100%	100%
int indexOf(Object,int)	**20%**	10%		void clear(int)	**90%**	80%
void clear()	100%	100%		boolean equals(Object)	0%	0%
boolean equals(Object)	100%	100%		String toString()	100%	100%
String toString()	100%	100%		boolean isEmpty()	100%	100%
boolean contains(Object)	**50%**	40%		int length()	**30%**	0%
boolean isEmpty()	100%	100%		int size()	100%	100%
int lastIndexOf(Object,int)	0%	0%		void set(int)	**100%**	70%
int lastIndexOf(Object)	100%	100%		void set(int, boolean)	100%	100%
boolean addAll(Collection)	**90%**	70%		void set(int, int)	**70%**	40%
boolean addAll(int,Collection)	**30%**	20%		void set(int, int, boolean)	40%	**70%**
int size()	100%	100%		void flip(int, int)	**60%**	20%
Object[] toArray()	100%	100%		void flip(int)	**90%**	50%
Object[] toArray(Object[])	40%	40%		void and(BitSet)	0%	0%
void addElement(Object)	100%	100%		void andNot(BitSet)	**60%**	30%
Object elementAt(int)	100%	100%		int cardinality()	100%	100%
Object remove(int)	**20%**	0%		boolean intersects(BitSet)	**20%**	0%
boolean remove(Object)	100%	100%		int nextClearBit(int)	0%	0%
Enumeration elements()	100%	100%		int nextSetBit(int)	10%	10%
Object set(int,Object)	**100%**	80%		void or(BitSet)	0%	0%
int capacity()	100%	100%		void xor(BitSet)	**90%**	30%
boolean containsAll(Collection)	100%	100%				
void copyInto(Object[])	100%	100%				
void ensureCapacity(int)	100%	100%				
Object firstElement()	100%	100%				
void insertElementAt(Object,int)	80%	**90%**				
Object lastElement()	100%	100%				
boolean removeAll(Collection)	100%	100%				
void removeAllElements()	100%	100%				
boolean removeElement(Object)	30%	**40%**				
void removeElementAt(int)	**20%**	10%				
boolean retainAll(Collection)	100%	100%				
void setElementAt(Object,int)	**100%**	70%				
void setSize(int)	100%	100%				
List subList(int,int)	**30%**	0%				
void trimToSize()	100%	100%				

In terms of speed, the overhead introduced by embedding the adaptive strategy into the evolutionary algorithm was negligible; each generation took, on average, 23.25 seconds using the adaptive methodology, and 23.21 seconds using the static approach. The time overhead introduced by the adaptation procedure was a mere 0.19%.

5 Conclusions

Recent research on Evolutionary Testing has relied heavily on Genetic Programming for representing and evolving test data for Object-Oriented software. The main contribution of this work is that of proposing a dynamic adaptation strategy for promoting the introduction of relevant Method Call Sequences into the generated test cases by means of Mutation.

The Adaptive Evolutionary Testing strategy proposed obtains feedback from the individuals produced and evaluated in the preceding generation, and dynamically updates the selection probability of the constraints defined in the Function Set so as to encourage the selection of interesting genetic material and promote diversity and test case feasibility. The experimental studies implemented indicate a considerable improvement in the algorithm's efficiency when compared to its static counterpart, while introducing a negligible overhead.

References

[1] Angeline, P.J.: Adaptive and self-adaptive evolutionary computations. In: Computational Intelligence: A Dynamic Systems Perspective, pp. 152–163. IEEE Press, Los Alamitos (1995)

[2] Arcuri, A., Yao, X.: A memetic algorithm for test data generation of object-oriented software. In: Proceedings of the 2007 IEEE Congress on Evolutionary Computation (CEC), pp. 2048–2055. IEEE, Los Alamitos (2007)

[3] Arcuri, A., Yao, X.: On test data generation of object-oriented software. In: TAICPART-MUTATION 2007: Proceedings of the Testing: Academic and Industrial Conference Practice and Research Techniques - MUTATION, pp. 72–76. IEEE Computer Society, Washington (2007)

[4] Ferrer, J., Chicano, F., Alba, E.: Dealing with inheritance in oo evolutionary testing. In: GECCO 2009: Proceedings of the 11th Annual conference on Genetic and evolutionary computation, pp. 1665–1672. ACM, New York (2009)

[5] Hinterding, R., Michalewicz, Z., Eiben, A.E.: Adaptation in evolutionary computation: A survey. In: Proceedings of the Fourth International Conference on Evolutionary Computation (ICEC 1997), pp. 65–69. IEEE Press, Los Alamitos (1997)

[6] Inkumsah, K., Xie, T.: Evacon: A framework for integrating evolutionary and concolic testing for object-oriented programs. In: Proc. 22nd IEEE/ACM International Conference on Automated Software Engineering (ASE 2007), pp. 425–428 (2007)

[7] Koza, J.R.: Genetic Programming: On the Programming of Computers by Means of Natural Selection (Complex Adaptive Systems). The MIT Press, Cambridge (1992)

[8] Luke, S.: Issues in scaling genetic programming: Breeding strategies, tree generation, and code bloat. PhD thesis, Department of Computer Science, University of Maryland, A. V. Williams Building, University of Maryland, College Park, MD 20742 USA (2000)

[9] Luke, S.: Two fast tree-creation algorithms for genetic programming. IEEE Transactions on Evolutionary Computation 4(3), 274–283 (2000)

[10] McMinn, P.: Search-based software test data generation: A survey. Software Testing, Verification and Reliability 14(2), 105–156 (2004)

[11] Montana, D.J.: Strongly typed genetic programming. Evolutionary Computation 3(2), 199–230 (1995)

[12] Ribeiro, J.C.B., Zenha-Rela, M., de Vega, F.F.: An evolutionary approach for performing structural unit-testing on third-party object-oriented java software. In: Krasnogor, N., Nicosia, G., Pavone, M., Pelta, D.A. (eds.) NICSO. Studies in Computational Intelligence, vol. 129, pp. 379–388. Springer, Heidelberg (2007)

[13] Ribeiro, J.C.B., Zenha-Rela, M., de Vega, F.F.: Test case evaluation and input domain reduction strategies for the evolutionary testing of object-oriented software. Information & Software Technology 51(11), 1534–1548 (2009)

[14] Seesing, A., Gross, H.G.: A genetic programming approach to automated test generation for object-oriented software. ITSSA 1(2), 127–134 (2006)

[15] Tonella, P.: Evolutionary testing of classes. In: ISSTA 2004: Proceedings of the 2004 ACM SIGSOFT international symposium on Software testing and analysis, pp. 119–128. ACM Press, New York (2004)

[16] Wappler, S.: Automatic generation of object-oriented unit tests using genetic programming. PhD thesis, Technischen Universitat Berlin (2007)

[17] Wappler, S., Wegener, J.: Evolutionary Unit Testing Of Object-Oriented Software Using A Hybrid Evolutionary Algorithm. In: CEC 2006: Proceedings of the 2006 IEEE Congress on Evolutionary Computation, pp. 851–858. IEEE, Los Alamitos (2006)

[18] Wappler, S., Wegener, J.: Evolutionary unit testing of object-oriented software using strongly-typed genetic programming. In: GECCO 2006: Proceedings of the 8th annual conference on Genetic and evolutionary computation, pp. 1925–1932. ACM Press, New York (2006)

Evolutionary Algorithms for Planar MEMS Design Optimisation: A Comparative Study

Elhadj Benkhelifa, Michael Farnsworth*, Ashutosh Tiwari, and Meiling Zhu

Abstract. The evolutionary approach in the design optimisation of MEMS is a novel and promising research area. The problem is of a multi-objective nature; hence, multi-objective evolutionary algorithms (MOEA) are used. The literature shows that two main classes of MOEA have been used in MEMS evolutionary design Optimisation, NSGA-II and MOGA-II. However, no one has provided a justification for using either NSGA-II or MOGA-II. This paper presents a comparative investigation into the performance of these two MOEA on a number of MEMS design optimisation case studies. MOGA-II proved to be superior to NSGA-II. Experiments are, herein, described and results are discussed.

1 Introduction

Micro-electro-mechanical systems (MEMS) or micro-machines [1] are a field grown out of the integrated circuit (IC) industry, utilizing fabrication techniques from the technology of Very-Large-Scale-Integration (VLSI). The goal is to develop smart micro devices which can interact with their environment in some form.

Elhadj Benkhelifa · Michael Farnsworth · Ashutosh Tiwari
Decision Engineering Centre, Building 50, School of Applied Sciences, Cranfield Campus,
Cranfield University College Road, Cranfield Bedfordshire, MK43 0AL
e-mail: {e.benkhelifa,m.j.farnsworth,a.tiwari}@cranfield.ac.uk

Meiling Zhu
Microsystems and Nanotechnology, Building 40, School of Applied Sciences,
Cranfield Campus, Cranfield University College Road, Cranfield Bedfordshire, MK43 0AL
e-mail: m.zhu@cranfield.ac.uk

* Corresponding author.

J.R. González et al. (Eds.): NICSO 2010, SCI 284, pp. 199–210, 2010.
springerlink.com   © Springer-Verlag Berlin Heidelberg 2010

The paradigm of MEMS is well established within both the commercial and academic fields. At present encompassing more than just the mechanical and electrical [2], MEMS devices now cover a broad range of domains, including the fluidic, thermal, chemical, biological and magnetic systems. This has resulted in a host of applications to arise, from micro-resonators and actuators, gyroscopes, micro-fluidic devices [3], and biological lab on chip devices [4], to name but a few.

Developing MEMS by silicon micromachining fabrication techniques [5] requires both many prototypes and a long line of experimentation (design process). The process of MEMS design itself is broken down into many levels into which a designer may provide input and ultimately model, analyse and optimise a device. The process itself has been outlined by both fedder [6] and also senturia [7]. Normally, designs are produced in a trial and error approach dependant on user experience and naturally an antithesis to the goal of allowing designers the ability to focus on device and system design. This approach, nominally coined a "Build and Break" iterative, is both time-consuming and expensive.

A number of Computer Aided Design (CAD) tools and simulators have been developed and used to facilitate an improvement in the design process; however, this does not solve a fundamental problem with the current approach to MEMS design optimisation. The development of a design optimisation environment, which can allow MEMS designers to automate the process of modelling, simulation and optimisation at all levels of the MEMS design process, is fundamental to the eventual progress in MEMS Industry. Such an environment reduces the burden put on the designer and providing mediums that will potentially produce optimal devices within design constraints [20]. Work in design automation and optimisation can be seen to fall into two distinct areas; firstly the more traditional approaches found within numerical methods such as gradient-based search [8] [9]; and secondly the use of more powerful stochastic methods such as simulated annealing [10] and/or Evolutionary Algorithms (EAs) [11][12][16]. The current work has employed the latter to evolve and optimise new MEMS devices. Different researchers have used different classes of EAs in this subject domain [10],[12],[13],[14],[15],[17], however it is not clear which particular EA approach is the most appropriate and efficient in MEMS design synthesis and optimisation. This paper presents a comparative investigation into the performance of, particularly, two well known and widely used EAs (Multi-Objective Genetic Algorithm: MOGA-II [18] and NSGA-II [19]) on a number of MEMS design optimisation case studies. MOGA-II proved to be superior to NSGA-II. Experiments are, herein, described and results are discussed.This study also allows the validation of a design optimisation framework, by coupling both areas of MEMS simulation and analysis with optimization routines.

The next section describes the evolutionary design optimisation environment for MEMS. The subsequent section presents the experimental setup for three case studies of increasing complexity, followed by results and discussions in sections 4 and 5, respectively. Finally, a conclusion of the findings is presented in section 6.

2 MEMS Design Optimisation Framework

It is important as a designer to be able to undertake automated design optimisation whenever possible in order to speed up the design process. In response to this we establish a design optimization framework which links a powerful optimization environment tool based on EAs, with MEMS simulator SUGAR Fig 1. The MOEAs follow an iterative process, selecting designs based upon their performance in respect to the objectives set out, evolving them using powerful operators. Analysis is then undertaken by the simulator which is passed a parameters structure which overrides a default model design. Finally analysis is retrieved and designs are evaluated and ranked and finally replacement operators tune out worse designs by replacing them with better offspring.

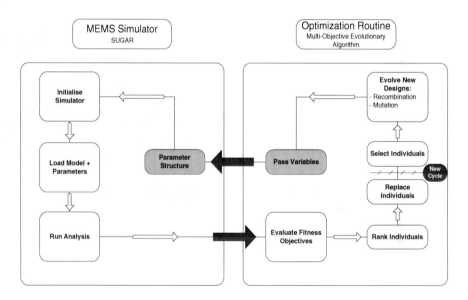

Fig. 1 An Evolutionary Design Optimisation Framework for MEMS

3 Experiments Set Up

Drawing on previous work undertaken in the field [15][10][21], planar MEMS devices form the basis for our evaluation of our design optimisation approach. A set of three case studies of increasing complexity have been implemented within our design optimisation environment, which forms a suitable strategy to evaluate the performance of the algorithms in question. The experiments investigate the performance of MOGA-II and NSGA-II for the design and optimisation of MEMS through these case studies. For each case study five experimental runs of each algorithm are conducted. MOGA-II is an improved version of MOGA by Poloni [22],

Table 1 Experimental parameter settings for MOGAII and NSGAII

MOGA-II		NSGA-II	
Probability of directional crossover	80%	Probability of SBX crossover	80%
Probability of classical crossover	14%	Probability of Mutation	1%
Probability of Mutation	1%	Distribution Index for crossover	20
DNA string Mutation ratio	5%	Distribution Index for mutation	20
Population	100	Population	100
Generations	100	Generations	100

utilizing a smart multi search elitism, and a triad of operators (classical one-point crossover, directional crossover and bit flip mutation) each with their own probability of invocation. As with classical MOGA, the representation is a binary string and in order to simulate continuous variables a sufficiently high base value must be used to divide between upper and lower bounds the possible variable values. NSGA-II is an elite preserving multi objective genetic algorithm, which also includes a diversity heuristic to maintain a uniform spread on the Pareto front. Unlike the standard MOGA, NSGA-II uses a real-valued representation, and therefore both recombination and mutation operators revolve around these real values. Both algorithms use some form of elitism based generational evolution and a breakdown of each is found below. The algorithms' parameters are fixed as shown in Table 1.

Algorithm 1: MOGA-II Pseudo Code

1. Initialize population

   a. Generate random population of size N and elite set $E = \theta$

2. Evaluate objective values
3. Assign rank based on Pareto dominance - 'Sort'
4. Generate offspring population

   a. Combine both population and elite sets P' = P ∪E
   b. If the cardinality of P' is greater than the cardinality of P reduce P' removing randomly the exceeding points.
   c. Compute the evolution from P' to P" applying MOGA operators
      i. Randomly assign one operator (Local tournament selection, directional crossover, one point crossover or bit flip mutation) based upon probability of invocation.

5. Evaluate objective values of population P"
6. Assign rank to P" individuals based on Pareto dominance - 'Sort'
7. Copy all non-dominated designs of P" to E - 'Sort'
8. Update E by removing duplicated or dominated designs
9. Resize the elite set E if it is bigger than the generation size N removing randomly the exceeding individuals
10. Return to step 2 considering P" as the new P until termination

---

Algorithm 2: NSGA-II Pseudo Code

1. Initialize population.
   a. Generate random population P of size N.
2. Evaluate objective values.
3. Assign rank based on Pareto dominance - 'Fast-Sort'.
4. Generate offspring population.
   a. Create population P' using tournament selection and apply variation operators (Simulated binary crossover and mutation).
5. Evaluate objective values of population P'.
6. Combine both population sets P and P' to give set of size 2N P''.
7. Assign rank to P'' individuals based on Pareto dominance - 'Fast-Sort'.
   a. Fill new P set with non-dominated fronts until cardinality is reached from set P''
   b. If the cardinality of new set P is greater than the size N reduce P by computing the crowding distance of the last front set to be added and fill remaining slots using crowded-comparison operator.
8. Return to step 4 until termination.

The case studies experimented with are; a simple meandering spring, a meandering resonator and finally a real world example of an ADXL150 accelerometer. For each case study, the performance of MOGA-II and NSGA-II is evaluated.

3.1 Case Study: Meandering Spring

The core topology of a large class of MEMS, such as micro-resonators and accelerometers consists generally of a spring + mass system, where a mass is suspended by a spring like structure anchored to a substrate, and the shape and topology of which effects the behaviour of the device. Therefore the ability to evolve spring like structures which match certain behaviour is important for the eventual design optimisation of more complex spring + mass systems such as a micro-resonator. Following previous work [15] we look to synthesize a simple meandering spring, composed of several beams, each of which has three variables, length, width and angle. The variable design parameters can be seen in Table 2.

Table 2 Variables design parameters used in spring case study, taken from [15]

Min Width	Max Width	Min Length	Max Length	Angle Min	Angle Max	Min Beam No	Max Beam No
2E-06	2E-05	2E-05	4E-04	-90	90	1	6

The objectives chosen for the experiment were to evolve designs that matched a certain behaviour in this instance each spring was to have a stiffness in the x direction Kx = 2N/m, and a stiffness in the y direction Ky = 2N/m following a force applied deflection. In this instance the objectives simply become the minimization of error from the design goal of 2 N/m.

3.2 Case Study: Meandering Resonator

It is important to be able to design micro resonators to match a certain frequency which can be integrated into a band-pass filter device. Following previous work [12] we look to evolve a MEMS resonator in order to match certain behaviour and design objectives. For this case study a set of four meandering springs are evolved each of which consists of several beams. The same variable parameters as described in table 2 are used and the central mass shape is fixed as in [10]. In order to reduce the search space complexity, a symmetry constraint to the design is applied, where one spring is evolved and then mirrored in both the x and y directions. The objectives for each design remain the same as for the spring, but also a third objective of having a first mode resonance frequency of 93723 Rad/s as taken from [12] are shown in Table 3.

Table 3 Design Objectives for Meandering Resonator

Objective	Target
Stiffness Kx N/m	2.0 (Minimize Error)
Stiffness Kx N/m	2.0 (Minimize Error)
Frequency Rad/s	93723 Rad/s (Minimize Error)

3.3 Case Study: ADXL150 Accelerometer

The goal to produce devices that mimic already viable real world macro designs but at a much smaller and more energy efficient way is a possibility with MEMS technology. The ADXL accelerometer series is a device which has been fabricated and tested in real world applications and seen it replacing its macro counterpart. This device can detect acceleration, as a result of force and gravity. This is crucial in one of the applications of this device that of car airbag deployment. Upon impact with another vehicle, acceleration as a result of the force occurs, this is then detected via the accelerometer device and if over a given threshold the signal can trigger the deployment of the airbag and thus save lives. The design variables follow that of previous work undertaken in [21] and are summarised in table 4. They consist of a central mass and a special case spring known as a "serpentine" spring, along with the sensing comb that runs alongside the mass. In this particular case study a symmetry constraint is applied to the serpentine springs, as a result only one spring is evolved and then mirrored in the x and y directions. The design objectives for these experiments are shown in Table 5.

Table 4 Design Objectives for ADXL150 Accelerometer

Variable	Lower Bounds	Upper Bounds
Mass Length	300μm	$600\ \mu$m
Mass Width	$50\ \mu$m	$150\ \mu$m
Finger Length	$30\ \mu$m	$130\ \mu$m
Finger Width	$2\ \mu$m	$4\ \mu$m
Short Beam Length	$10\ \mu$m	$10\ \mu$m
Short Beam Width	$2\ \mu$m	$2\ \mu$m
Long Beam Length	$10\ \mu$m	$100\ \mu$m
Long Beam Width	$2\ \mu$m	$2\ \mu$m
Crenulations	1	6

Table 5 ADXL150 Case Study design Objectives

Objective	Target
Frequency Rad/s	150,796 Rad/s (Minimize Error)
Total Area μm^2	Minimize
Sense Capacitance fF	Maximize

4   Results

For each case study results are represented in four sets of values; firstly the number of pareto solutions that were present at the end for each experiment (exp) run; secondly the number of pareto solutions from a particular experiment that remained when all five sets were combined, thirdly the number that remained when constraints on objective values were added and finally near the bottom we compare the number of pareto solutions from these sets that remain for each algorithm when MOGA-II and NSGA-II pareto individuals are combined.

Table 6 Table 7 and Table 9 shows the number of Pareto optimal solutions found within each experimental run for case studies described in sections 3.1, 3.2, and 3.3,

Table 6 MOGAII v NSGAII Experimental Results for the Meandering Spring

	MOGA-II				NSGA-II		
Exp	No of Pareto Sol in Exp	No of Pareto Sol Collated	No Sol < 1% Error per Obj	Exp	No of Pareto Sol in Exp	No of Pareto Sol Collated	No Sol < 1% Error per Obj
1	4316	2299	0	1	2944	1	0
2	26	0	0	2	2322	0	0
3	910	910	910	3	2866	0	0
4	2919	2873	1	4	2886	0	0
5	1920	0	0	5	209	209	209
Total	10091	6082	911	Total	11227	210	209
Total MOGAII v NSGAII	-	-	1	Total MOGAII v NSGAII	-	-	209

Table 7 MOGAII v NSGAII Experimental Results for the Meandering Resonator

	MOGA-II				NSGA-II		
Exp	No of Pareto Sol in Exp	No of Pareto Sol Collated	No Sol < 1% Error per Obj	Exp	No of Pareto Sol in Exp	No of Pareto Sol Collated	No Sol < 1% Error per Obj
1	66	22	7	1	220	10	0
2	31	18	12	2	16	0	0
3	49	1	0	3	87	82	8
4	215	1	0	4	182	31	0
5	42	20	10	5	164	1	0
Total	403	62	29	Total	669	124	8
Total MOGAII v NSGAII	-	-	29	Total MOGAII v NSGAII	-	-	2

Table 8 MOGAII v NSGAII Top 10 Frequency Results for culled < 1% set for the Meandering Resonator

	MOGA-II					NSGA-II			
Exp	ID	Freq Error Rad/s	Kx Error N/m	Ky Error N/m	Exp	ID	Freq Error Rad/s	Kx Error N/m	Ky Error N/m
1	9630	1.07	1.253E-02	6.911E-03	3	29624	20.15	1.851E-02	8.094E-03
5	47112	2.75	1.644E-02	4.814E-03	3	29693	30.08	1.445E-02	7.766E-03
5	47555	3.93	5.284E-03	3.878E-04	3	29734	33.24	6.372E-03	6.175E-03
1	8608	4.04	3.722E-04	9.670E-03	3	29806	72.17	4.353E-03	2.785E-03
1	9053	4.04	3.722E-04	9.670E-03	3	29660	122.65	4.875E-04	3.309E-03
1	9172	4.04	3.722E-04	9.670E-03	3	28978	245.77	5.755E-03	2.047E-03
1	9880	4.04	3.722E-04	9.670E-03	3	29563	279.95	8.898E-03	1.072E-03
2	19326	10.45	3.079E-03	8.302E-03	3	28733	668.50	1.678E-03	4.429E-05
2	18276	13.29	1.664E-05	6.495E-03	-	-	-	-	-
2	19258	13.91	3.165E-03	3.322E-03	-	-	-	-	-

Table 9 MOGAII v NSGAII Experimental Results for the ADXL150 Accelerometer

	MOGA-II				NSGA-II		
Exp	No of Pareto Sol in Exp	No of Pareto Sol Collated	No Sol < 1% Error per Obj	Exp	No of Pareto Sol in Exp	No of Pareto Sol Collated	No Sol < 1% Error per Obj
1	1525	551	47	1	1741	684	36
2	1389	646	69	2	1781	289	34
3	1613	547	88	3	1298	382	7
4	1494	940	146	4	1325	857	19
5	1464	695	134	5	1229	449	22
Total	7485	3379	484	Total	7374	2661	118
Total MOGAII v NSGAII	-	-	484	Total MOGAII v NSGAII	-	-	18

Table 10 MOGAII v NSGAII Top 10 Total Area Results for culled < 1% set for the ADXL150 Accelerometer

MOGA-II					NSGA-II				
Exp	ID	Freq Error Rad/s	Kx Error N/m	Ky Error N/m	Exp	ID	Freq Error Rad/s	Kx Error N/m	Ky Error N/m
1	9630	1.07	1.253E-02	6.911E-03	3	29624	20.15	1.851E-02	8.094E-03
5	47112	2.75	1.644E-02	4.814E-03	3	29693	30.08	1.445E-02	7.766E-03
5	47555	3.93	5.284E-03	3.878E-04	3	29734	33.24	6.372E-03	6.175E-03
1	8608	4.04	3.722E-04	9.670E-03	3	29806	72.17	4.353E-03	2.785E-03
1	9053	4.04	3.722E-04	9.670E-03	3	29660	122.65	4.875E-04	3.309E-03
1	9172	4.04	3.722E-04	9.670E-03	3	28978	245.77	5.755E-03	2.047E-03
1	9880	4.04	3.722E-04	9.670E-03	3	29563	279.95	8.898E-03	1.072E-03
2	19326	10.45	3.079E-03	8.302E-03	3	28733	668.50	1.678E-03	4.429E-05
2	18276	13.29	1.664E-05	6.495E-03	-	-	-	-	-
2	19258	13.91	3.165E-03	3.322E-03	-	-	-	-	-

respectively. For the constrained set all individuals which did not have an error value within 1% of the target for each objective were removed and in the case of the ADXL150 accelerometer an additional constraint of designs with a minimum sensitivity of 133fF was applied.

Table 8 highlights the top ten results from the culled 1% set, ranked by frequency error objective for the Meandering Resonator case study. Table 10 highlights the top ten results from our culled 1% set, ranked by total area objective for the ADXL150 Accelerometer case study.

5 Comparison and Discussion

From the above presented results one can begin to paint a picture into the performance of the two selected algorithms on this particular subset of case studies for MEMS design optimisation. To begin with, it seems that both algorithms are robust enough to provide similar sets of Pareto fronts from each experimental run when they are collated. However of the two algorithms, MOGA-II provides results which fair better, with NSGA-II falling down somewhat with the meandering spring case study. Of the number of Pareto solutions found within the target constraints for each case study, MOGA-II outperforms NSGA-II, generally producing two thirds more solutions for all three case studies. A direct comparison between the final Pareto sets for each algorithm provides a similar result, with MOGA-II providing more individuals within the Pareto front for the ADXL150 Accelerometer and the Meandering Resonator case-studies, with only NSGA-II providing better results on the Meandering Spring example.

Reasons behind such discrepancies could fall into the differences found within each algorithm; these can lay either in, naturally, the choice of representation, the role of each algorithms variation operators, or some of the diversity heuristics used. If one picks up on the third case study, the ADXL150 Accelerometer, from Table 10 one can see a deviation in terms of behaviour between MOGA-II and NSGA-II. For this case study, the constraints focused upon designs which had as small a total area as possible while maintaining a sensitivity value above 133 fF and a minimum frequency error below 1% of the target goal. Though both MOGA-II and NSGA-II were able to provide individuals which lied within these constraints, NSGA-II designs seem to lie heavily towards an increased sensitivity, rather than focusing upon reduced total area.

From the outset this seems to cast a shadow on the performance of NSGA-II in this example; however it may be an unfair assessment and hence requires further analysis. NSGA-II employs a diversity heuristic in the form of a crowded distance operator to enforce a uniform spread of Pareto solutions while MOGA-II does not in any specific way emulate this behavior. The ADXL150 example contains two particular objectives which somewhat work in tandem, that being total area and sensitivity. In this instance changing the mass length of the device can either result in a decrease in total area and subsequent decrease in sensitivity or provide the opposite effect. In parallel, increasing finger length can increase sensitivity and the total area and vice versa. These two variable changes seem to have the most accessible influence in objective function performance, and as such most likely drive our algorithms for fitter individuals. As NSGA-II looks to find a suitable spread it will produce designs which lie upon the whole gradient between these two competing objectives, while MOGA-II does not feel this selective pressure and can perhaps begin to concentrate on designs which target improved frequency objectives. As a result MOGA-II can possibly produce designs which target all three objectives more easily than NSGA-II. Given the final constraint where we want to focus on one particular area of a front, something NSGA-II looks to avoid, the MOGA-II is not encumbered by this and as a result it seems able to produce superior designs. This is only a speculative explanation and requires further investigation into what effect NSGA-II's crowded diversity heuristic has in terms of performance and the reasons why.

Finally, it is a possibility that each algorithm local search approach, be it MOGA-II's single bit flip operator or NSGA-II's real valued polynomial mutation may provide a profound difference in performance when it comes to local search performance at near optimal design spaces. In the field of MEMS design it is important for operators to cope with such small scales and is therefore something for further investigation.

6   Conclusions

The paper presents an important study that compares the performance of two widely known and used evolutionary algorithms for the design optimisation of MEMS devices, namely, NSGA-II and MOGA-II. Experiments are conducted on three MEMS case studies with increasing complexity. Initial results clearly show the superiority of MOGA-II over NSGA-II. Speculative explanations are discussed in section 5, however, further work is needed to evaluate the reasons why the performance of the two algorithms differed, and essentially the role of various heuristics and operators in the evolutionary design optimisation of this application domain.

References

[1] Fujita, H.: Two Decades of MEMS– from Surprise to Enterprise. In: Proceedings of MEMS, Kobe, Japan, January 21-25 (2007)

[2] Hsu, T.R.: MEMS and Microsystems, 2nd edn. Wiley, Chichester (2008)

[3] Isoda, T., Ishida, Y.: Seperation of Cells using Fluidic MEMS Device and a Quantitative Analysis of Cell Movement. Transactions of the Institute of Electrical Engineering of Japan 126(11), 583–589 (2006)

[4] Hostis, F.l., Green, N.G., Morgan, H., Akaisi, M.: Solid state AC electroosmosis micro pump on a Chip. In: International Conference on Nanoscience and Nanotechnology, ICONN, Brisbane, Qld, July 2006, pp. 282–285 (2006)

[5] Hao, Y., Zhang, D.: Silicon-based MEMS process and standardization. In: Proceedings of the 7th International Conference on Solid-State and Integrated Circuits Technology 2004, vol. 3, pp. 1835–1838 (2004)

[6] Fedder, G.: Structured Design of Integrated MEM. In: Twelfth IEEE International Conference on Micro Electro Mechanical Systems, MEMS 1999, Orlando, FL, USA, pp. 1–8 (1999)

[7] Senturia, S.D.: Microsystem Design. Kluwer Academic Publishers, Dordrecht (2001)

[8] Haronain, D.: Maximizing microelectromechanical sensor and actuator sensitivity by optimizing geometry. Sensors and Actuators A 50, 223–236 (1995)

[9] Iyer, S., Mukherjee, T., Fedder, G.: Automated Optimal Synthesis of Microresonators. In: Solid-State Sensors and Actuators, Chicago, IL, pp. 12–19 (1997)

[10] Kamalian, R., Zhou, N., Agogino, A.M.: A Comparison of MEMS Synthesis Techniques. In: Proceedings of the 1st Pacific Rim Workshop on Transducers and Micro/Nano Technologies, Xiamen, China, July 22-24, pp. 239–242 (2002)

[11] Li, H., Antonsson, E.K.: Evolutionary Techniques in MEMS Synthesis. In: Proc. DETC 1998, 1998 ASME Design Engineering Technical Conferences, Atlanta, GA (1998)

[12] Zhou, N., Agogino, A.M., Pister, K.S.: Automated Design Synthesis for Micro-Electro-Mechanical Systems (MEMS). In: Proceedings of the ASME Design Automation Conference, ASME CD ROM, Montreal, Canada, September 29-October 2 (2002)

[13] Kamalian, R.H., Takagi, H., Agogino, A.M.: Optimized Design of MEMS by Evolutionary Multi-objective Optimization with Interactive Evolutionary Computation. In: Deb, K., et al. (eds.) GECCO 2004. LNCS, vol. 3103, pp. 1030–1041. Springer, Heidelberg (2004)

[14] Zhang, Y., Kamalian, R., Agogino, A.M., Sequin, C.: Hierarchical MEMS Synthesis and Optimization. In: Varadan, V.K. (ed.) Proceedings of SPIE, Smart Structures and Materials 2005: Smart Electronics, MEMS, BioMEMS, and Nanotechnology. International Society for Optical Engineering, CD ROM. Paper 5763-12, vol. 5763, pp. 96–106 (2005)

[15] Zhou, N., Zhu, B., Agogino, A.M., Pister, K.: Evolutionary Synthesis of MEMS (Microelectronic Mechanical Systems) Design. In: Proceedings of ANNIE 2001, IEEE Neural Networks Council and Smart Engineering Systems Laboratory, Marriott Pavilion Hotel, St. Louis, Missouri, November 4-7, vol. 11, pp. 197–202. ASME Press (2001)

[16] Benkhelifa, E., Farnsworth, M., Tiwari, A., Zhu, M.: An Integrated Framework for MEMS Design Optimisation using modeFrontier. In: EnginSoft International Conference 2009, CAE Technologies For Industry and ANSYS Italian Conference 2009 (2009)

[17] Lohn, J.D., Kraus, W.F., Hornby, G.S.: Automated Design of a MEMS Resonator. In: Proceedings of the Congress on Evolutionary Computation, pp. 3486–3491 (2007)

[18] Poles, S.: MOGA-II An Improved Multi-Objective Genetic Algorithm. Technical report 2003-006, Esteco, Trieste (2003)

[19] Deb, K., Agrawal, S., Pratap, A., Meyarivan, T.: A fast elitist non-dominated sorting genetic algorithm for multi-objective optimization: NSGA-II. In: Schonauer, M., et al. (eds.) PPSN 2000. LNCS, vol. 1917, pp. 849–858. Springer, Heidelberg (2000)

[20] Benkhelifa, E., Farnsworth, M., Bandi, G., Tiwari, A., Zhu, M., Ramsden, J.: Design and Optimisation of Microelectromechanical Systems: A Review of the State-of-the-Art. International Journal of Design Engineering, Special Issue Evolutionary Computing for Engineering Design (2009) (accepted to be published)

[21] Zhang, Y., Kamalian, R., Agogino, A.M., Séquin, C.H.: Design Synthesis of Microelectromechanical Systems Using Genetic Algorithms with Component-Based Genotype Representation. In: Proc. of GECCO 2006 (Genetic and Evolutionary Computation Conference), Seattle, July 8-12, vol. 1, pp. 731–738 (2006) ISBN 1-59593 187-2

[22] Poloni, C., Pediroda, V.: GA coupled with computationally expensive simulations: tools to improve efficiency. In: Genetic Algorithms and Evolution Strategies in Engineering and Computer Science, pp. 267–288. John Wiley and Sons, England (1997)

A Distributed Service Oriented Framework for Metaheuristics Using a Public Standard

P. García-Sánchez, J. González, P.A. Castillo, J.J. Merelo, A.M. Mora,
J.L.J. Laredo, and M.G. Arenas

Abstract. This work presents a Java-based environment that facilitates the development of distributed algorithms using the OSGi standard. OSGi is a plug-in oriented development platform that enables the installation, support and deployment of components that expose and use services dynamically. Using OSGi in a large research area, like the Heuristic Algorithms, facilitate the creation or modification of algorithms, operators or problems using its features: event administration, easy service implementation, transparent service distribution and lifecycle management. In this work, a framework based in OSGi is presented, and as an example two heuristics have been developed: a Tabu Search and a Distributed Genetic Algorithm.

1 Introduction

Nowadays the Metaheuristics Research Area has a wide number of algorithms and problems. There are many implementations of them, using several programming languages, frameworks and architectures, but without using a well-defined plug-in specification.

When building quality software systems it is necessary to design them with a high level of modularity. Besides the benefits that classic modularization paradigms can offer (like object-oriented modelling) and the improvements in test, reusability, availability and maintainability, it is necessary to explore another modelling techniques, like the plug-in based development [21]. This kind of development simplifies aspects such as the complexity, personalization, configuration, development and cost of the software systems. In the optimization heuristics software area, the benefits the usage of this kind of development can offer are concreted in the development

P. García-Sánchez · J. González · P.A. Castillo · J.J. Merelo · A.M. Mora ·
J.L.J. Laredo · M.G. Arenas
Dept. of Computer Architecture and Computer Technology
e-mail: pgarcia@atc.ugr.es, jesus@atc.ugr.es, pedro@atc.ugr.es,
jmerelo@geneura.ugr.es, amorag@geneura.ugr.es,
juanlu@geneura.ugr.es, maribel@atc.ugr.es

J.R. González et al. (Eds.): NICSO 2010, SCI 284, pp. 211–222, 2010.
springerlink.com   © Springer-Verlag Berlin Heidelberg 2010

of algorithms, experimental evaluation, and combination of different optimization paradigms [21].

On the other hand, other patterns for integration, like SOA, have emerged. SOA (Service Oriented Architecture) [18] is a paradigm for organizing and utilizing distributed capabilities, called *services*. A service is an interaction depicted in Figure 1.

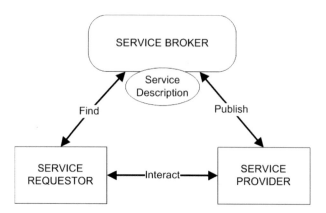

Fig. 1 Service interaction schema. The service provider publish a service description that is used by the requester to find and use services

The service provider publishes service descriptions (or interfaces) in the service registry, so the service requesters can discover services and bind to the service providers.

Distributed computing offers the possibility of taking advantage of parallel processing in order to obtain a higher computing power than other multiprocessor architectures. Two clear examples are the research lines centred in clusters [5] and GRID [9] for parallel processing. SOA it is also used in this area, using platforms based in Web Services [18], and new standards for this paradigm have emerged, like OSGi.

OSGi (Open Service Gateway Initiative) [2] was proposed by a consortium of more than eighty companies in order to develop an infrastructure for the deployment of service in heterogeneous network of devices, mainly oriented to domotic [15]. Nowadays it defines a specification for a Service Oriented Architecture for virtual machines (VMs). It provides very desiderable features, like packet abstraction, lifecycle management, packaging or versioning, allowing significant reduction of the building, support and deployment complexity of the applications.

OSGi technology allows the components to be dynamically discovered among them to increase the collaboration to minimize and manage the coupling among modules. Moreover, the OSGi Alliance has developed several standard component interfaces for common usage patterns, like HTTP servers, configuration, logs, security, management or XML management among others, whose implementations can be obtained by third-parties. Nowadays there are some challenges in the OSGi development [12], but they only affect to the creation of very complex applications.

Therefore, the objective of the proposed environment is to facilitate the development of distributed computing applications by using the OSGi standard, taking advantage of the plug-in software development and SOA that can compete with existing distributed applications in easy of use, compatibility and development.

The rest of this work is structured as follows: first the state of the art in similar applications is described (section 2). Section 3 introduces the technologies used in the development of this work. Then, we present (section 4) the design of the proposed architecture (called OSGiLiath) and the development of two computing applications using a Distributed Genetic Algorithm and a Tabu Search. Experiments and yielded results are shown in section 6. Finally the conclusions and future work are presented.

2 State of the Art

Nowadays there are many works about heuristic frameworks. Most of them have the lack of low generality, because they are focused in an specific field, like EasyLocal++ [10] (focused in Local Search) or SIGMA [11] (in the field of optimization-based decision support systems). Another common problem is that they are just libraries (like ECJ [14], Evolutionary Computation in Java), they have no GUIs, or they are complicated to install and require many programming skills. Another issue could be the lack of comfort, for example, C++ has a more complicate sintaxis than other languages.

Among this great number of frameworks we want to focus in the most widely accepted distributed algorithms frameworks. MALLBA [1] is based in software skelletons with a common and public interface. Every skeleton implements a resolution technique for optimization in the fields of exact, heuristic or hybrid optimization. It provides LAN and WAN capacity distribution with MPI . However, it is not based in the plug-in development, so it can not take advantage of features like the life-cycle management, versioning, or dynamic service binding, as OSGi proposes.

Another important platform is DREAM [3], which is an open source framework for Evolutionary Algorithms based on Java that defines an island model and uses the Gossip protocol and TCP/IP sockets for communication. It can be deployed in P2P platforms and it is divided in five modules. Every module provides an user interface and different interaction and abstraction level, but adding new functionalities is not so easy, due to the system must be stopped before adding new modules and the implementation of interfaces must be defined in the source code, so a new compilation is needed. OSGi lets the addition of new functionalities only compiling the new features, not the existing ones.

ParadiseEO [6] allows the design of Evolutionary Algorithms and Local Search with hybridization, providing a variety of operators and evaluation functions. It also implement the most common parallel and distributed models, and it is based in standard libraries like MPI, PVM and Pthreads. But it has the same problems that the previous frameworks, not lifecycle managment or service oriented programming. GAlib [23] is very similar and share the same characteristics and problems.

In the field of the plug-in based frameworks, HeuristicLab [20] is the most important example. It also allows the distributed programming using Web Services and a centralized database, instead using their own plug-in design for this distributed communication. Moreover, the used plug-in system does not uses a public specification like OSGi. And also it is a proprietary software, like their execution environment, the .NET platform [7].

Finally, METCO framework [13] also have the same problems, it not uses a standard plug-in system or SOA, but let the implementation of existing interfaces, and lets the user configure its existing functionalities.

In summary, the previous works present a number of shortcomings when designing and adding new features: they need to modify source code or be stopped in order to add new features and they are not based in a public plug-in specification. Also they not have an event administration mechanism and they are not service-oriented, so they not take advantage of this paradigm.

3   Used Technologies

OSGi features can be useful in the development of distributed algorithms, so this section describes the tools and communication protocols employed within the presented framework.

3.1   OSGi

OSGi implements a dynamic component model, unlike normal Java environments. Applications or components (also called *bundles*) can be remotely installed, started, stopped, updated or uninstalled on the fly; moreover, the classes and packaging management is specified in detail. The framework provides APIs for the management of services that are exposed or used by the bundles.

A **bundle** is a file that contains compiled and packaged classes and a configuration file. This file indicates which classes imports or exports the bundle.

The most important concept in OSGi is the **service**. The services allow to connect *bundles* in a dynamic way, offering a publication-search-connection model. That is, a *bundle* exposes a service by a Java interface, and another bundle (or itself) implements that interface. A third *bundle* can access this service using the exposed interface without having any knowledge of how it is implemented, using the *Service Registry*. The Figure 2 shows an example of the OSGi architecture.

It would be useful if this connection could be done out of the source code, so the OSGi also provides **components**. A component is a class inside a *bundle* together with an XML description. This description is interpreted in execution time to create and remove services depending the availability of other services, other components or configuration data. The main difference between a component and a normal class inside a *bundle* is that in the second the association between interface and implementation of the service must be defined in the source code, and also the dependency management and the service state detection, being this a tedious work for the

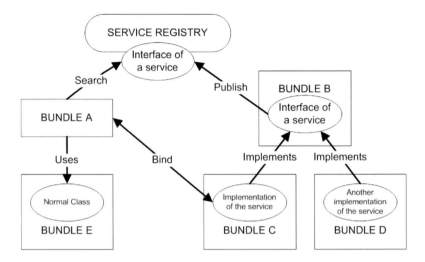

Fig. 2 In OSGi a service can be implemented by several bundles. Other bundles may chose among this implementations using the Service Registry

programmer. To facilitate this task in OSGi the *Declarative Services* specification [17] arises. It lets that, for example, we could create a class that is not activated until an specific and required service is detected. When this service is active, the class can use it with a *bind* method. It is important to note that implementation will be injected in execution time, not in compilation time.

OSGi also provides event handling with an implementation of the event broker pattern, the **Event Admin**. It is an interbundle communication mechanism based on a publish-and-subscribe model. Some bundles publish events and some other bundles can read this events, being this task transparent for the programmer: the sender does not need to know who are listening their events, and the listener can filter among the events.

3.2 R-OSGi

One of the problems of OSGi today is its inability to invoke remote services and its lack of a distributed module management, so other protocols adapters have been created, like JINI [22] and UPnP [16]. Nevertheless these approximations can be considered invasive, due to their requirement of re-structuring the application. This is the reason that R-OSGi arises [19]. R-OSGi is a middleware layer inside OSGi that lets a more transparent distribution of the application parts simply distributing its software modules. Inside the OSGi framework the remote and local services are indistinguishable, so the existent OSGi applications can be distributed without modification using R-OSGi. Moreover, this middleware does not imposes client-server assignation because the modules relationship is symmetric. The authors have demonstrated that the R-OSGi is similar to the highly optimized Java 5 RMI implementation and two times faster than UPnP.

R-OSGi creates client *proxies*. For the client of a service, this proxies behave as local services and they also are provided by locally instantiated bundles. However a *proxy bundle* redirects all received calls to the original service that resides in the remote machine, and propagates the result of the call back to the client. An example of this architecture is shown in Figure 3. The events used in the previously explained Event Admin are also transmitted in a transparent manner: the senders and the receivers of the events do not need to add anything to the program code in order to receive the events among distributed nodes, because they do not need to know where the nodes are.

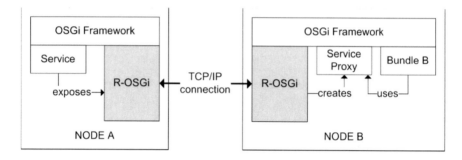

Fig. 3 Architectural overview of R-OSGi. The node B uses *Service Proxy* as a normal service

4 OSGiLiath Platform

This section dives in the functionality and design of the proposed environment, called OSGiLiath (*OSGi Laboratory for Implementation and Testing of Heuristics*). This environment is a framework for the development of heuristic optimization applications, not centred on a concrete paradigm, and whose main objective is to promote the OSGi usage and offer to programmers the next features:

- Easy interfaces
- Asynchronous data sending/receiving
- Component Oriented Programming
- Client/Server or Distributed Model
- Paradigm independent
- Declarative Services
- Remote event handling

The source code is available in http://atc.ugr.es/~pgarcia, under a GPL license. The environment presented in this work lets defining implementations for specific problems using the OSGi benefits. Its architecture is composed by three levels or layers: *Interface*, *Heuristic* and *Problem* (see Figure 4).

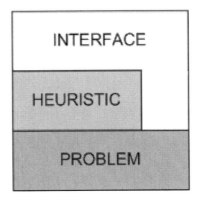

Fig. 4 Defined layers in OSGiLiath

The *Interface* layer provides a hierarchy of interfaces defined to develop distributed heuristics. Some examples are *Algorithm*, *Distributed Algorithm*, *Solution*, *Problem*, *Input Data* or *Parameters*. It also provides interfaces and objects for distributed programming, like *Server*, *Node* or *Task*. This class hierarchy, exported as a *bundle* is well-defined, because it will be the basis to construct the full application. As every *bundle*, it can export these interfaces to be used by another bundles. These interfaces must be implemented in the next framework level, the *Heuristic* layer. Using the OSGi Declarative Services Specification [17], the instances of these implementations will be activated when they are necessary and accessed among them. Finally, (*Problem*) layer defines what problems will be executed in the framework.

Furthermore, using the R-OSGi functionality we can add the feature of distributed applications in an undetermined number of nodes. In this case, we have to implement several *Tasks*, whose implementation can be in different nodes. Given the platform architecture the *Heuristic* or *Problem* layers could be in remote nodes, so the user could define new problems or heuristics and automatically bind with the necessary elements to execute.

5   Development Example Using OSGiLiath

As an example of usage of the presented framework, a tabu search and a distributed genetic algorithm have been developed to solve the *Vehicle Routing Problem* (VRP) and the capabilities of the framework have been tested. Due to space restriction we refer the reader to [8], which explains the implemented Tabu Search and a more formal problem approach. The Tabu Search is a sequential algorithm, while the Genetic Algorithm uses a distributed island model: every node executes a separate algorithm and swaps individuals with the other nodes.

5.1   Specifying an Application

The first step to develop in OSGiLiath is to implement the interfaces defined in previous sections to build specific implementations. For example, *TabuSearch* and *DistributedGeneticAlgorithm* are implementations of *Algorithm* and *DistributedAlgorithm*. The implementation of each algorithm must be as general as possible, due to the implementation of the problem to solve its developed in the next level. So, in this layer more interfaces are defined, like *StopCriterion*, *TabuList*, *Mutation*, *Fitness* or *IndividualInitialization*. This level uses the feature of Declarative Services in order to obtain automatically the implementation of that interfaces.

5.2   Specifying the Problem

Finally in this level the problem to solve is specified in more detail. For example we have implemented the interfaces *Problem*, *Individual*, *Crossover* or *TabuList* with the *ProblemVRP*, *IndividualVRP*, *CrossoverVRP* and *TabuListVRP* classes. Due to they have been exposed as declarative services, when they are activated, the services defined in the previous level also will be activated.

All work developed in this level can be added to the base platform, since all component are clearly differentiated, and other developers could implement their own problems to apply the Genetic Algorithm or Tabu Search, or add new algorithms to solve the VRP problem.

5.3   Adding Distributed Capacity

Using declarative services implementation of *Task* interfaces are created. In the Tabu Search example, remote nodes could search the best neighbourhood of the current solution, receiving a movement list and the Tabu List, but due to the canonical Tabu Search is difficult to parallelize because of the latency we only have tested the sequential algorithm.

In the case of the Genetic Algorithm, every certain number of iterations each node receives one of the best individuals of the other nodes, randomly selected. Thanks to the OSGi features, every service can be distributed in a transparent way (operators, algorithms, initers, schedulers). The programmer does not need how the communication is performed or where the implementation is, he only needs to know the interface of the service.

All the nodes have knowledge of what the other are doing, thanks to the OSGi event handling mechanism. Whenever an iteration or algorithm over, events are published and read by the others, so the algorithms can synchronize or inform to others about their results.

Along with the challenges of OSGi [12], there exists the issue of the loss of abstraction in the development of the interfaces of our framework, so a study to find balance between cohesive and loose coupled hierarchy will be performed in future. In problem-specific algorithms, where exist a tightly coupled association, the usage

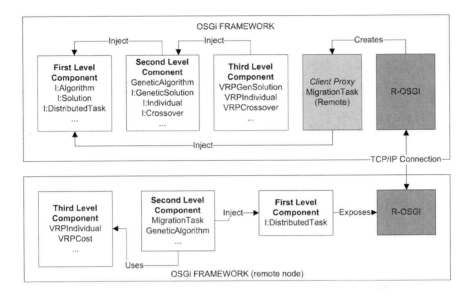

Fig. 5 OSGiLiath architecture. The user can implement heuristics and problems interfaces

of events and automatic communication mechanisms will be helpful if they are used properly.

6 Experiments

Once the algorithms development have been explained we present the obtained results. We have to say that the presented work is a proof-of-concept, so these results are shown as example. We have used a 4 nodes cluster, each one of them with a 1.6 GHz, 4 GB RAM and Java version 1.5. The common parameters for the algorithms are a stop criterion of 60 iterations without improve the best solution and random initial solutions. In Tabu Search the Tabu List have 30 moves. The Genetic Algorithm parameters are: 200 individual population with elitism, migration of one of the 10 best individuals, randomly selected every 10 iterations; mutation probability is 0.5 and a tournament selection for crossover of the 50 best individuals. The instances of the two problems have been extracted from [4].

The obtained results are shown in Table 1. As can be seen, the Genetic Algorithm results outperforms the Tabu Search, due to the used crossover swaps complete routes, unlike the Tabu Movements, that moves an unique shop in the routes. The time taken in the sequential Genetic Algorithm also is lower than the sequential Tabu Search.

However, the purpose of this work is not perform an analysis of the presented algorithms, but show the ease of using this framework in the distributed algorithm development.

Table 1 Result table for the experiments (average ± standard deviation)

Nodes	Cost	Iterations	Time (s)
	Tabu Search		
1	2330.18 ± 86.41	312.13 ± 20.26	224.70 ± 11.61
	Genetic Algorithm		
1	2318.83 ± 72.89	4222.60 ± 435.04	100.43 ± 10.51
2	2268.11 ± 74.53	4759.52 ± 798.54	113.33 ± 87.80
3	2223.18 ± 54.13	4903.66 ± 338.40	128.94 ± 65.67
4	2212.24 ± 29.85	4740.20 ± 278.45	124.85 ± 98.20

Every experiment was executed 10 times.

7 Conclusions and Future Work

This work presents an environment for the development of distributed algorithms extensible via plug-ins and based in a wide-accepted software specification (OSGi). OSGi features (declarative services, dynamic life-cycle management, or package abstraction) are used to easily create algorithms in a layered way. Moreover, it uses R-OSGi to develop distributed services. We have shown the Tabu Search and the Genetic Algorithm implementation as an example.

As future work an automatic generated GUI will be developed to dynamically control which problems, algorithms or parameters to use. A study about scalability using other algorithms (like GRASP, Scatter Search, Ant Colony Optimization and others) will be performed. Also, we are going to increase the usage of the OSGi capabilities, like the Event Administration or automatic service management in a deeper way. Additionally we intend to create a web portal to centralize all new implementations of problems and algorithms to let the distribution within the base platform, so the users just have to write the level 3 classes to solve particular problems. An study of porting existing software to our framework (especially those works that are written in Java, like DREAM or ECJ) will be performed. Moreover, due to the ease of implementations binding with their interfaces it is planned to develop the functionality of chosing one implementation or another depending on several parameters or, for example, using Genetic Programming to evolve and hybridize algorithms.

Acknowledgements. Supported by projects AmIVital (CENIT2007-1010) and EvOrq (TIC-3903).

References

[1] Alba, E., Almeida, F., Blesa, M., Cotta, C., Díaz, M., Dorta, I., Gabarró, J., León, C., Luque, G., Petit, J., Rodríguez, C., Rojas, A., Xhafa, F.: Efficient parallel LAN/WAN algorithms for optimization, the MALLBA project. Parallel Computing 32(5-6), 415–440 (2006)

[2] Alliance, O.: OSGi alliance (2004), http://www.osgi.org/
[3] Arenas, M., Collet, P., Eiben, A., Jelasity, M., Merelo, J.J., Paechter, B., Preuß, M., Schoenauer, M.: A framework for distributed evolutionary algorithms. In: Guervós, J.J.M., Adamidis, P.A., Beyer, H.-G., Fernández-Villacañas, J.-L., Schwefel, H.-P. (eds.) PPSN 2002. LNCS, vol. 2439, pp. 665–675. Springer, Heidelberg (2002)
[4] BranchAndCutorg. Vehicle routing data sets (2003),
     http://branchandcut.org/VRP/data/
[5] Buyya, R.: High Performance Cluster Computing: Architectures and Systems. Prentice-Hall, Englewood Cliffs (1999)
[6] Cahon, S., Melab, N., Talbi, E.: ParadisEO: A framework for the reusable design of parallel and distributed metaheuristics. Journal of Heuristics 10(3), 357–380 (2004)
[7] Escoffier, C., Donsez, D., Hall, R.S.: Developing an OSGi-like Service Platform for .NET. In: 3rd IEEE Consumer Communications and Networking Conference, vol. 1-3, pp. 213–217 (2006)
[8] Esparcia-Alcázar, A.I., Cardós, M., Merelo, J.J., Martínez-García, A., García-Sánchez, P., Alfaro-Cid, E., Sharman, K.: EVITA: An integral evolutionary methodology for the inventory and transportation problem. Studies in Computational Intelligence 161, 151–172 (2009)
[9] Foster, I.: The Grid: A new infrastructure for 21st Century Science. Phisics Today 55, 42–47 (2002)
[10] Gaspero, L., Schaerf, A.: Easylocal++: an object-oriented framework for the flexible desgin of local search algorithms and metaheuristics. In: Proceedings of 4th Metaheuristics International Conference (MIC 2001), pp. 287–292 (2001)
[11] González, J.R., Pelta, D.A., Masegosa, A.D.: A framework for developing optimization-based decision support systems. Expert Systems with Applications 36(3, Part 1), 4581–4588 (2009)
[12] Kriens, P.: Research challenges for OSGi (2008),
     http://www.osgi.org/blog/2008/02/
     research-challenges-for-osgi.html
[13] León, C., Miranda, G., Segura, C.: Metco: A parallel plugin-based framework for multi-objective optimization. International Journal on Artificial Intelligence Tools 18(4), 569–588 (2009)
[14] Luke, S., et al.: ECJ: A Java-based Evolutionary Computation and Genetic Programming Research System (2009),
     http://www.cs.umd.edu/projects/plus/ec/ecj
[15] Marples, D., Kriens, P.: The Open Services Gateway Initiative: An introductory overview. IEEE Communications Magazine 39(12), 110–114 (2001)
[16] Miller, B.A., Nixon, T., Tai, C., Wood, M.D.: Home networking with universal plug and play. IEEE Communications Magazine 39(12), 104–109 (2001)
[17] OSGi Alliance. Declarative services specification, pp. 281–314 (2007),
     http://www.osgi.org/download/
     r4-v4.2-cmpn-draft-20090310.pdf
[18] Papazoglou, M.P., Van Den Heuvel, W.: Service oriented architectures: Approaches, technologies and research issues. VLDB Journal 16(3), 389–415 (2007)
[19] Rellermeyer, J.S., Alonso, G., Roscoe, T.: R-osgi: Distributed applications through software modularization. In: Cerqueira, R., Campbell, R.H. (eds.) Middleware 2007. LNCS, vol. 4834, pp. 1–20. Springer, Heidelberg (2007)

[20] Wagner, S., Affenzeller, M.: Heuristiclab grid - a flexible and extensible environment for parallel heuristic optimization. In: Proceedings of the International Conference on Systems Science, vol. 1, pp. 289–296 (2004)

[21] Wagner, S., Winkler, S., Pitzer, E., Kronberger, G., Beham, A., Braune, R., Affenzeller, M.: Benefits of plugin-based heuristic optimization software systems. In: Moreno Díaz, R., Pichler, F., Quesada Arencibia, A. (eds.) EUROCAST 2007. LNCS, vol. 4739, pp. 747–754. Springer, Heidelberg (2007)

[22] Waldo, J.: The Jini architecture for network-centric computing. Communications of the ACM 42(7), 76–82 (1999)

[23] Wall, B.: A genetic algorithm for resource-constrained scheduling, Ph.D. thesis. MIT, Cambridge (1996), http://lancet.mit.edu/ga

Cellular Genetic Algorithm on Graphic Processing Units

Pablo Vidal and Enrique Alba

Abstract. The availability of low cost powerful parallel graphic cards has estimulated a trend to implement diverse algorithms on Graphic Processing Units (GPUs). In this paper we describe the design of a parallel Cellular Genetic Algorithm (cGA) on a GPU and then evaluate its performance. Beyond the existing works on master-slave for fitness evaluation, we here implement a cGA exploiting data and instructions parallelism at the population level. Using the CUDA language on a GTX-285 GPU hardware, we show how a cGA can profit from it to create an algorithm of improved physical efficiency and numerical efficacy with respect to a CPU implementation. Our approach stores individuals and their fitness values in the global memory of the GPU. Both, fitness evaluation and genetic operators are implemented entirely on GPU (i.e. no CPU is used). The presented approach allows us benefit from the numerical advantages of cGAs and the efficiency of a low-cost but powerful platform.

Keywords: Cellular Genetic Algorithm, Parallellism, GPGPU, CUDA.

1 Introduction

Cellular Genetic Algorithms (cGAs) are effective optimization techniques solving many practical problems in science and engineering [1]. The basic algorithm (cGA)

Pablo Vidal
LabTEm - Laboratorio de Tecnologías Emergentes, Unidad Académica Caleta Olivia,
Universidad Nacional de La Patagonía Austral,
Ruta 3 Acceso Norte s/n, (9011) Caleta Olivia Sta. Cruz - Argentina
Phone/Fax: +54 0297 4854888
e-mail: `pablo.vidal.20@gmail.com`

Enrique Alba
Dept. de Lenguajes y Ciencias de la Computación, University of Málaga, ETSI Informática,
Campus de Teatinos, Málaga - 29071, Spain
e-mail: `eat@lcc.uma.es`

J.R. González et al. (Eds.): NICSO 2010, SCI 284, pp. 223–232, 2010.
springerlink.com   © Springer-Verlag Berlin Heidelberg 2010

is selected here because of its high performance and because of its swarm intelligence structure (i.e. emergent behavior and decentralized control flow). By evolving this kind of algorithm is able of keeping a high diversity in the population until reaching the region containing the global optimum. This kind of algorithms may benefit from parallelism as a way of speeding up its operations [2] when the instance of the problem is complex.

Graphic Processing Units (GPUs) are well-known hardware cards with a fixed function, being traditionally used for visualization purposes. However, the new generations of GPUs have also unleashed a promising potential for scientific computing, seen as a new hardware allowing the use of high arithmetic capacity and high performance.

Thus, researchers and developers have begun to harness GPUs for general purpose computation [4] [7]. In addition to their low cost and ubiquitous availability, GPUs have a superior processing architecture when compared to modern CPUs, and thus present a tremendous opportunity for developing lines of research in optimization algorithms especially targeted for GPUs, this its shown in present works such as [3] [5].

Therefore, we work here with a parallel cGA running entirely on GPU (i.e. no CPU is needed only to start and stop the algorithm), and demonstrate that the proposed optimization technique (called cGA GPU) is quite amenable for massive parallelism to obtain larger performances with reduction of times and improvements of the speedup. This approach offers the possibility to solve larges problem instances with the improved computing capacity of a GPU. All this will be shown on a benchmark of discrete and continuous problem to claim not only for time reductions but also for numerical advantages of this swarm intelligence algorithm.

The paper is structured as follows, The next section contains some background about the parallelism, we explain the Cellular Genetic Algorithm and its implementation in GPU. Section 3 describes the experimental setup, while Section 4 explains the test problems used, details of the cGA parameters, and the statistical tests performed.

Finally, Section 5 provides the obtained results and Section 6 offers our conclusions, as well as some comments on the future work.

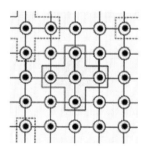

Fig. 1 Toroidal structure of a cGA population

2   Description of a Cellular Genetic Algorithm

Cellular GAs (cGAs) are a subclass of Genetic Algorithm (GAs) in which the population is structured in a specified topology defined as a connected graph, 2D toroidal grid, in which each vertex is an individual that communicates with its nearest neighbours (e.g, North, South, East, West) and use these individuals for crossover and mutation. Algorithm 1 (and Figure 1) presents the structure of a cGA.

Each individual interacts only with their neighbours. The resulting overlapped small neighbourhoods help in exploring the search space because the induced slow diffusion of solutions through the population provides a kind of exploration, while exploitation takes place inside each neighbourhood by genetic operations. The reader can find a deeper estudy on cGAs in [1].

Algorithm 4. Pseudocode of Canonical Cellular GA

```
 1:  pop ← initializePopulation(pop)
 2:  pop ← evaluatePopulation(pop)
 3:  while not stop criterion do do
 4:      for each individual do do
 5:          neighbours ← calculateNeigbourhood(individual)
 6:          parents ← selection (neighbours)
 7:          offspring ← Recombination(parents,prob_Recombination);
 8:          offspring ← Mutation(offspring,prob_Mutation);
 9:          pop' ← evaluate(offspring);
10:          replacement(pop',individual,offspring);
11:      end for
12:  end while
```

2.1   The Proposal

The basic idea behind most parallel programs is to divide a task into subtasks and solve the subtasks simultaneously using multiple processors. This divide and conquer approach can be applied to GAs in many different ways, and the literature contains many examples of successful parallel implementations [2]. Some parallelization methods use a single population, while others divide the population into several relatively isolated subpopulations. Some methods exploit massively parallel computer architectures, while others are better suited to multicomputers with fewer and more powerful processing elements.

In the case of NVIDIA GPUs have (currently) up to 30 Streaming Multiprocessors (SM); each SM has eight parallel thread processors called Streaming Processors (SP). The SPs run synchronously, meaning all eight SPs run a copy of the same program, and actually execute the same instruction at the same time by each thread created (see Figure 2). Different SMs run asynchronously, much like commodity multicore processors. For achieving this, the notion of kernel is defined. A kernel is a function callable from the host and executed on the specified GPU simultaneously by several SPs in parallel.

Algorithm 5. Pseudocode of Cellular GA on a GPU

 1: initialize_cGA(Input_param)
 2: generate_random_numbers(seeds)
 3: allocate problems, seeds for random numbers and data inputs on GPU device memory
 4: **for each** individual **in parallel do do**
 5:     individual ← **initializeOnGPU**(individual)
 6:     individual ← **evaluateOnGPU**(individual)
 7: **end for**
 8: **while** not Stop Criterion **do do**
 9:     neighbours ← **calculateNeigbourhoodOnGPU**(individual)
10:     parents ← **selectionOnGPU** (neighbours)
11:     offspring ← **RecombinationOnGPU**(parents,prob_Recombination);
12:     offspring ← **MutationOnGPU**(offspring,prob_Mutation);
13:     **evaluateOnGPU(offspring);**
14:     **replacementOnGPU(individual,offspring);**
15: **end while**

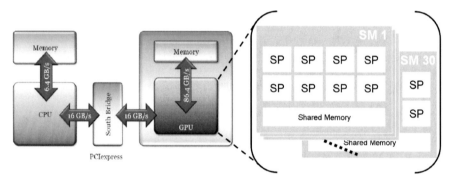

Fig. 2 Description of the architecture between CPU and GPU

In our present work the proposed algorithm exploits the inherent parallelism of a GPU using a direct mapping between the population structure and the threads of the GPU. First of all, at initialization stage, the memory allocations on GPU have to made. Input parameters for the algorithm (the population generated in the CPU and the configuration parameters for the algorithm), are stored in the global memory of the GPU. The population generated is transfered from the CPU to the device memory, this is a synchronous operation. Since we are not having a Pseudo Random Number Generator (PRNG) for GPUs, we used a PRNG that is provided by the SDK of CUDA named Merseinne Twister; the only condition for its use is to initially copy from CPU to GPU a group of seeds neccesary for execute the PRNG. Once the copies are done, we execute a series of subtasks implemented only in the GPU called through a kernel function (that allows to invoke functions implemented in the GPU) and these are executed for every thread. As a second step, for each individual, we need to identify its neighrbourhood. Third, we proceed to apply the GA operators on the solution neighbourhood in each thread. Now, we synchronize all threads for taking the fourth step: replacement of the individual with the offspring (if a condition is satisfied). Finally, this process is repeated until a stop condition is satisfied. This algorithm is synchronous, as the individuals of the population of the next generation are formally created all at the same time. We can see a general model of the proposal algorithm for GPU in the Figure 3.

The implementation for this algorithm was done with CUDA [6] for GPU .

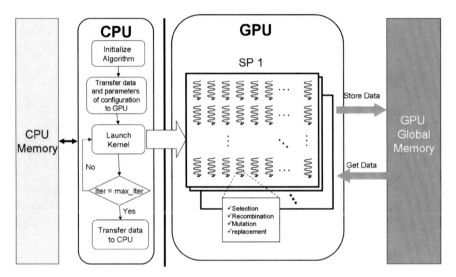

Fig. 3 Description of the architecture between CPU and GPU

3 Experimental Setup

This section is devoted to describing the methodology that we have used in the experiments carried out in this work. First, we present the benchmark problems used to compare the cGA GPU. In order to show the performance on a wide spectrum of problems we encompass tests both in discrete and continuous domains. Also, we try to use standard benchmarks as the ones reported in CEC 2005 [8] and 2008 [9] standards.

4 Methodology and Configurations Used

We have selected for our tests the following problems: Colville Minimization, ECC, MMDP (discrete optimization) and Shifted GriewankȦfs function, Shifted RastriginȦfs function and Shifted RosenbrockȦfs function (continuous optimization). These problems were selected because they are generally popular in GAs and/or used in previous works on GPUs [1] [12].

Our GPU-based implementation is compared against previous software implementations on a CPU implemented in JCell [1].

Now, we explain the statistical test that we have applied to ensure the confidence of the obtained results. Since we are dealing with stochastic algorithms and we want to provide the results with confidence, we have made 30 independent runs of each experiment, and the following statistical analysis has been performed throughout this work. Firstly, a Kolmogorov Sminorv test was performed in order to check whether the values of the result follow a normal (Gaussian) distribution or not. If the distribution is normal, we will apply Levene test for the homogeneity of the variances. If samples have equals variance (positive Levene test), an ANOVA test is performed, otherwise a Welch test is performed. For non Gaussian distributions, the non-parametric Kruskal-Wallis test is used to compare the medians of the algorithms.

We always consider in this work a confidence level of 95% (i.e., significance level of 5% or p-value under 0.05) in the statistical tests, which means that the differences are unlikely to have occurred by chance with a probability of 95%. Successful tests are marked with "+" symbols in the last column in the first table; conversely, "•" means that no statistical confidence was found (p-value 0:05).

In order to make a meaningful comparison among the algorithms, we have used a common parameterization. The details are described in Table 1, where we include the maximum number of generations as the stop condition for all the algorithms in each execution (500). The toroidal grid has different sizes for evaluate the behavior of the algorithms and compare that exist some advantage or not to use different population sizes for each problem. So, we define four population sizes: 32×32, 64×64, 256×256 and 512×512 individuals. The neighbourhood used is composed of five individuals: the considered individual plus those located at its North, East, West and South (see Fig. 1). One selection method have been used in this work: on parent is always the cosidered individual itself, while the other one is obtained by using Roulette Wheel (RW) selection in its 4-neighbourhood. For the recombination operator, we obtain just one offspring from two parents: the one having the largest portion of the best parent. The DPX recombination is applied always (probability $p_c = 1.0$), this operator is a crossover of two points, keeping the largest part of the best parent. The bit mutation probability is set to $p_m = 0.05$. We will replace the considered individual on each generation only if its offspring has a better fitness value, called Replace if Better [11]. All these parameters are selected after previous works [1] and an own initial setting study.

Table 1 SpeedUp in seconds obtained with different population size

Parameters	Value
Max. Number of Generations	500
Population Size	$\{32^2, 64^2, 256^2, 512^2\}$
Neighborhood	N - E - W - S
Selection of Parents	itself + Roulette Wheel (RW)
Recombination	DPX = 1.0
Mutation	0.05
Replacement	Replace if the new individual is better

Experiments were run on a machine equiped with a Intel R . CoreTM Quad processor running at 2.67GHz , under Windows XP operating system, and having 4 GB of memory. The GPU used is a nVIDIA GeForce GTX 285 equipped with 1GB of RAM. The environment used is Microsoft Visual C++ 2008 Express Edition together with the Toolkit SDK for CUDA v2.1 with the nVIDIA driver v180.49.

5 Results

In this section we present the results obtained when solving the problems selected with the proposed cGA GPU algorithm. We here describe the numeric and time performance of the cGA on GPU. In order to compare the time performance we use a sequential version of a cGA implemented in JCell [1] that is executed in the CPU.

The results of speedup are summarized in Table 2: for each problem, the average speedup of the 30 executions is shown. This value is the result of the average of the time for the algorithm in CPU divided by the average time of the algorithm on GPU. Thus, a value over 1.0 means a more efficient performance of the GPU versus the CPU.

The results of our tests show that the speedup ranges from 5 to 24. In general, as the population size increase we see that the GPU can achieve a better performance.

For a population space as 32×32, the CPU implementation still remains faster than those in a GPU; the reason is probably because the population are very small and the existency the some overhead between the CPU and GPU to call the kernel functions affects the time performance. We would like to point out that the efficiency showed for the GPU is equivalent to 24 proccessors, an a insteresting benefit drawn from a commodity computer.

Another interesting observation is that there is not significant difference between the speedup of the discrete and continuous domains. This indicates that the GPU is effective to evaluate problem instances of both domains.

The result of the statistical tests are in column Test of the Table 2, where the symbol "+" means that statiscally significant differences exist. In most of the instances of the problems, the existing statistically significant differences favor the cGA implemented in the GPU versus the CPU. As well, Table 3 gives for each problem the time (in seconds) of the algorithm executed in CPU and in GPU respectively (each column shows the time of CPU and the GPU time separated by a "−"). As expected, the time of the GPU is very small (between 0.14 and 0.35 seconds) while for the CPU the execution time range between 0.11 and 7.89 seconds. In most of the cases, the time of the GPU is shorter than the one on CPU (an exception occured just for the population of 32×32). Table 5 gives results about of the average of fitness solutions obtained for each problem. This table shows the average value and the standard deviation of the averaged best final fitness value for each problem and algorithm configuration. The values obtained show that the algorithm gets very frequently a near optimal value for every problem. Also, those values are competitive against other algorithms in the literature [10]. Table 4 show the results of the average of fitness solutions obtained in CPU. This table show that the values obtained

Table 2 SpeedUp in seconds obtained with different population sizes

Population	SpeedUp Discrete Problems			SpeedUp Continuos Problems			Test
	Colville Minimization	ECC	MMDP	Rastrigin	Rosenbrock	Griengwak	
32×32	0.561	0.660	0.784	0.494	0.539	0.826	•
64×64	5.441	5.450	5.645	5.417	5.688	5.783	+
256×256	16.433	16.830	14.485	15.463	17.830	16.964	+
512×512	23.593	22.419	22.789	20.810	20.982	20.421	+

Table 3 Average of time performance in seconds with different population sizes

Population	SpeedUp Discrete Problems			SpeedUp Continuos Problems		
	Colville Minimization	ECC	MMDP	Rastrigin	Rosenbrock	Griengwak
32×32	0.10-0.18	0.11-0.17	0.11-0.14	0.12-0.22	0.11-0.19	0.12-0.15
64×64	1.19-0.21	1.25-0.23	1.21-0.21	1.17-0.21	1.20-0.21	0.21-0.20
256×256	4.16-0.25	4.63-.0.27	4.09-0.28	4.31-0.27	4.66-0.26	4.53-0.26
512×512	7.46-0.32	7.89-0.35	7.66-0.33	7.21-0.34	7.12-0.33	7.35-0.35

Table 4 Average of solutions fitness obtained with different population sizes for CPU

Population	Average Solutions			Average Solutions Continuos Problem		
	Colville Minimization	ECC	MMDP	Rastrigin	Rosenbrock	Griengwak
32×32	$0.133_{\pm 5.176e\text{-}5}$	$0.066_{\pm 1.065e\text{-}3}$	$39.896_{\pm 8.709e\text{-}6}$	$4.637e\text{-}5_{\pm 5.512e\text{-}5}$	$2.600e\text{-}5_{\pm 3.075e\text{-}5}$	$3.733e\text{-}5_{\pm 3.750e\text{-}3}$
64×64	$0.111_{\pm 3.684e\text{-}6}$	$0.066_{\pm 0.633e\text{-}3}$	$39.900_{\pm 9.145e\text{-}6}$	$2.978e\text{-}5_{\pm 1.136e\text{-}6}$	$1.645e\text{-}6_{\pm 5.170e\text{-}6}$	$2.687e\text{-}5_{\pm 3.410e\text{-}3}$
256×256	$0.100_{\pm 1.033e\text{-}6}$	$0.067_{\pm 0.361e\text{-}6}$	$39.911_{\pm 1.365e\text{-}6}$	$1.218e\text{-}5_{\pm 7.872e\text{-}6}$	$1.639e\text{-}5_{\pm 2.816e\text{-}6}$	$2.350e\text{-}5_{\pm 3.590e\text{-}6}$
512×512	$0.010_{\pm 0.310e\text{-}6}$	$0.067_{\pm 0.003e\text{-}6}$	$39.999_{\pm 2.713e\text{-}5}$	$1.749e\text{-}6_{\pm 4.350e\text{-}6}$	$3.311e\text{-}5_{\pm 4.997e\text{-}6}$	$1.356e\text{-}6_{\pm 1.450e\text{-}6}$

Table 5 Average of solutions fitness obtained with different population sizes for GPU

Population	Average Solutions			Average Solutions Continuos Problem		
	Colville Minimization	ECC	MMDP	Rastrigin	Rosenbrock	Griengwak
32×32	$0.330_{\pm 9.660e\text{-}2}$	$0.065_{\pm 0.865e\text{-}3}$	$39.590_{\pm 1.070}$	$4.850e\text{-}5_{\pm 2.970e\text{-}5}$	$2.600e\text{-}5_{\pm 9.330e\text{-}5}$	$3.733e\text{-}5_{\pm 3.750e\text{-}3}$
64×64	$0.330_{\pm 3.122e\text{-}2}$	$0.066_{\pm 0.633e\text{-}3}$	$39.720_{\pm 0.080}$	$4.560e\text{-}5_{\pm 6.810e\text{-}5}$	$1.645e\text{-}5_{\pm 7.360e\text{-}5}$	$2.687e\text{-}5_{\pm 3.410e\text{-}3}$
256×256	$0.130_{\pm 1.030e\text{-}3}$	$0.066_{\pm 0.361e\text{-}6}$	$39.860_{\pm 0.080}$	$4.540e\text{-}5_{\pm 6.330e\text{-}5}$	$1.639e\text{-}5_{\pm 3.870e\text{-}5}$	$2.391e\text{-}5_{\pm 1.830e\text{-}3}$
512×512	$0.100_{\pm 1.000e\text{-}3}$	$0.067_{\pm 0.003e\text{-}5}$	$39.940_{\pm 8.000e\text{-}3}$	$4.210e\text{-}5_{\pm 1.090e\text{-}5}$	$1.500e\text{-}5_{\pm 3.600e\text{-}5}$	$2.375e\text{-}5_{\pm 1.050e\text{-}3}$

for the CPU and GPU are very similar with a approximation very similar. As a conclusion, the algorithm implemented in GPU presents a robust numerical behavior because the values are very near or they reached the optimal. So, we can conclude that in general the cGA GPU is better than the sequential cGA, bothnumerically and in time.

6 Conclusions

In this work we have presented a novel implementation of a cGA running on a GPU. All operators have been implemented directly in the GPU. We test the performance of the algorithm with 6 different problems in continuous and discrete domain, and we compare against a standard cGA. We showed that the inherent parallelism of the GPU can be exploited to accelerate a cGA.

In the future, we will apply the presented approach to other complex real-world problems. Especially those that remains open because at their large dimensions, as well as to applications in industry. Another future work will be to implement other families of evolutionary algorithms and evaluate its performance in multiGPU architectures.

Acknowledgements. Authors acknowledge funds from the Spanish Ministry of Sciences and Innovation European FEDER under contract TIN2008-06491-C04-01 (M* project http://mstar.lcc.uma.es) and CICE, Junta de Andalucía under contract P07-TIC-03044 (DIRICOM project http://diricom.lcc.uma.es).

References

[1] Alba, E., Dorronsoro, B.: Cellular Genetic Algorithms. Operations Research / Computer Science, vol. 42. Springer, Heidelberg (2008)

[2] Alba, E.: Parallel metaheuristics: A new class of algorithms (August 2005)

[3] Lewis, T.E., Magoulas, G.D.: Strategies to minimise the total run time of cyclic graph based genetic programming with gpus (2009)

[4] Luebke, D., Harris, M., Krüger, J., Purcell, T., Govindaraju, N., Buck, I., Woolley, C., Lefohn, A.: Gpgpu: general purpose computation on graphics hardware. In: SIGGRAPH 2004: ACM SIGGRAPH 2004 Course Notes, vol. 33. ACM, New York (2004)

[5] Maitre, O., Baumes, L.A., Lachiche, N., Corma, A., Collet, P.: Coarse grain parallelization of evolutionary algorithms on gpgpu cards with easea (2009)

[6] Nickolls, J., Buck, I., Garland, M., Skadron, K.: Scalable parallel programming with cuda. In: SIGGRAPH 2008: ACM SIGGRAPH 2008 classes, pp. 1–14. ACM, New York (2008)

[7] Owens, J.D., Luebke, D., Govindaraju, N., Harris, M., Kruger, J., Lefohn, A.E., Purcell, T.J.: A survey of general-purpose computation on graphics hardware. Computer Graphics Forum 26(1), 80–113 (2007)

[8] Suganthan, P.N., Hansen, N., Liang, J.J., Deb, K., Chen, Y.-P., Auger, A., Tiwari, S.: Problem definitions and evaluation criteria for the CEC 2005 special session on real-parameter optimization (2005)

[9] Tang, K., Yao, X., Suganthan, P.N., MacNish, C., Chen, Y.P., Chen, C.M., Yang, Z.: Benchmark functions for the CEC 2008 special session and competition on large scale global optimization (November 2007)

[10] Tseng, L.-Y., Chen, C.: Multiple trajectory search for large scale global optimization. In: Evolutionary computation, CEC 2008 (IEEE World Congress on Computational Intelligence). IEEE Congress (2008)

[11] Whitley, D.L.: The genitor algorithm and selection pressure: Why rank-based allocation of reproductive trials is best. In: Proceedings of the 3rd international conference on genetic algorithms (1989)

[12] Yu, Q., Chen, C., Pan, Z.: Parallel genetic algorithms on programmable graphics hardware. In: Wang, L., Chen, K., S. Ong, Y. (eds.) ICNC 2005, part III. LNCS, vol. 3612, pp. 1051–1059. Springer, Heidelberg (2005)

Evolutionary Approaches to Joint Nash – Pareto Equilibria

D. Dumitrescu, Rodica Ioana Lung, and Tudor Dan Mihoc

Abstract. A new type of equilibrium incorporating different rationality types for finite non cooperative games with perfect information is introduced. The concept of strategic game is generalized in order to admit players with different rationalities. Generative relations are used to characterize several types of equilibria with respect to players rationality. An evolutionary technique for detecting it is considered. Numerical experiments show the potential of the method.

1 Introduction

Equilibrium concepts are the most common solutions proposed in game theory. In a particular game it is usually considered that players interact according to a unique equilibrium concept, i.e. only players guided by the same kind of equilibrium are allowed to interact. This restriction induces unrealistic predictions. For example, the concept of Nash equilibrium sometimes can lead to deceptive results [5].

In real life players (agents) can be more or less cooperative, more or less competitive and more or less rational. In order to cope with more complex situations a concept of generalized game is presented. Players are allowed to have different behaviors/rationality types considering an adequate meta-strategy concept.

According to [3] game equilibria can be characterized using appropriate generative relations. Thus Nash equilibrium is characterized by the ascendancy relation [8] and Pareto equilibrium by the Pareto domination. Combining the two relations may lead to different types of joint Nash–Pareto equilibria.

An evolutionary technique for detecting the joint Nash–Pareto equilibrium for the generalized game is used.

D. Dumitrescu · Rodica Ioana Lung · Tudor Dan Mihoc
Babes Bolyai University
e-mail: ddumitr@cs.ubbcluj.ro, rodica.lung@econ.ubbcluj.ro,
        mihoct@cs.ubbcluj.ro

J.R. González et al. (Eds.): NICSO 2010, SCI 284, pp. 233–243, 2010.
springerlink.com   © Springer-Verlag Berlin Heidelberg 2010

2 Generalized Games

In order to cope with different rationality types the concept of generalized game is defined [3].

Definition 2.1. A finite strategic *generalized game* is defined as a system by $G = (N, M, U)$ where:

- $N = \{1, ..., n\}$, represents the set of players, n is the number of players;
- for each player $i \in N$, S_i represents the set of actions available to him, $S_i = \{s_{i_1}, s_{i_2}, ..., s_{i_{m_i}}\}$; $S = S_1 \times S_2 \times ... \times S_N$ is the set of all possible situations of the game;
- for each player $i \in N$, M_i represents the set of available meta-strategies, a meta-strategy is a system $(s_i|r_i)$ where $s_i \in S_i$ and r_i is the ith player rationality type;
- $M = M_1 \times M_2 \times ... \times M_N$ is the set of all possible situations of the generalized game and $(s_1|r_1, s_2|r_2, ..., s_n|r_n) \in M$ is a meta-strategy profile.
- for each player $i \in N$, $u_i : S \to \mathbf{R}$ represents the payoff function.

$$U = \{u_1, ..., u_n\}.$$

Remark 2.1. In a generalized game the set of all possible meta-strategies represents the meta-strategy search space.

3 Generative Relations for Generalized Games

Three generative relations are considered in this section. Two of them correspond to Pareto and Nash equilibria. The third induces a new type of joint Nash–Pareto equilibrium.

3.1 n_P–Strict Pareto Domination

We introduce the n_P–strict Pareto domination in order to be able to combine the concepts of Nash and Pareto domination.

In a finite strategic generalized game consider the set of players Pareto biased

$$I_P = \{j \in \{1, ..., n\} | r_j = \text{Pareto}\}$$

and $n_P = card\ I_P$, where $card\ A$ denotes the number of elements in the set A.

Let us consider two meta strategy profiles x and y from M.

Definition 3.1. The meta strategy profile x n_P–strict Pareto dominates the meta strategy profile y if the payoff of each Pareto biased player from I_P using meta strategy x is strictly greater than the payoff associated to the meta strategy y, i.e.

$$u_i(x) > u_i(y), \forall i \in I_P.$$

Remark 3.1. The set of non dominated meta strategies with respect to the n_P–strict Pareto domination relation when $n_P = n$ is a subset of the Pareto front.

3.2 Nash – Ascendancy

Similar to Pareto equilibrium a particular relation between strategy profiles can be used in order to describe Nash rationality. This relation is called Nash-ascendancy (NA).

A strategy is called Nash equilibrium [7] if each player has no incentive to unilaterally deviate i.e. it can not improve the payoff by modifying its strategy while the others do not modify theirs.

We denote by (s_{i_j}, s_{-i}^*) the strategy profile obtained from s^* by replacing the strategy of player i with s_{i_j} i.e.

$$(s_{i_j}, s_{-i}^*) = (s_1^*, s_2^*, ..., s_{i-1}^*, s_{i_j}, s_{i+1}^*, ..., s_1^*).$$

Definition 3.2. The strategy profile x Nash-ascends the strategy profile y, and we write $x <_{NA} y$ if there are less players i that can increase their payoffs by switching their strategy from x_i to y_i then vice versa.

In [8] is introduced an operator

$$k : S \times S \to \mathbf{N},$$

$$k(y,x) = card\{i \in \{1,...,n\} | u_i(x_i, y_{-i}) \geq u_i(y), x_i \neq y_i\}.$$

$k(y,x)$ denotes the number of players which benefit by switching from y to x.

Proposition 3.1. *The strategy x Nash-ascends y (x is NA-preferred to y), and we write $x <_{NA} y$, if the inequality*

$$k(x,y) < k(y,x),$$

holds.

According to [8] the set of all strategies from S non-dominated by respect of Nash ascendancy relation equals the set of Nash equilibria.

This result proves that the Nash ascendancy is the generative relation for the Nash equilibrium.

3.3 Joint Nash–Pareto Domination

Let us consider two meta-strategies

$$x = (x_1 | r_1, x_2 | r_2, ..., x_n | r_n) \text{ and } y = (y_1 | r_1, y_2 | r_2, ..., y_n | r_n).$$

Let us denote by I_N the set of Nash biased players (N-players) and by I_P the set of Pareto biased players (P-players). Therefore we have

$$I_N = \{i \in \{1, ..., n\} | r_i = \text{Nash}\}.$$

We consider the operators k_P and k_N defined as:

$$k_P(x,y) = card\{j \in I_P | u_j(x) > u_j(y), x \neq y\}$$

and respectively

$$k_N(x,y) = card\{i \in I_N | u_i(y_i, x_{-i}) \geq u_i(x), x_i \neq y_i\}.$$

Remark 3.2. $k_P(x,y)$ measures the *relative efficiency* of the meta strategies x and y with respect to Pareto rationality and $k_N(x,y)$ measures the *relative efficiency* of the meta strategies x and y with respect to Nash rationality.

Definition 3.3. The meta strategy x N–P dominates the meta strategy y if and only if the following statements hold

1. $k_P(x,y) = n_P$
2. $k_N(x,y) < k_N(y,x)$

In what follows we consider that efficiency relation induces a new type of equilibrium called *joint Nash-Pareto equilibrium*.

Remark 3.3. Joint Nash-Pareto equilibrium defined in this section is a concept completely different from the existing concept of Pareto-Nash equilibria [10].

4   Detecting Joint N–P Equilibria in Generalized Games

Consider a three player non-cooperative game. Let r_i be the rationality type of player i.

If $r_1 = r_2 = r_3 = $ Nash then all players are Nash biased and the corresponding solution concept is the Nash equilibrium.

If $r_1 = r_2 = r_3 = $ Pareto then all players are Pareto biased and the corresponding equilibria are described by the set of strictly non dominated strategies (Pareto front).

We also intend to explore the joint cases where one of the players is Nash biased and others are Pareto and the one where one is Pareto and the others are Nash biased.

In order to detect the joint Nash–Pareto equilibria of the generalized game an evolutionary approach is used.

Let us consider an initial population $P(0)$ of p meta strategies for the generalized three player game. Each member of the population has the form

$$x = (s_1 | r_1, s_2 | r_2, s_3 | r_3).$$

Pairs of meta-strategies are randomly chosen from the current population $P(t)$. For each pair a binary tournament is considered. The meta strategies are compared by means of the domination relation. An arbitrary tie breaking is used if the two meta strategies have the same efficiency. The winers of two binary tournaments are re-combined using the simulated binary crossover (SBX) operator [11] resulting two offspring. Offspring population is mutated using real polynomial mutation [2], re-sulting an intermediate population P'. Population $P(t)$ and P' are merged.

The resulting set of meta strategies is sorted with respect to the efficiency relation using a fast non dominant sorting approach [2]. For each meta strategy M' the num-ber expressing how many meta strategies in the merged population are less efficient then M' is computed. On this basis the first p meta strategies are selected from the merged population. Selected meta strategies represent the new population $P(t+1)$.

Let us remark that in the proposed technique selection for recombination and sur-vival is driven by the efficiency relation. Therefore the population of meta-strategies is expected to converge toward the joint Nash–Pareto front. According to the pro-posed approach the members of this front represent the joint N–P equilibria of the generalized game.

5   Numerical Experiments

In order to illustrate the proposed concepts the oligopoly Cournot model is conside-red (see for instance [6]).

Let q_1, q_2 and q_3 denote the quantities of a product. This unique product is pro-duced by three companies. The market price denoted by $P(Q)$ is given by

$$P(Q) = \begin{cases} a - Q, & \text{for } Q < a, \\ 0, & \text{for } Q \geq a. \end{cases}$$

where

$$Q = q_1 + q_2 + q_3,$$

is the aggregate quantity on the market and $a > 0$ is a constant characterizing the market.

The cost for the company i of producing q_i units is $C_i(q_i)$

$$C_i(q_i) = c_i q_i,$$

where $c_i < a$. Suppose that the companies choose their quantities simultaneously. The payoff for the company i is its profit, which can be expressed as:

$$\pi_i(q_1, q_2, q_3) = q_i P(Q) - C_i(q_i)$$
$$= q_i [a - (q_1 + q_2 + q_3) - c_i], i = 1, 2, 3.$$

A game strategy is a triple

$$s = (q_1, q_2, q_3).$$

Several experiments have been performed for this game by using RED tech-nique [3].

5.1 Symmetric Games

The symmetric Cournot model with parameters $a = 24$ and $c_1 = c_2 = c_3 = 9$ is considered.The payoff corresponding to Nash equilibrium is (14.00, 14.00, 14.00).

Table 1 Average payoff and standard deviation (St, Dev.) of the final populations in 30 runs with 100 meta-strategies after 30 generations for the symmetric Cournot model where all three players are Nash biased

N-N-N	Average payoff			St. dev.			Maximum payoff			Minimum payoff		
player	p1	p2	p3	p1	p2	p3	p1	p2	p3	p1	p2	p3
Average	14,05	14,06	14,05	0,03	0,04	0,04	14,85	15,57	15,00	12,25	12,49	12,45
St. Dev.	0,02	0,02	0,02	0,08	0,09	0,08	1,39	2,80	1,83	3,25	3,00	3,05

According to the data from the Table 1 in less than 30 generations the algorithm converges to the Nash equilibrium point.

Table 2 Average payoff and standard deviation (St. Dev.) of the final populations in 30 runs with 100 meta-strategies after 30 generations for the symmetric Cournot model where two player are Nash biased and one is Pareto

N-N-P	Average payoff			St. dev.			Maximum payoff			Minimum payoff		
player	p1	p2	p3	p1	p2	p3	p1	p2	p3	p1	p2	p3
Average	10,99	11,01	29,80	52,81	53,02	182,28	25,92	25,71	56,24	0,00	0,00	0,49
St. Dev.	0,36	0,33	0,78	1,75	2,33	17,62	0,92	0,88	0,00	0,00	0,00	1,67

The resulting front in the Nash-Nash-Pareto case spreads from the standard Nash equilibrium corresponding to the two player–Cournot game (25.00, 25.00) to the Nash equilibrium corresponding to the three player–Cournot game, and from there to the edges of Pareto front. The equilibrium set is depicted in Figure 1 and Figure 2 from two angles, for a better view. The numerical results are presented in Table 2.

As we can see in the Figure 3 in the Nash-Pareto-Pareto case the result is similar to the Pareto front, an result that is determined by the strength of the Pareto component in the generative relation for the joint Nash–Pareto equilibrium. As we can see in Table 3 the minimum values for all three players are 0.00 and the maximum are 56.24, the same like the ones for the Pareto front.

5.2 Asymmetric Games

First, let us consider the two player asymmetric Cournot game with parameters $a = 24$, $c_1 = 9$ and $c_2 = 12$.

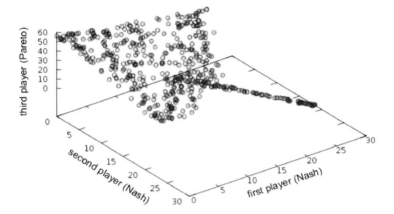

Fig. 1 The payoffs for the Nash-Nash-Pareto front detected in less than 30 iterations for the symmetric Cournot game

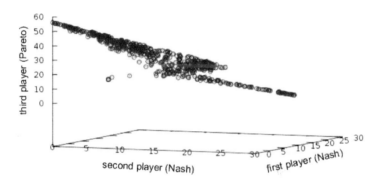

Fig. 2 The payoffs for the Nash-Nash-Pareto front detected in less than 30 iterations for the symmetric Cournot game

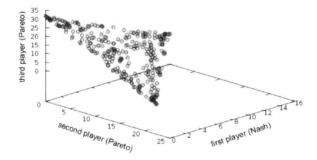

Fig. 3 The payoffs for the Nash-Pareto-Pareto front detected in less than 30 iterations for the symmetric Cournot game

Table 3 Average payoff and standard deviation (St. Dev.) of the final populations in 30 runs with 100 meta-strategies after 30 generations for the symmetric Cournot model where one player is Nash biased and the other two Pareto

N-P-P	Average payoff			St. dev.			Maximum payoff			Minimum payoff		
player	p1	p2	p3	p1	p2	p3	p1	p2	p3	p1	p2	p3
Average	17,74	18,52	18,44	242,42	247,83	247,97	56,23	56,24	56,24	0,00	0,00	0,00
St. Dev.	0,40	0,36	0,42	7,36	6,98	6,82	0,04	0,00	0,00	0,00	0,00	0,00

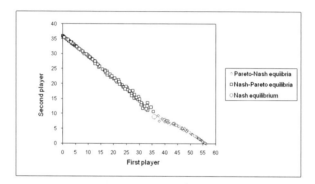

Fig. 4 The payoffs for the Nash-Pareto, Pareto-Nash and Nash-Nash equilibria approximations detected after 30 iterations for Cournot's model with $a = 24$, $c_1 = 9$, and $c_2 = 12$

Table 4 Numerical results for the two asymmetric games with a Nash-Nash-Pareto rationality in final population in 30 runs, for 100 meta strategies and after 30 iterations

N-N-P	First game			Second game		
	$c_1 = 9$	$c_2 = 12$	$c_3 = 9$	$c_1 = 9$	$c_2 = 12$	$c_3 = 5$
Player	p1	p2	p3	p1	p2	p3
Average payoff	17,19	2,72	46,86	16,12	3,93	29,19
St. Dev.	10,59	2,49	25,90	11,29	2,67	15,45
Minimum payoff	0,00	0,00	0,18	0,00	0,00	0,11
Maximum payoff	36,72	9,24	90,25	36,61	9,43	56,25

In Figure 4 are depicted the payoff functions for two players. The results are in concordance with those obtained in [3]. The difference between c_1 and c_2 determines an asymmetry for the represented detected equilibria.

The asymmetric three player Cournot games with parameters $a = 24$, $c_1 = 9$, $c_2 = 12$, $c_3 = 5$ and respectively $a = 24$, $c_1 = 9$, $c_2 = 12$, $c_3 = 9$ are considered. The asymmetries allow us to better understand the players behavior in the joint Nash–Pareto equilibrium.

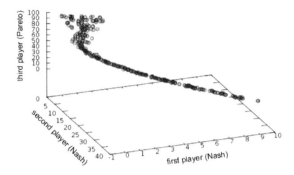

Fig. 5 The payoffs for the Nash-Nash-Pareto front detected in less than 30 iterations for the asymmetric Cournot game with parameters $a = 24$, $c_1 = 9$, $c_2 = 12$, $c_3 = 5$

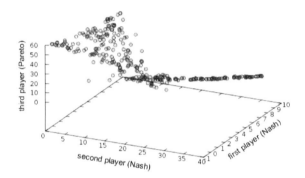

Fig. 6 The payoffs for the Nash-Nash-Pareto front detected in less than 30 iterations for the symmetric Cournot game with parameters $a = 24$, $c_1 = 9$, $c_2 = 12$, $c_3 = 9$

Comparing Figure 5 and Figure 6 one can observe the influence of the Pareto component in to the final front. If parameter c_3 is close to parameters c_1 and c_2 the Pareto influence determines the set so spread out in a plane similar to the pure Pareto front. The distribution between the Nash equilibrium for two players (as the third gains nothing) and three players Nash equilibrium remains also for the asymmetric game.

Analyzing the results one can observe that for these particular cases of joint Nash–Pareto rationalities, symmetric or not, there is no minimum guaranteed payoff for any rationality.

As regarding the maximum payoffs an interesting feature appears if we construct a new game based on the maximum payoffs. The players strategies will be their rationality (Nash or Pareto) and their payoffs the maximum gains in the joint Nash-Pareto equilibria. Solving this game in GAMBIT the pure strategies Nash equilibrium is the Pareto-Pareto-Pareto rationality.

6   Conclusions and Future Work

A concept of generalized game is used in order to capture the behavior of players with several types of rationalities. A new generative relation between meta strategies induces a new solution concept called joint Nash–Pareto equilibrium. Proposed method allows the combination of different types of equilibria in a game.

An evolutionary technique for detecting an approximation of the generalized equilibria is used. The idea are exemplified for Cournot games with three players and two types of rationality.

Results indicate the potential of the proposed technique. Different analyses prove that for the presented games there is no minimal payoff but the possibility of a maximum gain is obtained for the Pareto rationality. These experimental results offer an inside view of the problems arising when two different type of equilibria are considered in the same game.

Future work will address generalized games having other rationality types then Nash and Pareto and other methods of combining them.

Acknowledgements. This research is supported partially by the CNCSIS Grant ID508 *"New Computational paradigms for dynamic complex problems"* funded by the MEC and from the SECTORAL OPERATIONAL PROGRAMME HUMAN RESOURCES DEVELOPMENT, Contract POSDRU 6/1.5/S/3 "Doctoral studies: through science towards society", Babeş - Bolyai University, Cluj - Napoca, România.

References

[1] Bade, S., Haeringer, G., Renou, L.: More strategies, more Nash equilibria, Working Paper 2004-15, School of Economics University of Adelaide University (2004)
[2] Deb, K., Agrawal, S., Pratab, A., Meyarivan, T.: A Fast Elitist Non-Dominated Sorting Genetic Algorithm for Multi-Objective Optimization: NSGA-II. In: Schoenauer, M., Deb, K., Rudolph, G., Yao, X., Lutton, E., Merelo, J.J., Schwefel, H.-P. (eds.) PPSN 2000. LNCS, vol. 1917, pp. 849–858. Springer, Heidelberg (2000)
[3] Dumitrescu, D., Lung, R.I., Mihoc, T.D.: Evolutionary Equilibria Detection in Non-cooperative Games. In: Giacobini, M., Brabazon, A., Cagnoni, S., Di Caro, G.A., Ekárt, A., Esparcia-Alcázar, A.I., Farooq, M., Fink, A., Machado, P. (eds.) EvoCOMNET. LNCS, vol. 5484, pp. 253–262. Springer, Heidelberg (2009)
[4] McKelvey, R.D., McLennan, A.: Computation of equilibria in finite games. In: Amman, H.M., Kendrick, D.A., Rust, J. (eds.) Handbook of Computational Economics. Elsevier, Amsterdam (1996)
[5] McKelvey, R.D., Palfrey, T.: An experimental study of the centipede game. Econometrica 60(4), 803–836 (1992)
[6] Lung, R.I., Muresan, A.S., Filip, D.A.: Solving multi-objective optimization problems by means of natural computing with application in finance. In: Aplimat 2006, Bratislava, February 2006, pp. 445–452 (2006)
[7] Nash, J.: Non-cooperative games. Annals of Mathematics 54, 286–295 (1951)
[8] Lung, R.I., Dumitrescu, D.: Computing Nash Equilibria by Means of Evolutionary Computation. Int. J. of Computers, Communications & Control, 364–368 (2008)

[9] Osborne, M.J., Rubinstein, A.: A Course in Game Theory. MIT Press, Cambridge (1994)

[10] Maskin, E.: The theory of implementation in Nash equilibrium:A survey. In: Hurwicz, L., Schmeidler, D., Sonnenschein, H. (eds.) Social Goals and Social Organization, pp. 173–204. Cambridge University Press, Cambridge (1985)

[11] Deb, K., Beyer, H.: Self-adaptive genetic algorithms with simulated binary crossover. Complex Systems 9, 431–454 (1995)

Accelerated Genetic Algorithms with Markov Chains

Guan Wang, Chen Chen, and K.Y. Szeto

Abstract. Based on the mutation matrix formalism and past statistics of genetic algorithm, a Markov Chain transition probability matrix is introduced to provide a guided search for complex problem optimization. The important input for this guided search is the ranking scheme of the chromosomes. It is found that the effect of mutation using the transition matrix yields faster convergence as well as overall higher fitness in the search for optimal solutions for the 0-1 Knapsack problem, when compared with the mutation-only-genetic-algorithm, which include the traditional genetic algorithm as a special case. The accelerated genetic algorithm with Markov Chain provides a theoretical basis for further mathematical analysis of evolutionary computation, specifically in the context of adaptive parameter control.

1 Introduction

Successful applications of genetic algorithm using the Darwinian principle of survival of the fittest have been implemented in many areas [1, 2], such as in solving the crypto-arithmetic problem [3], time series forecasting [4], traveling salesman problem [5], function optimization [6], adaptive agents in stock markets [7, 8], and airport scheduling [9]. A drawback for the practitioners of genetic algorithm is the need for expertise in the specific application, as its efficiency depends very much on the parameters chosen in the evolutionary process. An example of this drawback can be found in the ad-hoc manner in choosing the selection mechanism, where

Guan Wang
School of Physics, Peking University, Beijing, China

Chen Chen
Columbia University, New York, USA

K.Y. Szeto
Department of Physics, the Hong Kong University of Science and Technology Hong Kong, China
e-mail: phszeto@ust.hk

J.R. González et al. (Eds.): NICSO 2010, SCI 284, pp. 245–254, 2010.
springerlink.com

different percentages of the population for survival for different problems are used. Since the parameters used in a specific application will generally be suboptimal in a different application, the traditional usage of genetic algorithm is more like an artisan approach. To tackle this shortcoming, we attempt an adaptive approach where the number of parameters requiring input from experts is reduced to a minimum. The basic aim is to make use of past data collected in the evolutionary process of genetic algorithm to tune the parameters such as the survival probability of a chromosome. This idea is similar to the Estimation Distribution Algorithm or its variants [10–13]. We can collect the statistics of the chromosomes in past generations to decide the probability of mutation of each locus [14, 15]. A general formalism using this idea has been constructed through the time dependent mutation matrix, and the algorithm called MOGA (mutation only genetic algorithm) has been shown to be very successful in solving various practical problems [3–9]. Generalization of this formalism to include crossover has also been developed in a recent publication [16]. This approach is reviewed in Section 2. The objective of this paper is to present a more systematic way of making use of the statistic of the locus during evolution to provide a guided search in optimization. Similar to MOGA discussed in Section 2, the mutation probability is time dependent. However, the dependence is not based on the tensor product of the chromosome mutation probability and locus mutation probability. We make use a Markov chain to describe the evolution of the probability of mutation, or more generally, the transition probability between 0 and 1 state in a binary encoding of the chromosomes in the genetic algorithm. The entire approach is based on the existence of a target state provided by the fit chromosomes, with proper weighting. In this approach, which we called MCGA (Markov Chain Genetic Algorithm), the really important input of this algorithm is the weighting function applied for the chromosomes, which is defined by the fitness ranking scheme. We will develop this algorithm in section 3. We compare this MCGA with our previous MOGA algorithm using the standard 0-1 knapsack problem in section 4. Even though this comparison is limited to the knapsack problem, we expect this approach of MCGA is more flexible for further development. We will discuss these issues at the end of the paper.

2   Mutation Matrix and MOGA

In traditional simple genetic algorithm, the mutation/crossover operators are processed on the chromosome indiscriminately over the loci without making use of the loci statistics, which has been demonstrated to provide useful information on mutation operator [16]. In our mutation matrix formalism, the traditional genetic algorithm can be treated as a special case. Let's consider a population of N chromosomes, each of length L and binary encoded. We describe the population by a $N \times L$ matrix, with with entry $A_{ij}(t)$, $i = 1,...,N$; $j = 1,...,L$ denoting the value of the jth locus of the ith chromosome. The convention is to order the rows of A by the fitness of the chromosomes, $f_i(t) \leq f_k(t) \ for \ i \geq k$. Traditionally we divide the population of N chromosomes into three groups: (1) Survivors who are the fit ones. They form the first rows of the population matrix A(t+1). Here $N_1 = c_1 N$ with the survival selection ratio $0 < c_1 < 1$. (2) The number of children is $N_2 = c_2 N$ and is generated from the fit chromosomes by genetic operators such as mutation. Here $0 < c_2 < 1 - c_1$ is the second parameter of the model. We replace the next N_2 rows in the population matrix A(t+1). (3) The remaining $N_3 = N - N_1 - N_2$ rows are the randomly generated chromosomes to ensure the diversity of the population so that the genetic algorithm continuously explores the solution space. In our formalism, we introduce a mutation matrix with elements $M_{ij}(t) \equiv a_i(t)b_j(t)$, $i = 1, ..., N$; $j = 1, ..., L$; $0 \leq a_i(t), b_j(t) \leq 1$ where $a_i(t)$ and $b_j(t)$ are called the row mutation probability and column mutation probability respectively. Traditional genetic algorithm with mutation as the only genetic operator corresponds to a time independent mutation matrix with elements $M_{ij}(t) \equiv 0$ for $i = 1, ..., N_1$, $M_{ij}(t) \equiv m \in (0, 1)$ for $i = N_1 + 1, ..., N_2$, and finally we have $M_{ij}(t) \equiv 1$ for $i = N_2 + 1, ..., N$. Here m is the time independent mutation rate. We see that traditional genetic algorithm with mutation as the only genetic operator requires at least three parameters: N_1, N_2, and m. We first consider the case of mutation on a fit chromosome. We expect to mutate only a few loci so that it keeps most of the information unchanged. This corresponds to "exploitation" of the features of fit chromosomes. On the other hand, when an unfit chromosome undergoes mutation, it should change many of its loci so that it can explore more regions of the solution space. This corresponds "exploration". Therefore, we require that $M_{ij}(t)$ should be a monotonic increasing function of the row index i since we order the population in descending order of fitness. There are many ways to introduce the row mutation probability. One simple solution is to use $a_i(t) = (i - 1)/(N - 1)$. Next, we must decide on the choice of loci for mutation once we have selected a chromosome to undergo mutation. This is accomplished by computing the locus mutation probability of changing to X (X=0 or 1) at locus j as p_{jX} by

$$ p_{jX} = \sum_{k=1}^{N} (N + 1 - k)\delta_{kj}(X) \bigg/ \sum_{m=1}^{N} m \qquad (1) $$

Here k is the rank of the chromosome in the population. $\delta_{kj}(X) = 1$ if the j-th locus of the k-th chromosome assume the value X, and zero otherwise. The factor in the denominator is for normalization. Note that p_{jX} contains information of both locus and row and the locus statistics is biased so that heavier weight for chromosomes with high fitness is assumed. This is in general better than the original method of Ma and Szeto[14] where there is no bias on the row. After defining p_{jX}, we define the column mutation rate as

$$ b_j = \left(1 - |p_{j0} - 0.5| - |p_{j1} - 0.5|\right) \bigg/ \sum_{j'=1}^{L} b_{j'} \qquad (2) $$

For example, if 0 and 1 are randomly distributed, we have $p_{j0} = p_{j1} = 0.5$. There will be no useful information about the locus, so we should mutate this locus, and $b_j = 1$. When there is definitive information, such as when $p_{j0} = 1 - p_{j1} = 0$ or 1, we should not mutate this column and we have $b_j = 0$. Once the mutation matrix M is obtained, we are ready to discuss the strategy of using M to evolve A. There are two ways to do Mutation Only Genetic Algorithm (MOGA). We can first decide which row (chromosome) to mutate, then which column (locus) to mutate, we call this particular method the Mutation Only Genetic Algorithm by Row or abbreviated as MOGAR. Alternatively, we can first select the column and then the row to perform mutation. We call this the Mutation Only Genetic Algorithm by Column or abbreviated as MOGAC. For MOGAR, we go through the population matrix A(t) by row first. The first step is to order the set of locus mutation probability $b_j(t)$ in descending order. This ordered set will be used for the determining of the set of column position (locus) in the mutation process. Now, for a given row i, we generate a random number x. If $x < a_i(t)$, then we perform mutation on this row, otherwise we proceed to the next row and $A_{ij}(t + 1) = A_{ij}(t)$, $j = 1,...,L$. If row i is to be mutated, we determine the set $R_i(t)$ of loci in row i to be changed by choosing the loci with $b_j(t)$ in descending order, till we obtain $K_i(t) = a_i(t) * L$ members. Once the set $R_i(t)$ has been constructed, mutation will be performed on these columns of the i-th row of the A(t) matrix to obtain the matrix elements $A_{ij}(t + 1)$, $j = 1,...,L$. We then go through all N rows, so that in one generation, we need to sort a list of L probabilities $b_j(t)$ and generate N random numbers for the rows. After we obtained A(t+1), we need to compute the $M_{ij}(t + 1) = a_i b_j(t + 1)$ and proceed to the next generation. For MOGAC, the operation is similar to MOGAR mathematically except now we rotate the matrix A by 90 degrees. Now, for a given column j we generate a random number y. If $y < b_j(t)$, then we mutate this column, otherwise we proceed to the next column and $A_{ij}(t + 1) = A_{ij}(t)$, $i = 1,...,N$. If column j is to be mutated, we determine the set $S_j(t)$ of chromosomes in column j to be changed by choosing the rows with the $a_i(t)$ in descending order, till we obtain $W_j(t) = b_j(t) * N$ members. Since our matrix A is assumed to be row ordered by fitness, we simply need to choose the $N, N - 1,...,N - W_j + 1$ rows to have the j-th column in these row mutated to obtain the matrix elements $A_{ij}(t + 1)$, $i = 1,...,N$. We then go through all L columns, so that in one generation, we need to sort a list of N fitness values and generate L random numbers for the columns. For a controlled comparison between MOGA and MCGA, we first choose which row (chromosome) to mutate, then which column (locus) to mutate. In this way, we can will compare MCGA with MOGAR on fair ground, since in our MCGA, we also decide which row to mutate before deciding which locus to change. The difference then is in the change of the locus for a given chromosome. In MOGAR, this change of the locus is governed by the probability b(j), while in MCGA, this change is based on the transition matrix P for that particular locus j at time t.

3   Markov-Chain Accelerated Genetic Algorithms

In MOGA, the mutation matrix is obtained through the assignment of the row and column mutation probability, $M_{ij}(t) \equiv a_i(t)b_j(t)$, $i = 1,...,N$; $j = 1,...,L$. In the implementation of the mutation, there are two different ways or ordering, MOGA by row or MOGA by column. However, both methods of implementation are rather ad-hoc. It will be desirable to implement the mutation on a more theoretical platform. Since we impose no memory effect, we should be able to describe the evolution process of the mutation probability through a Markov chain. Let us de-note the population matrix by A. There are N rows each representing a chromo-some and L columns each representing a locus. We assume that A is a binary matrix where every entry is either 0 or 1. Let's assume that there is a transition probability between the population A(t) with A(t+1) in the next generation. This transition should involve the basic features of genetic operators, which in the present paper concerns only the mutation process. In the proposed genetic algorithm, we consider a mutation-only updating scheme between two generations. During each generation, chromosomes are sorted from the fittest one to the least fit one. The chromosome at the i-th fittest place is assigned a row mutation probability a(i) according to some monotonic in-creasing function of its ranking: r(i)=(i-1)/N', if i-1<N' and r(i)=1, otherwise. Here we use N'=N/2. This choice of r(i) defines the ranking scheme we used. We also define the survival probability as s(i)=1-r(i). In this way, a(i) and s(i) together decide the probability to mutate to the other 0-1 state or to remain in the current 0-1 state for the i-th fittest chromosome. In the spirit of "survival of the fittest", we can use s(i) as the statistical weight of importance for the i-th chromosomes: w(i)=s(i). After discussing the row mutation, let's now turn to the column mutation, which addresses the relevance of the statistical importance of the locus. For each locus j, we define C0(j) and C1(j) which count in the current generation the numbers of chromosomes whose j-th entries are "0"s and "1"s. Next we normalize these counts to obtain

$$n_o(j,t) = \frac{C_o(j)}{C_o(j) + C_1(j)}; \qquad n_1(j,t) = \frac{C_1(j)}{C_o(j) + C_1(j)} \qquad (3)$$

The normalized vector $\mathbf{n}(j,t) = (n_o(j), n_1(j))$ characterizes the state distribution of the j-th locus among all the chromosomes in the current generation at time t. In order to direct the current population to a preferred state distribution for locus j, we first look at those rows of the population matrix A that has the j-th locus assuming the value u(=0 or 1). Among these rows with j-th locus equal to u, let's assume that the row that has maximum weight is the i*-th row, we should then follow the i* row as it is the fittest and the weight w(i*) is largest among the N rows. We can then rewrite

$$C'_0(j) = \max\{W(i)|A_{ij} = 0; i = 1,...,N\};$$
$$C'_1(j) = \max\{W(i)|A_{ij} = 1; i = 1,...,N\} \qquad (4)$$

with their normalized forms

$$n_o(j,t+1) = \frac{C'_o(j)}{C'_o(j) + C'_1(j)}; \qquad n_1(j,t+1) = \frac{C'_1(j)}{C'_o(j) + C'_1(j)}. \qquad (5)$$

The vector $\mathbf{n}(j,t+1) = (n_o(j,t+1), n_1(j,t+1))$ provides a direction that the popu-lation should evolve. This vector characterizes the target state distribution of the locus j among all the chromosomes in the next generation. Note that in this definition of the target state, we make use of the weight w(i) of the rows, which we intuitively set to be the survival probability of the chromosomes s(i). In turn, the survival probability is (1-r(i)), which is the type of ranking r(i) used in the formulation. A different scheme of fitness ranking of the chromosomes will pro-duce a different set of w(i), thereby a different direction vector $\mathbf{n}(j,t+1)$. Therefore, the ultimate tunable quantity in MCGA is our choice of ranking scheme for the chromosomes.

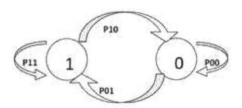

Fig. 1 Markov-Chain with transition probability P_{ab} from the current state a to the state b

Assuming that we have chosen a monotonic ranking scheme for the fitness, and have worked out the direction vector for locus j, we can then apply Markov Chain theory to compute the transition probability of the various states as defined below

$$\begin{pmatrix} n_o(j,t+1) \\ n_1(j,t+1) \end{pmatrix} = \begin{pmatrix} P_{00}(j,t) & P_{10}(j,t) \\ P_{01}(j,t) & P_{11}(j,t) \end{pmatrix} \begin{pmatrix} n_o(j,t) \\ n_1(j,t) \end{pmatrix} \qquad (6)$$

In the Markov Chain described in Fig. 1, state 1 changes to state 0 with probability $P_{10}(j,t)$, and remains to be in state 1 with probability $P_{11}(j,t) = 1 - P_{10}(j,t)$. Similarly, state 0 changes to state 1 with probability $P_{01}(j,t)$, and remains to be state 0 with probability $P_{00}(j,t) = 1 - P_{01}(j,t)$. These conditions allow us to solve for the transition probability $P_{10}(j,t)$ explicitly in terms of the single variable $P_{00}(j,t)$ as

$$\begin{aligned} P_{10}(j,t) &= \tfrac{n_o(j,t+1)-P_{00}(j,t)n_o(j,t)}{n_1(j,t)} \\ P_{11}(j,t) &= 1 - P_{10}(j,t) \text{and} P_{01}(j,t) = 1 - P_{00}(j,t). \end{aligned} \qquad (7)$$

Thus, we need only to know the probability $P_{00}(j,t)$ to compute all the remaining probabilities. At the beginning, we can set $P_{00}(j,t_{ini}) = 0.5$ since our guess of the solution is random. In the next time step, we need to make an assumption on $P_{00}(j,t)$. An intuitive assignment $P_{00}(j,t) = n_0(j,t)$ gives very good result. Unlike MOGA, where a chromosome mutates using the mutation matrix, here our Markov Chain Genetic Algorithm (MCGA) makes use of the transition probability matrix $P_{ab}(j,t)$ to move from one state to the next. The entire evolutionary computation of MCGA depends mainly on the ranking scheme used for the chromosomes.

4 Experiments

In order to evaluate the effectiveness of MCGA, we compare it with MOGAR (Mutation Only Genetic Algorithm by Row). We first choose a row to mutate ac-cording to the row mutation probability, same as MOGAR, but in the next step when we perform the locus mutation in this row we use the Markov Chain method. This gives a time dependent transition probability between the 0 and 1 states. We have performed numerous experiments to compare MCGA with MOGAR using mutation matrix. Note that this is a control comparison since we fix the first step to be the same.

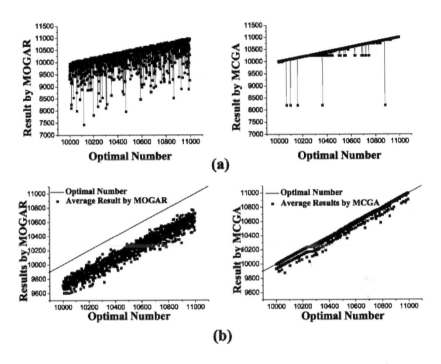

Fig. 2 MOGAR and MCGA for the Knapsack Problem after 40 generations. (a) for one run (b) the average for 30 runs

Based on tests performed when comparing the locus dependent mutation rate genetic algorithm [14] with other standard methods in solving the knapsack problems [17], we continue to use this standard problem for the comparison of MOGAR and MCGA, which are two more sophisticated version of locus dependent mutation genetic algorithm. The knapsack has a maximum weight volume C, called the optimal number. We consider the allocation of n items in the knapsack. When item i is in the knapsack, its loading is given by the weight w_i while the profit for carrying it is p_i. Given $\{w_1, w_2, ..., w_n\}$ and $\{p_1, p_2, ..., p_n\}$, our goal is to maximize $\sum_{i=1}^{n} p_i x_i$ subject to the constraint $\sum_{i=1}^{n} w_i x_i \leq C$ .In our experiment we have n=16 items with weight $w_i = 2^{i-1}$ and profit $p_i = 1$ for all i. Our goal is to find configuration of item occupancy $\{x_i\}$ so that we get as close to the limit defined by the maximum knapsack weight volume as possible. Here x_i is a binary variable which is 1 if item i is in the knapsack, and 0 if it is not. We can now use the string of $\{x_i\}$ to denote a chromosome. We set the optimal number (C) between 10000 and 11000, and try to find the best chromosome that fulfills our goal using MOGAR and MCGA. The results are shown in Fig. 2 (a). Each algorithm runs for 40 generations. Here we can see that MCGA has a higher probability of reaching the optimum than MOGAR, which very often get trapped in local optima. We also run the program for 30 times and the results are shown in Fig. 2 (b). We can see that the results of MCGA converge to the optimal number much better than that of MOGAR.

Next we define a performance measure to compare these two algorithms. The first one is the "hit rate". For the "hit rate", we first introduce a "hit scale" by which we say that the result "hits" the optimum when it surpasses a critical thre-shold value. This threshold can be defined as a given percentage of the optimum. For example, if we set the threshold to be within 30% of the optimum, then in our case with an optimal number 10000, we will say a particular solution of the knap-sack problem has "hit" the optimum when its value lies between 7000 and 10000. The "hit rate" is then defined as the number of hit result over the total number of tests performed and is between 0 for "never hitting" and 1 for "always hitting". In the context of our numerical work, we set the optimal number of the knapsack in the range between 10000 and 11000, with a total number of tests being 1000. We change the hit scale and the result is shown in Fig. 3 (a). We can see that the hit rate of MCGA is closer to 1.0 than MOGAR with small hit scale. This indicates that MCGA can obtain more accurate results than MOGAR. The second performance measure used in the comparison of MOGA with MCGA address the number of generations in each algorithm. If we run these algorithms for a given number (g) of generations, we can compute the average best fitness, which we call the "average fit rate", defined as

$$\overline{F}(g) = \frac{1}{1000} \sum_{h=10000}^{11000} \frac{B(h,g)}{h} \qquad (8)$$

Here B(h,g) is the fitness of the best chromosome in the g-th generation and with given optimum value of h. In a given test with g generation and optimal number h, the fitness of the best chromosome will be h if it hits the optimum value, otherwise it will be less. Therefore, the average fit rate will be smaller than one, unless the algorithm is perfect and find optimum every time for given g and h. We have performed test for generation g less than 81. The results are shown in Fig. 3 (b). We can see that MCGA has a better fit rate than MOGAR in general, especially in the beginning generations. This is consistent with the result for MCGA in Fig. 3 (a) where we see that MCGA is more accurate than MOGAR. Furthermore, the result on fit rate indicates that MCGA can achieve more accurate results with less number of generations than MOGAR.

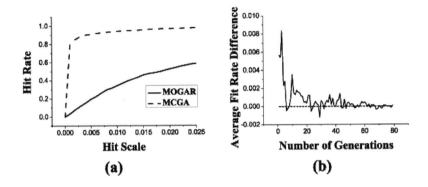

Fig. 3 (a) Hit Rate and (b) Fit Rate for MOGAR and MCGA

5   Conclusion

We test our new genetic algorithm using a Markov Chain formulation of the mutation matrix, so that the direction of search is a guided search using a transition matrix based on a given ranking scheme of the chromosomes. This new algorithm (MCGA) makes transparent the fundamental assumption of the evolutionary process lies in the choice of ranking scheme. The important input for this guided search is the ranking scheme of the chromosomes. The simple ranking scheme used in this paper suggests that MCGA is better than the original formulation of MOGA, which makes use of past statistics to construct a mutation matrix. We compare a particular choice of MOGA using row first (MOGAR) and conclude that MCGA is in general better than MOGAR both in terms of accuracy (measured by hit rate) and in terms of speed (measured by fit rate) for the knapsack problem. We expect this superiority of MCGA over MOGA remains valid in other standard tests of the algorithms. Various combinations of these algorithms involve resource allocation, which have been addressed for MOGA, but not yet for MCGA. In future works, we will present the results on managing these combinations using intelligent resource allocation techniques [9, 15, 18].

Acknowledgements. K.Y. Szeto acknowledges the support of CERG grant 602506 and 602507.

References

[1] Holland, J.H.: Adaptation in Natural and Artificial Systems. University of Michigan Press, Ann Arbor (1975)

[2] Goldberg, D.E.: Genetic Algorithms in Search, Optimization, and Machine Learning. Addison-Wesley, Reading (1989)

[3] Li, S.P., Szeto, K.Y.: Crytoarithmetic problem using parallel Genetic Algorithms. In: Mendl 1999, Brno, Czech (1999)

[4] Szeto, K.Y., Cheung, K.H.: Multiple time series prediction using genetic algorithms optimizer. In: Proceedings of the International Symposium on Intelligent Data Engineering and Learning, IDEAL 1998, Hong Kong, pp. 127–133 (1998)

[5] Jiang, R., Szeto, K.Y., Luo, Y.P., Hu, D.C.: Distributed parallel genetic algorithm with path splitting scheme for the large traveling salesman problems. In: Shi, Z., Faltings, B., Musen, M. (eds.) Proceedings of Conference on Intelligent Information Processing, 16th World Computer Congress 2000, Beijing, August 21-25, pp. 478–485. Publishing House of Electronic Industry (2000)

[6] Szeto, K.Y., Cheung, K.H., Li, S.P.: Effects of dimensionality on parallel genetic algorithms. In: Proceedings of the 4th International Conference on Information System, Analysis and Synthesis, Orlando, Florida, USA, vol. 2, pp. 322–325 (1998)

[7] Szeto, K.Y., Fong, L.Y.: How adaptive agents in stock market perform in the presence of random news: a genetic algorithm approach. In: Leung, K.-S., Chan, L., Meng, H. (eds.) IDEAL 2000. LNCS(LNAI), vol. 1983, pp. 505–510. Springer, Heidelberg (2000)

[8] Fong, A.L.Y., Szeto, K.Y.: Rule Extraction in Short Memory Time Series using Genetic Algorithms. European Physical Journal B 20, 569–572 (2001)

[9] Shiu, K.L., Szeto, K.Y.: Self-adaptive Mutation Only Genetic Algorithm: An Application on the Optimization of Airport Capacity Utilization. In: Fyfe, C., Kim, D., Lee, S.-Y., Yin, H. (eds.) IDEAL 2008. LNCS, vol. 5326, pp. 428–435. Springer, Heidelberg (2008)

[10] Harik, G.R., Lobo, F.G., Goldberg, D.E.: The compact genetic algorithm. IEEE Transactions on Evolutionary Computation, 523–528 (1998)

[11] Baluja, S.: Population-based incremental learning: A method for integrating genetic search based function optimization and competitive learning (1994)

[12] Pelikan, M., Goldberg, D.E., Cantu-Paz, E.: BOA: The Bayesian optimization algorithm. In: Proc. of the Genetic and Evolutionary Computation Conference GECCO, pp. 525–532 (1999)

[13] Pelikan, M., Goldberg, D.E.: Scalable optimization via probabilistic modeling: From algorithms to applications, pp. 63–90 (2006)

[14] Ma, C.W., Szeto, K.Y.: Locus Oriented Adaptive Genetic Algorithm: Application to the Zero/One Knapsack Problem. In: Proceeding of The 5th International Conference on Recent Advances in Soft Computing, RASC 2004, Nottingham, UK, pp. 410–415 (2004)

[15] Szeto, K.Y., Zhang, J.: Adaptive genetic algorithm and quasi-parallel genetic algorithm: Application to knapsack problem. In: Lirkov, I., Margenov, S., Waśniewski, J. (eds.) LSSC 2005. LNCS, vol. 3743, pp. 189–196. Springer, Heidelberg (2006)

[16] Law, N.L., Szeto, K.Y.: Adaptive Genetic Algorithm with Mutation and Crossover Matrices. In: Proceeding of the 12th International Joint Conference on Artificial Intelligence (IJCAI 2007), January 6 - 12. Theme: AI and Its Benefits to Society, pp. 2330–2333. International Joint Conferences on Artificial Intelligence, Hyderabad (2007)

[17] Gordon, V., Bohm, A., Whitley, D.: A Note on the Performance of Genetic Algorithms on Zero-One Knapsack Problems. In: Proceedings of the 9th Symposium on Applied Computing (SAC 1994), Genetic Algorithms and Combinatorial Optimization, Phoenix, Az, pp. 194–195 (1994)

[18] Szeto, K.Y., Rui, J.: A quasi-parallel realization of the Investment Frontier in Computer Resource Allocation Using Simple Genetic Algorithm on a Single Computer. In: Fagerholm, J., Haataja, J., Järvinen, J., Lyly, M., Råback, P., Savolainen, V. (eds.) PARA 2002. LNCS, vol. 2367, pp. 116–126. Springer, Heidelberg (2002)

Adapting Heuristic Mastermind Strategies to Evolutionary Algorithms

Thomas Philip Runarsson and Juan J. Merelo-Guervós

Abstract. The art of solving the Mastermind puzzle was initiated by Donald Knuth and is already more than thirty years old; despite that, it still receives much attention in operational research and computer games journals, not to mention the nature-inspired stochastic algorithm literature. In this paper we try to suggest a strategy that will allow nature-inspired algorithms to obtain results as good as those based on exhaustive search strategies; in order to do that, we first review, compare and im-prove current approaches to solving the puzzle; then we test one of these strategies with an estimation of distribution algorithm. Finally, we try to find a strategy that falls short of being exhaustive, and is then amenable for inclusion in nature inspired algorithms (such as evolutionary of particle swarm algorithms). This paper proves that by the incorporation of what we call *local entropy* into the fitness function of the evolutionary algorithm it becomes a better player than a random one, and gives a rule of thumb on how to incorporate the best heuristic strategies to evolutionary algorithms without incurring in an excessive computational cost.

Keywords: puzzles, games, Mastermind, bulls and cows, search strategies, oracle games, evolutionary algorithms, estimation of distribution algorithms.

Thomas Philip Runarsson
School of Engineering and Natural Sciences,
University of Iceland,
Reykjavik, Iceland
e-mail: tpr@hi.is

Juan J. Merelo-Guervós
Department of Architecture and Computer Technology, ETSIIT, University of Granada, Spain
e-mail: jmerelo@geneura.ugr.es

J.R. González et al. (Eds.): NICSO 2010, SCI 284, pp. 255–267, 2010.
springerlink.com   © Springer-Verlag Berlin Heidelberg 2010

1 Introduction

Mastermind in its current version is a board game that was introduced by the telecommunications expert Mordecai Merowitz [15] and sold to the company In-victa Plastics, who renamed it to its actual name; in fact, Mastermind is a version of a traditional puzzle called *bulls and cows* that dates back to the Middle Ages. In any case, Mastermind is a puzzle (rather than a game) in which two persons, the *codemaker* and *codebreaker* try to outsmart each other in the following way:

- The codemaker sets a length ℓ combination of κ symbols. In the classical version, $\ell = 4$ and $\kappa = 6$, and color pegs are used as symbols over a board with rows of $\ell = 4$ holes; however, in this paper we will use uppercase letters starting with A instead of colours.
- The codebreaker then tries to guess this secret code by producing a combination.
- The codemaker gives a response consisting on the number of symbols guessed in the right position (usually represented as black pegs) and the number of symbols in an incorrect position(usually represented as white pegs).
- The codebreaker then, using that information as a hint, produces a new combination until the secret code is found.

For instance, a game could go like this: The codemaker sets the secret code *ABBC*. The rest of the game is shown in Table 1.

Table 1 Progress in a Mastermind game that tries to guess the secret combination *ABBC*. The player here is not particularly clueful, playing a third combination that is not *consistent* with the first one, not coinciding in two positions and one color (corresponding to the 2 black/1 white response given by the codemaker) with it

Combination	Response
AABB	2 black, 1 white
ACDE	1 black, 1 white
FFDA	1 white
ABBE	3 black
ABBC	4 black

Different variations of the game include giving information on which position has been guessed correctly, avoiding repeated symbols in the secret combination (*bulls and cows* is actually this way), or allowing the codemaker to change the code during the game (but only if this does not make responses made so far false).

In any case, the codebreaker is allowed to make a maximum number of combinations (usually fifteen, or more for larger values of κ and ℓ), and score corresponds to the number of combinations needed to find the secret code; after repeating the game a number of times with codemaker and codebreaker changing sides, the one with the lower score wins.

Since Mastermind is asymmetric, in the sense that the position of one of the players after setting the secret code is almost completely passive, and limited to give hints as a response to the guesses of the codebreaker, it is rather a puzzle than a game, since the codebreaker is not really matching his skills against the codemaker, but facing a problem that must be solved with the help of hints, the implication being that playing Mastermind is more similar to solving a Sudoku than to a game of chess; thus, the solution to Mastermind, unless in a very particular situation (always playing with an opponent who has a particular bias for choosing codes, or maybe playing the dynamic code version), is a search problem with constraints.

What makes this problem interesting is its relation to other, generally called *oracle* problems such as circuit and program testing, differential cryptanalysis and other puzzle games (these similarities were reviewed in our previous paper [10]) is the fact that it has been proved to be NP-complete [5, 14] and that there are several open issues, namely, what is the lowest average number of guesses you can achieve, how to minimize the number of evaluations needed to find them (and thus the runtime of the algorithm), and obviously, how it scales when increasing κ and ℓ. This paper will concentrate on the first issue.

This NP completeness implies that it is difficult to find algorithms that solve the problem in a reasonable amount of time, and that is why in our previous work [2, 9, 10] we introduced stochastic evolutionary and simulated annealing algorithms that solved the Mastermind puzzle in the general case, finding solutions in a reasonable amount of time that scaled roughly logarithmically with problem size. The strategy followed to play the game was optimal in the sense that is was guaranteed to find a solution after a finite number of combinations; however, there was no additional selection on the combination played other than the fact that it was consistent with the responses given so far.

In this paper, after reviewing how the state of the art in solving this puzzle has evolved in the last few years, we examine how we could improve the code-breaking skills of an evolutionary algorithm by using different techniques, and how these techniques can be further optimized. In order to do that we examine different ways of scoring combinations in the search space, how to choose one combination out of a set of combinations that have exactly the same score, and how all that can be applied to a simple estimation of distribution algorithm to improve results over a standard one. This paper presents for the first time an evolutionary algorithm that biases search so that combinations played have a better chance of reducing the size of the remaining search space, and adapt to an stochastic environment deterministic techniques that had been previously published; all techniques, unlike our former papers, have been tested over the whole code space, instead of a random sample, so that they can be compared and yield significant results.

The rest of the paper is organized as follows: next we establish terminology and examine the state of the art; then heuristic strategies for Mastermind are examined in Section 3; the way they could be adapted to an evolutionary algorithm is presented in Section 5, and finally, conclusions are drawn in the closing section 6.

2   State of the Art

Before presenting the state of the art, a few definitions are needed. We will use the term *response* for the return code of the codemaker to a played combination, c_{played}. A response is therefore a function of the combination, c_{played} and the secret combination c_{secret}, let the response be denoted by $h(c_{played}, c_{secret})$. A combination c is *consistent with* c_{played} iff

$$h(c_{played}, c_{secret}) = h(c_{played}, c) \qquad (1)$$

that is, if the combination has as many black and white pins with respect to the played combination as the played combination with respect to the secret combination. Furthermore, a combination is *consistent* iff

$$h(c_i, c) = h(c_i, c_{secret}) \text{ for } i = 1..n \qquad (2)$$

where n is the number of combinations, c_i, played so far; that is, c is *consistent with* all guesses made so far. A combination that is consistent is a candidate solution. The concept of consistent combination will be important for characterizing different approaches to the game of Mastermind.

One of the earliest strategies, by Knuth [6], is perhaps the most intuitive for Mastermind. In this strategy the player selects the guess that reduces the number of remaining consistent guesses and the opponent the return code leading to the maximum number of guesses. Using a complete minimax search Knuth shows that a maximum of 5 guesses are needed to solve the game using this strategy. This type of strategy is still the most widely used today: most algorithms for Mastermind start by searching for a consistent combination to play.

In some cases once a single consistent guess is found it is immediately played, in which case the object is to find a consistent guess as fast as possible. For example, in [10] an evolutionary algorithm is described for this purpose. These strategies are fast and do not need to examine a big part of the space. Playing a consistent combinations eventually produces a number of guesses that uniquely determine the code. However, the maximum, and average, number of combinations needed is usually high. Hence, some bias must be introduced in the way combinations are searched. If not, the guesses will be no better than a purely random approach, as solutions found (and played) are a random sample of the space of consistent guesses.

The alternative to discovering a single consistent guess is to collect a set of consistent guesses and select among them the best alternative. For this a number of heuristics have been developed over the years. Typically these heuristics require all consistent guesses to be first found. The algorithms then use some kind of search over the space of consistent combinations, so that only the guess that extracts the most information from the secret code is issued, or else the one that reduces as much as possible the set of remaining consistent combinations. However, this is obviously not known in advance. To each combination corresponds a partition of the rest of the space, according to their match (the number of blacks and white pegs that would be the response when matched with each other). Let us consider the first combination:

if the combination considered is AABB, there will be 256 combinations whose response will be 0b, 0w (those with other colors), 256 with 0b, 1w (those with either an A or a B), etc. Some partitions may also be empty, or contain a single element (4b, 0w will contain just AABB, obviously). For a more exhaustive explanation see [7]. Each combination is thus characterized by the features of these partitions: the number of non-empty ones, the average number of combinations in them, the maximum, and other characteristics one may think of.

The path leading to the most successful strategies to date include using the *worst case, expected case, entropy* [3, 13] and *most parts* [7] strategies. The *entropy* strategy selects the guess with the highest entropy. The entropy is computed as follows: for each possible response i for a particular consistent guess, the number of remaining consistent guesses is found. The ratio of reduction in the number of guesses is also the *a priori* probability, p_i, of the secret code being in the corresponding partition. The entropy is then computed as $\sum_{i=1}^{n} p_i \log_2(1/p_i)$, where $\log_2(1/p_i)$ is the information in bit(s) per partition, and can be used to select the next combination to play in Mastermind [13]. The *worst case* is a one-ply version of Knuth's approach, but Irving [4] suggested using the *expected case* rather than the worst case. Kooi [7] noted, however, that the size of the partitions is irrelevant and that rather the number of non empty partitions created, n, was important. This strategy is called *most parts*. The strategies above require one-ply look-ahead and either determining the size of resulting partitions and/or the number of them. Computing the number of them is, however, faster than determining their size. For this reason the *most parts* strategy has a computational advantage.

The heuristic strategies described above use some form of look-ahead which is computationally expensive. If no look-ahead is used to guide the search a guess is selected purely at *random*. However, it may be possible to discriminate by using local information. If this were possible one could even dismiss searching for all consistent guesses and search for a single consistent guess with the bias. In section 3 these heuristic strategies are compared. In section 4 an EDA using only local information is compared with those that need to examine all consistent guessed in order to select the best one.

3   Comparison of Heuristic Strategies

As has been mentioned before, there have been a number of different strategies proposed over the years for selecting among consistent guesses in Mastermind. These heuristics do not consider an exhaustive minimax search, but rather one-ply search. What is, however, not clear in these research papers is how ties are broken, which probably implies that a *first come, first served* approach is taken, using the first combination in lexicographical order out of all tied combinations. For this reason we propose to perform a comparison of the heuristic methods here where the ties are broken randomly. Each strategy is, therefore, used on all possible secret combinations (they are $6^4 = 1296$) using ten independent runs.

The heuristics compared are the *entropy*, *most parts* and *worst case* strategy, as performed by Bestavros and Belal [3]. The worst case refers to the fact that for each possible return code for a particular guess the smallest reduction in assumed, i.e. the worst case. The actual consistent guess chosen is the one which maximizes the worst case. Finally, the *expected size* strategy, [4] is also tested; in this strategy the expected case is used instead of the worst case. These strategies are compared with the *random* strategy.

Table 2 A comparison of the mean number of games played using all 6^4 colour combinations and breaking ties randomly, ranked from best to worst average number of guesses needed. Statistics are given for 10 independent experiments. The maximum number of moves used for the 10×6^4 games is also presented in the final column. Horizontal separators are given for statistically independent results

Strategy	min	mean	median	max	st.dev.	max guesses
Entropy	4.383	4.408	4.408	4.424	0.012	6
Most parts	4.383	4.410	4.412	4.430	0.013	7
Expected size	4.447	4.470	4.468	4.490	0.015	7
Worst case	4.461	4.479	4.473	4.506	0.016	6
Random	4.566	4.608	4.608	4.646	0.026	8

The results of the experiments are given in table 2. The first combination played is always AABC, as proposed by [4]. The Wilcoxon rank sum (used instead of t-test since the variable does not follow a normal distribution) with a 0.05 significance level is used to determine which results are statistically different form another. The horizontal lines are used to group together heuristics that are not statistically different from the other. From these results we can gather that there is no statistical difference between the *entropy* and *most parts* strategies. However, out of all games played the maximum number of guesses needed by the Entropy strategy was only 6 while for most parts it was 7. These strategies are also better than the *worst* and *expected* case, which are statistically equivalent. For the worst case strategy used, nevertheless, only a maximum of 6 guesses, unlike the expected case with 7. The worst performer is the *random* strategy which also required a maximum of 8 guesses. Finally, note that the optimal expected result on playing all secrets is 4.340 [8].

4 Estimation of Distribution Algorithm Using Local Entropy

One of the common approaches to using evolutionary algorithms for Mastermind, is simply to search for a single consistent guess which is then immediately played. This is especially true for the generalized version of the game, for $N > 6$ and $L > 4$, where the task of just finding a consistent guess can be difficult. The result of such an approach is likely to do as well as the random strategy discussed in the previous sections if there are no major biases in searching for these consistent solutions. For steady state evolutionary algorithms it may, however, be the case that the consecutive consistent guesses found may be similar to others played before; that is, the strategy of play may not necessarily be purely random. In any case it is highly likely that evolutionary algorithms of this type will not do better than the random strategy, as seen above, since consistent combinations found are a random sample of the set of consistent combinations.

In this section we investigate the performance of strategies that find a single consistent guess and play it immediately. In this case we use an estimation of distribution algorithm [12] *EDA* included with the `Algorithm::Evolutionary` Perl module [11], with the whole EDA-solving algorithm available as `Algorithm::-MasterMind::EDA` from CPAN (the comprehensive Perl Archive Network). This is an standard EDA that uses a population of 200 individuals and a replacement rate of 0.5; each generation, half the population is generated from the previously generated distribution. The first combination played was AABB, since it was not found significantly different from using AABC, as before.

The fitness function used previously [10] to find consistent guesses is as follows,

$$f(c_{guess}) = \sum_{i=1}^{n} |h(c_i, c_{guess}) - h(c_i, c_{secret})|$$

that is, the sum of the absolute difference of the number of white and black pegs needed to make the guess consistent. However, this approach is likely to perform as well as the random strategy discussed in the previous section. When finding a single consistent guess we cannot apply the heuristic strategies from the previous section. For this reason we introduce now a local entropy measure, which can be applied to non-consistent guesses and so bias our search. The local entropy assumes that the fact that some combinations are better than others depends on its informational content, and that in turn depends on the entropy of the combination along with the rest of the combinations played so far. To compute *local entropy*, the combination is concatenated with n combinations played so far and its Shannon entropy computed:

$$s(c_{guess}) = \sum_{g \in \{A,...,F\}} \frac{\#g}{(n+1)\ell} \log \left(\frac{(n+1)\ell}{\#g} \right) \qquad (3)$$

with g being a symbol in the alphabet and # denotes the number of them. Thus, the fitness function which includes the local entropy is defined as,

$$f_{\ell}(c_{guess}) = \frac{s(c_{guess})}{1 + f(c_{guess})}$$

In this way a bias is introduced to the fitness to as to select the guess with the highest local entropy. When a consistent combination is found, the combination with the highest entropy found in the generation is played (which might be the only one or one among several; however, no special provision is done to generate several).

The result of ten independent runs of the EDA over the whole search space are now compared with the results of the previous section. These results may be seen in table 3. Two EDA experiments are shown, one using the fitness function designed to find a consistent guess only (f) and ones using local entropy f_ℓ. The EDA using local entropy is statistically better than playing pure random, whereas the other EDA is not. In order to confirm the usefulness of the local entropy, an additional experiment was performed. This time, as in the previous sections, all consistent guesses are found and the one with the highest local entropy played. This results is labelled *LocalEntropy* in table 3. The results are not statistically different from the EDA results using fitness function f_ℓ.

Table 3 A comparison of the mean number of games played using all 6^4 colour combinations and breaking ties randomly, ranked from best to worse mean number of combinations. Statistics are given for 10 independent experiments. The maximum number of moves used for the 10×6^4 games is also presented in the final column. Horizontal separators are given for statistically independent results

Strategy	min	mean	median	max	st.dev.	max guesses
Entropy	4.383	4.408	4.408	4.424	0.012	6
Most parts	4.383	4.410	4.412	4.430	0.013	7
Expected size	4.447	4.470	4.468	4.490	0.015	7
Worst case	4.461	4.479	4.473	4.506	0.016	6
LocalEntropy	4.529	4.569	4.568	4.613	0.021	7
EDA+f_ℓ	4.524	4.571	4.580	4.600	0.026	7
EDA+f	4.562	4.616	4.619	4.665	0.032	7
Random	4.566	4.608	4.608	4.646	0.026	8

As a local conclusion, the *Entropy* method seemed to perform the best on average, but the estimation of distribution algorithm is not statistically different from (admittedly naive) exhaustive search strategies such as LocalEntropy and performs significantly better than the Random algorithm on average.

We should remark that the objective of this paper is not to show which strategy is the best runtime-wise, or which one offers the best algorithmic performance/runtime trade-off; but in any case we should note that the algorithm with the least number of evaluations and lowest runtime is the EDA. However, its average performance as a player is not as good as the rest, so some improvement might be obtained by creating a set of possible solutions. It remains to be seen how many solutions would be needed, but that will be investigated in the next section.

5   Heuristics Based on a Subset of Consistent Guesses

Following a tip in one of our former papers, recently Berghman et al. [1] proposed an evolutionary algorithm which finds a number of consistent guesses and then uses a strategy to select which one of these should be played. The strategy they apply is not unlike the *expected size* strategy. However, it differs in some fundamental ways.

In their approach each consistent guess is assumed to be the secret in turn and each guess played against every different secret. The return codes are then used to compute the size of the set of remaining consistent guesses in the set. An average is then taken over the size of these sets. Here, the key difference between the *expected size* method is that only a subset of all possible consistent guesses is used and some return codes may not be considered or considered more frequently than once, which might lead to a bias in the result. Indeed they remark that their approach is computationally intensive which leads them to reduce the size of this subset further. Note that Berghman et al. only present the result of a single evolutionary run and so their results cannot be compared with those here.

Their approach is, however, interesting, and lead us to consider the case where an evolutionary algorithms has been designed to find a maximum of μ consistent guesses within some finite time. It will be assumed that this subset is sampled uniformly and randomly from all possible consistent guesses. The question is, how do the heuristic strategies discussed above work on a randomly sampled subset of consistent guesses? The experiment performed in the previous sections are now repeated, but this time only using the four best one-ply look-ahead heuristic strategies on a random subset of guesses, bounded by size μ. If there are many guesses that give the same number of partitions or similar entropy then perhaps taking a random subset would be a good representation for all guesses. This has implications not only with respect to the application of EAs but also to the common strategies discussed here.

The size of the subsets are fixed at 10, 20, 30, 40, and 50, in order to investigate the influence of the subset size. The results for these experiments and their statistics are presented in table 4. The results are presented are as expected better as the subset size, μ, gets bigger. Noticeable is the fact that the *entropy* and *most parts* strategies perform the best as before, however, at $\mu = 40$ and 50 the entropy strategy is better.

Is there a statistical difference between the different subset sizes? To answer this we look at only the two best strategies in more detail, *entropy* and *most parts*, and compare their performances for the different subset sizes, μ, and using the complete set, case when $\mu = \infty$, as presented in table 3. These results are given in table 5 and 6. From this analysis it may be concluded that a set size of $\mu = 20$ is sufficiently large and not statistically different from using the entire set of consistent guesses. This is actually quite a large reduction is the set size, which is about 250 on average after the first guess, then 55, followed by 12 [1].

This implies that, at least in this case, using a subset of the combination pool that is around 1/10th of the total size potentially yields a result that is as good as using the whole set; even as algorithmically finding 20 tentative solutions is harder than finding a single one, using this in stochastic search algorithms such as the EDA mentioned above or an evolutionary algorithm holds the promise of combining the accuracy of exhaustive search algorithms with the speed of an EDA or an EA. In any case, for spaces bigger than $\kappa = 6, \ell = 4$ there is no other option, and this 1/10 gives at least a rule of thumb. How this proportion grows with search space size is still an open question.

Table 4 Statistics for the average number of guesses for different maximum sizes μ of subsets of consistent guesses. The horizontal lines are used as before to indicate statistical independent, with the exception of one case: for $\mu = 10$ the expected size and worst case are not independent

Strategy	min	mean	median	max	st.dev.	max guesses
$\mu = 10$						
Most parts	4.429	4.454	4.454	4.477	0.016	7
Entropy	4.438	4.468	4.476	4.483	0.016	7
Expected size	4.450	4.472	4.474	4.493	0.014	7
Worst case	4.447	4.486	4.487	4.519	0.020	7
$\mu = 20$						
Entropy	4.394	4.423	4.426	4.455	0.021	7
Most parts	4.424	4.431	4.427	4.451	0.009	7
Expected size	4.427	4.454	4.455	4.481	0.017	7
Worst case	4.429	4.453	4.451	4.486	0.017	7
$\mu = 30$						
Entropy	4.380	4.413	4.410	4.443	0.020	6
Most parts	4.393	4.416	4.416	4.435	0.015	7
Expected size	4.426	4.453	4.456	4.491	0.019	7
Worst case	4.434	4.459	4.461	4.477	0.013	7
$\mu = 40$						
Entropy	4.372	4.398	4.399	4.426	0.017	7
Most parts	4.383	4.424	4.427	4.448	0.020	7
Expected size	4.418	4.457	4.455	4.491	0.023	7
Worst case	4.424	4.458	4.457	4.490	0.022	7
$\mu = 50$						
Entropy	4.365	4.397	4.393	4.438	0.020	6
Most parts	4.400	4.424	4.422	4.454	0.017	7
Expected size	4.419	4.453	4.453	4.495	0.022	7
Worst case	4.431	4.456	4.457	4.474	0.012	6

Table 5 No statistical advantage is gained when using a set size larger than $\mu = 30$ when using the *entropy* strategy. However, there is also no statistically difference between $\mu = 20$ and both $\mu = 30$ and $\mu = \infty$ (the only cases not indicated by the horizontal lines)

$\mu =$	min	mean	median	max	st.dev.
10	4.438	4.468	4.476	4.483	0.016
20	4.394	4.423	4.426	4.455	0.021
30	4.380	4.413	4.410	4.443	0.020
40	4.372	4.398	4.399	4.426	0.017
50	4.365	4.397	4.393	4.438	0.020
∞	4.383	4.408	4.408	4.424	0.012

Table 6 No statistical advantage is gained when using a set size larger than $\mu = 20$ for the *most parts* strategy. However, there is a statistical difference between $\mu = 20$ and $\mu = \infty$ (the only case not indicated by the horizontal lines

$\mu =$	min	mean	median	max	st.dev.
10	4.429	4.454	4.454	4.477	0.016
20	4.424	4.431	4.427	4.451	0.009
30	4.393	4.416	4.416	4.435	0.015
40	4.383	4.424	4.427	4.448	0.020
50	4.400	4.424	4.422	4.454	0.017
∞	4.383	4.410	4.412	4.430	0.013

6   Discussion and Conclusion

In this paper we have tried to study and compare the different heuristic strategies for the simplest version of Mastermind in order to come up with a nature-inspired algorithm that is able to beat them in terms of running time and scalability. The main problem with heuristic strategies is that they need to have the whole search space in memory; even the most advanced ones that run over it only once will become unwieldy as soon as ℓ or κ increase. However, evolutionary algorithms have already been proved [10] to scale much better, the only problem being that their performance as players is no better than a random player.

In this paper, after improving (or maybe just clarifying) heuristic and deterministic algorithms with an random choice of a combination to play, we have incorporated the simplest of those strategies to an estimation of distribution algorithm (the so-called *local entropy*, which takes into account the amount of *surprise* the new combination implies); results are promising, but still fall short of the best heuristic strategies, which take into account the partition of search space created by each combination. That is why we have tried to compute the subset that would be able to obtain results that are indistinguishable, in the statistical sense, from those obtained with the whole set, coming up with a subset whose size is around 10% of the whole one, being thus less computational intensive and easily incorporated into an evolutionary algorithm.

However, how this is incorporated within the evolutionary algorithm remains to be seen, and will be one of our future lines of work. So far, distance to consistency and entropy are combined in an aggregative fitness function; the quality of partitions induced will also have to be taken into account; however, there are several ways of doing this: putting consistent solutions in an *archive*, in the same fashion that multiobjective optimization algorithms do, leave them into the population and take the quality of partitions as another objective, not to mention the evolutionary parameter issues themselves: population size, operator rate. Our objective, in this sense, will be not only to try and minimize the number of average/median games played, but also to minimize the proportion of the search space examined to find the final solution.

All the tests and algorithms have been implemented using the Matlab package, and are available as open source source software with a GPL licence from the authors. The evolutionary algorithm and several mastermind strategies are also available from CPAN; most results and configuration files needed to compute them are available from the group's CVS server (at http://sl.ugr.es/algmm).

Acknowledgements

This paper has been funded in part by the Spanish MICYT projects NoHNES (Spanish Ministerio de Educación y Ciencia - TIN2007-68083) and TIN2008-06491-C04-01 and the Junta de Andalucía P06-TIC-02025 and P07-TIC-03044. The authors are also very grateful to the traffic jams in Granada, which allowed limitless moments of discussion and interaction over this problem.

References

[1] Berghman, L., Goossens, D., Leus, R.: Efficient solutions for Mastermind using genetic algorithms. Computers and Operations Research 36(6), 1880–1885 (2009),
http://www.scopus.com/inward/
record.url?eid=2-s2.0-56549123376&partnerID=40
[2] Bernier, J.L., Herráiz, C.I., Merelo-Guervós, J.J., Olmeda, S., Prieto, A.: Solving *mastermind* using GAs and simulated annealing: a case of dynamic constraint optimization. In: Ebeling, W., Rechenberg, I., Voigt, H.-M., Schwefel, H.-P. (eds.) PPSN 1996. LNCS, vol. 1141, pp. 553–563. Springer, Heidelberg (1996),
http://www.springerlink.com/content/78j7430828t2867g
[3] Bestavros, A., Belal, A.: Mastermind, a game of diagnosis strategies. Bulletin of the Faculty of Engineering. Alexandria University (1986),
citeseer.ist.psu.edu/bestavros86mastermind.html,
http://www.cs.bu.edu/fac/best/res/papers/alybull86.ps
[4] Irving, R.W.: Towards an optimum mastermind strategy. Journal of Recreational Mathematics 11(2), 81–87 (1978-1979)
[5] Kendall, G., Parkes, A., Spoerer, K.: A survey of NP-complete puzzles. ICGA Journal 31(1), 13–34 (2008),
http://www.scopus.com/inward/
record.url?eid=2-s2.0-42949163946&partnerID=40
(Cited By (Since 1996) 1)
[6] Knuth, D.E.: The computer as Master Mind. J. Recreational Mathematics 9(1), 1–6 (1976-1977)
[7] Kooi, B.: Yet another Mastermind strategy. ICGA Journal 28(1), 13–20 (2005),
http://www.scopus.com/inward/
record.url?eid=2-s2.0-33646756877&partnerID=40
[8] Koyama, K., Lai, T.W.: An optimal Mastermind strategy. J. Recreational Mathematics 25(4) (1993/1994)
[9] Merelo-Guervós, J.J., Carpio, J., Castillo, P., Rivas, V.M., Romero, G.: Finding a needle in a haystack using hints and evolutionary computation: the case of genetic mastermind. In: Scott Brave, A.S.W. (ed.) Late breaking papers at the GECCO 1999, pp. 184–192 (1999)

[10] Merelo-Guervós, J., Castillo, P., Rivas, V.: Finding a needle in a haystack using hints and evolutionary computation: the case of evolutionary MasterMind. Applied Soft Computing 6(2), 170–179 (2006),
http://www.sciencedirect.com/science/article/
B6W86-4FH0D6P-1/2/40a99afa8e9c7734baae340abecc113a,
http://dx.doi.org/10.1016/j.asoc.2004.09.003

[11] Merelo-Guervós, J.J., Castillo, P.A., Alba, E.: ALGORITHM::EVOLUTIONARY, a flexible Perl module for evolutionary computation. Soft Computing - A Fusion of Foundations, Methodologies and Applications (2010), doi 10.1007/s00500-009-0504-3,
http://www.springerlink.com/content/8h025g83j0q68270

[12] Mühlenbein, H., Paass, G.: From recombination of genes to the estimation of distributions: I. binary parameters. In: Ebeling, W., Rechenberg, I., Voigt, H.-M., Schwefel, H.-P. (eds.) PPSN 1996. LNCS, vol. 1141, pp. 178–187. Springer, Heidelberg (1996)

[13] Neuwirth, E.: Some strategies for mastermind. Zeitschrift fur Operations Research. Serie B **26**(8), B257–B278 (1982)

[14] Stuckman, J., Zhang, G.Q.: Mastermind is np-complete. CoRR **abs/cs/0512049** (2005)

[15] Wikipedia: Mastermind (board game) — Wikipedia, The Free Encyclopedia (2009),
http://en.wikipedia.org/w/
index.php?title=Mastermind_board_game&oldid=317686771
(Online: accessed October 9, 2009)

Structural versus Evaluation Based Solutions Similarity in Genetic Programming Based System Identification

Stephan M. Winkler

Abstract. Estimating the similarity of solution candidates represented as structure trees is an important point in the context of many genetic programming (GP) applications. For example, when it comes to observing population diversity dynamics, solutions have to be compared to each other. In the context of GP based system identification, i.e., when mathematical expressions are evolved, solutions can be compared to each other with respect to their structure as well as to their evaluation. Obviously, structural similarity estimation of formula trees is not equivalent to evaluation based similarity estimation; we here want to see whether there is a significant correlation between the results calculated using these two approaches. In order to get an overview regarding this issue, we have analyzed a series of GP tests including both similarity estimation strategies; in this paper we describe the similarity estimation methods as well as the test data sets used in these tests, and we document the results of these tests. We see that in most cases there is a significant positive linear correlation for the results returned by the evaluation based and structural methods. Especially in some cases showing very low structural similarity there can be significantly different results when using the evaluation based similarity methods.

1 Solutions Similarity Estimation in GP Based System Identification

1.1 Related Work

Genetic diversity and population dynamics are very interesting aspects in the analysis of genetic programming (GP, [7, 8]) processes; several methods for measuring

Stephan M. Winkler
Department for Medical and Bioinformatics, Upper Austria University of Applied Sciences,
Heuristic and Evolutionary Algorithms Laboratory
e-mail: stephan.winkler@fh-hagenberg.at

J.R. González et al. (Eds.): NICSO 2010, SCI 284, pp. 269–282, 2010.
springerlink.com

the diversity of population using some kind of similarity measure can be found in the literature.

The entropy of a population of trees can be measured for example by considering the programs' scores (as is explained for example in [13]); in [10] the traditional fitness sharing concept from the work described in [4] is applied to test its feasibility in GP.

Several other approaches consider the programs' genotypes, i.e., their genetic make-up instead of their fitness values, the most common type of diversity measure being that of structural differences between programs. Koza [7] used the term variety to indicate the number of different programs in populations by comparing programs structurally and looking for exact matches. The Levenshtein distance [9] can be used for calculating the distance between trees, but it is considered rather far from ideal ([6], [12], [8]); in [5] an edit distance specific to genetic programming parse trees was presented which considered the cost of substituting between different node types.

A comprehensive overview of program tree similarity and diversity measures has been given for instance in [3]. The standard tree structures representation in GP makes it possible to use more fine grain structural measures that consider nodes, subtrees, and other graph theoretic properties (rather than just entire trees). In [6], for example, subtree variety is measured as the ratio of unique subtrees over total subtrees and program variety as a ratio of the number of unique individuals over the size of the population; [11] investigated diversity at the genetic level by assigning numerical tags to each node in the population.

1.2   Solutions Similarity Estimation Measures Used in This Work

In this section we describe measures which we have used for estimating the genetic diversity in GP populations as well as among populations of multi-population GP applications. What we use as basic measures for this are the following two functions that calculate the similarity of GP solution candidates or, a bit more specific, in our case formulas represented as structure trees:

- *Evaluation based* similarity estimation compares the subtrees of two GP formulas with respect to their evaluation on the given training or validation data. The more similar these evaluations are with respect to the squared errors or linear correlation, the higher is the similarity for these two formulas.
- *Structural* similarity estimation compares the genetic material of two solution candidates; we can so determine how similar the genetic make-up of formulas is without considering their evaluation.

As documented for example in [15] and [2], these similarity estimation measures can be used for monitoring population diversity in GP populations. We have analyzed the effects of the use of several different selection schemes as well as multi-population approaches. Please note that in these applications we use similarity estimation in the following way: The similarity measures used here are asymmetric, so when comparing structure trees T_1 and T_2 there might be a difference between the similarities $sim(T_1, T_2)$ and $sim(T_2, T_1)$. This is why we mostly use a symmetric variant of the measures described here: We calculate both similarity values and calculate their average as $sim_{avg}(T_1, T_2) = \frac{sim(T_1, T_2) + sim(T_2, T_1)}{2}$. This average similarity function (sim_{avg}) is used for estimating the similarities of GP individuals and monitoring the progress of genetic diversity in GP populations.

1.3 Evaluation Based Solutions Similarity Estimation

The main idea of our evaluation based similarity measures is that the building blocks of GP formulas are subtrees that are exchanged by crossover and so form new formulas. So, the evaluation of these branches of all individuals in a GP population can be used for measuring the similarity of two models m_1 and m_2:

For all sub-trees in the structure-tree of model m, collected in t, we collect the evaluation results by applying these sub-formulas to the given data collection *data* as

$$\forall (st_i \in t) \forall (j \in [1; N]) : e_i[j] = eval(st_i, data[j]) \tag{1}$$

where N is the number of samples included in the data collection, no matter if training or validation data are considered.

The evaluation based similarity of models m_1 and m_2, $es(m_1, m_2)$, is calculated by iterating over all subtrees of m_1 (collected in t_1) and, for each branch, picking that subtree of t_2 (containing all sub-trees of m_2) whose evaluation is most "similar" to the evaluation of that respective branch. So, for each branch b_a in t_1 we compare its evaluation e_a with the evaluation e_b of all branches b_b in t_2, and the "similarity" can be calculated using the sum of squared errors or the linear correlation coefficient:

- When using the sum of squared errors (*sse*) function, the sample-wise differences of the evaluations of the two given branches are calculated and their sum of squared differences is divided by the total sum of squares *tss* of the first branch's evaluation. This results in the similarity measure s for the given branches.

$$\overline{e_a} = \frac{1}{N} \sum_{j=1}^{N} e_a[j]; \ \overline{e_b} = \frac{1}{N} \sum_{j=1}^{N} e_b[j] \tag{2}$$

$$sse = \sum_{j=1}^{N} (e_a[j] - e_b[j])^2; \ tss = \sum_{j=1}^{N} (e_a[j] - \overline{e_a})^2; \ s_{sse}(b_a, b_b) = 1 - \frac{sse}{tss} \tag{3}$$

- Alternatively the linear correlation coefficient can be used:

$$s_{lc}(b_a, b_b) = \left| \frac{\frac{1}{n-1} \sum_{j=1}^{N} (e_a[j] - \overline{e_a})(e_b[j] - \overline{e_b})}{\sqrt{\frac{1}{n-1} \sum_{j=1}^{N} (e_a[j] - \overline{e_a})^2} \sqrt{\frac{1}{n-1} \sum_{j=1}^{N} (e_b[j] - \overline{e_b})^2}} \right| \tag{4}$$

No matter which approach is chosen, the calculated similarity measure for the branches b_a and b_b, $s(b_a, b_b)$, will always be in the interval $[0; 1]$; the higher this value becomes, the smaller is the difference between the evaluation results.

272   S.M. Winkler

As we can now quantify the similarity of evaluations of two given subtrees, for each branch b_a in t_a we can elicit that branch b_x in t_b with the highest similarity to b_a; the similarity values **s** are collected for all branches in t_a and their mean value finally gives us a measure for the evaluation based similarity of the models m_a and m_b, $es(m_a, m_b)$.

Optionally we can force the algorithm to select each branch in t_b not more than once as best match for a branch in t_a for preventing multiple contributions of certain parts of the models.

Finally, this similarity function can be parameterized by giving minimum and maximum bounds for the height and / or the level of the branches investigated. This is important since we can so control which branches are to be compared, be it the rather small ones, rather big ones or all of them.

Further details about this similarity measure can be found in [15].

1.4 Structural Solutions Similarity Estimation

Structural similarity estimation is, unlike the evaluation based method described before, independent of data; it is calculated on the basis of the genetic make-up of the models which are to be compared. When analyzing the structure of models we have to be aware of the fact that often structurally different models can be equivalent. This is why we have designed and implemented a method that systematically collects all pairs of ancestor and descendant nodes and information about the properties of these nodes. Additionally, for each pair we also document the distance (with respect to the level in the model tree) and the index of the ancestor's child tree containing the descendant node. The similarity of two models is then, in analogy to the method described in the previous section, calculated by comparing all pairs of ancestors and descendants in one model to all pairs of the other model and averaging the similarity of the respective best matches.

Figure 1 shows a simple formula and all pairs of ancestors and descendants included in the structure tree representing it; the input indices as well as the level differences ("level delta") are also given. Please note: The pairs given on the right side of Figure 1 are shown intentionally as they symbolize the pairs of nodes with level difference 0, i.e., nodes combined with themselves.

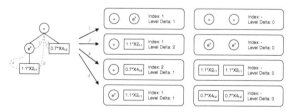

Fig. 1 Simple formula structure and all included pairs of ancestors and descendants (genetic information items)

We define a *genetic item* as a 6-tuple storing the following information about the ancestor node a and descendant node d: $type_a$ (the type of the ancestor a), $type_d$ (the type of the descendant d), δl (the level delta), *index* (the index of the child branch of a that includes d), np_a (the node parameters characterizing a), and np_d (the node parameters characterizing d); the parameters characterizing nodes are represented by tuples containing the following information: *var* (the variant (of functions)), *coeff* (the coefficient (of terminals)), *to* (the time offset (of terminals)), and *vi* (the variable index (of terminals)).

Now we can define the similarity of two *genetic items* gi_1 and gi_2, $s(gi_1, gi_2)$, as follows: Most important are the types of the definitions referenced by the nodes; if these are not equal, then the similarity is 0 regardless of all other parameters. If the types of the nodes correspond correctly, then the similarity of gi_1 and gi_2 is calculated using the similarity contributors $s_1 \ldots s_{10}$ of the parameters of gi_1 and gi_2 weighted with coefficients $c_1 \ldots c_{10}$. The similarity contributors $s_1 \ldots s_{10}$, all ranging from 0.0 to 1.0, are calculated with respect to input indices, variants, variable indices, level differences, coefficients, and time offsets; details can be found in [15] and [2].

Finally, there are two possibilities how to calculate the structural similarity of gi_1 and gi_2, $sim(gi_1, gi_2)$: On the one hand this can be done in an *additive* way, on the other hand in a *multiplicative* way.

- When using the *additive* calculation, which is the obviously more simple way, $sim(gi_1, gi_2)$ is calculated as the sum of these similarity contributions $s_{1\ldots10}$ weighted using the factors $c_{1\ldots10}$ and, for the sake of normalization of results, divided by the sum of the weighting factors:

$$sim_{add}(gi_1, gi_2) = \frac{\sum_{i=1}^{10} s_i \cdot c_i}{\sum_{i=1}^{10} c_i}. \tag{5}$$

- Otherwise, when using the *multiplicative* calculation method, we first calculate a punishment factor p_i for each s_i (again using weighting factors c_i, $0 \le c_i \le$ for all $i \in [1; 10]$) as $\forall (i \in [1; 10]) : p_i = (1 - s_i) \cdot c_i$ and then get the temporary similarity result as $sim_{tmp}(gi_1, gi_2) = \prod_{i=1}^{10} (1 - p_i)$.

  In the worst case scenario we get $s_i = 0$ for all $i \in [1; 10]$ and therefore the worst possible sim_{tmp} is $sim_{worst} = \prod_{i=1}^{10} (1 - ((1 - s_i) \cdot c_i)) = \prod_{i=1}^{10} (1 - c_i)$. As sim_{worst} is surely greater than 0 we linearly scale the results to the interval $[0; 1]$:

$$sim_{mult}(gi_1, gi_2) = \frac{sim_{tmp}(gi_1, gi_2) - sim_{worst}}{1 - sim_{worst}}. \tag{6}$$

In fact, we prefer this *multiplicative* similarity calculation method since it allows more specific analysis: By setting a weighting coefficient c_j to a rather high value (i.e., near or even equal to 1.0) the total similarity will become very small for pairs of genetic items that do not correspond with respect to this specific aspect j, even if all other aspects would lead to a high similarity result.

Based on this similarity measure it is easy to formulate a similarity function that measures the structural similarity of two model structures. In analogy to the approach presented in the previous section, for comparing models m_1 and m_2 we collect all pairs of ancestors and descendants (up to a given maximum level difference) in m_1 and m_2 and look for the best matches in the respective opposite model's pool of genetic items, i.e., pairs of ancestor and descendant nodes. As we are able to quantify the similarity of genetic items, for each genetic item gi_1 in the structure tree of m_1 we can elicit exactly that genetic item gi_x in the model structure m_2 with the highest similarity to gi_1; the similarity values s are collected for all genetic items contained in m_1 and their mean value finally gives us a measure for the structure-based similarity of the models m_1 and m_2, $sim(m_1, m_2)$.

2 Test Setup

For comparing structural and evaluation based similarity values we executed GP based system identification experiments using the following two data sets:

- The NO_x data set contains the measurements taken from a 2 liter 4 cylinder BMW diesel engine at a dynamical test bench (simulated vehicle: BMW 320d Sedan). Several emissions (including NO_x, CO and CO_2) as well as several other engine parameters were recorded; for identifying formulas for the NO_x emissions we have only used parameters which are directly measured from the engine's control unit and not in any sense connected to emissions. We cordially thank members of the Institute for Design and Control of Mechatronical Systems at JKU, Linz[1] who provided and helped us with these data.
- The *Thyroid* data set is a widely used machine learning benchmark data set containing 21 attributes and 7200 samples representing the results of medical measurements which were recorded while investigating patients potentially suffering from hypotiroidism[2]. In short, the task is to determine whether a patient is hypothyroid or not; three classes are formed: normal (not hypothyroid), hyperfunction and subnormal functioning.

Detailed information about these two data collections can also be found in [15] as well as in [2].

For the target variables of both data collections we trained nonlinear models using a functional basis containing standard functions (such as for example addition, subtraction, multiplication, trigonometrics, conditionals, and others) as described in [16]; the maximum formula tree height was set to 6, the maximum number of nodes was set to 50. We have used the GP implementation for HeuristicLab [14] and applied two different training methods for training models for both data sets: Standard GP as well as GP using strict offspring selection (OS, [1]). In both cases the

---

[1] The homepage of the Institute for Design and Control of Mechatronical Systems at the Johannes Kepler University, Linz can be found at http://desreg.jku.at/

[2] Further information about the data set used can be found on the UCI homepage http://www.ics.uci.edu/~{}mlearn/

population size was set to 1000, we used single point crossover and 15% structural as well as parametric mutation as described in [15], e.g; in standard GP we applied tournament selection ($k = 3$), in GP with OS we applied gender specific parents selection combining random and proportional selection. For standard GP processes the number of iterations was set to 2000, GP runs with offspring selection were terminated as soon as the selection pressure reached 100.

All test cases were executed three times independently; the maximum tree height was set to 6, the maximum tree size to 50 (for NO_x as well as *Thyroid* tests). The similarity values among individuals were calculated in the context of population diversity estimation analysis executed after every 100th generation in standard GP runs and after each 5th generation in GP runs with offspring selection. We have thus collected the results of all similarity calculations; as this is done for 1,000 models we get 1,000,000 for each similarity function each time the population is analyzed. For each standard GP test we therefore eventually get 21 million similarity values for each function (because we also analyze after initializing the population), and for each GP test with OS we get a comparable amount of similarity values[3]. We will in the following not care whether standard or extended GP produced pairs of solutions are compared; in total we will use data of approximately 120 million solution comparisons for each function and each data set.

The following similarity estimation functions are used:

- Evaluation based similarity estimation: As described in Section 1.3, all subtrees are evaluated on training and validation data, and we can analyze the similarity of the values calculated by evaluating the subtrees of the formula trees which are to be compared. We here use validation data for this similarity estimation and the squared differences based approach.
- Additive structural similarity estimation: Structural components of structure trees are analyzed as described in Section 1.4 using the additive approach; we here weight all possible contributing aspects equally, i.e. the contributions' weighting factors $c_{1...10}$ are all set to 1.0, only the level difference is weighted stronger with factor 4.0.
- Multiplicative structural similarity estimation: Again, structural components of structure trees are analyzed as described in Section 1.4 using the multiplicative approach; again, we set all weighting factors equally, namely to 0.2, only the level difference is weighted stronger with factor 0.8.

3   Test Results

The NO_x test series are hereafter referred to as series *(n)*, the *Thyroid* runs as *(t)*. The similarity values calculated for the *(n)* series using evaluation based, additive structural and multiplicative structural comparison are hereafter denoted as $\mathbf{n_e}$, $\mathbf{n_{s1}}$ and $\mathbf{n_{s2}}$, respectively; in analogy to this, the similarity values for the *(t)* series are denoted as $\mathbf{t_e}$, $\mathbf{t_{s1}}$ and $\mathbf{t_{s2}}$, respectively.

---

[3] This number is not constant for extended GP with OS due to the fact that the selection pressure reaches its limit not at the same time in each test case execution.

Please note that for each index i the values $n_e(i)$, $n_{s1}(i)$ and $n_{s2}(i)$ belong to the same pair of models (structure trees) that have been compared; in analogy to this, for each index i also the corresponding comparison results $t_e(i)$, $t_{s1}(i)$ and $t_{s2}(i)$ are associated to the same pair of formulas.

All test runs were executed on Pentium$^{\copyright}$ 4 computers with 3.00 GHz CPU speed and 2 GB RAM.

First, several statistics are calculated for the similarity values collected in $\mathbf{n_e}$, $\mathbf{n_{s1}}$, $\mathbf{n_{s2}}$, $\mathbf{t_e}$, $\mathbf{t_{s1}}$ and $\mathbf{t_{s2}}$; N_n stands for the number of values in $\mathbf{n_e}$, $\mathbf{n_{s1}}$ and $\mathbf{n_{s2}}$, N_t for the number of values in $\mathbf{t_e}$, $\mathbf{t_{s1}}$ and $\mathbf{t_{s2}}$. The results are summarized in Table 1; std here stands for standard deviation ($std(\mathbf{x}) = \sqrt{\frac{1}{N}\sum_{i\in[1;N]}(x_i - \bar{x})^2}$, $\bar{x} = \frac{1}{N}\sum_{i\in[1;N]}x$, $N = |x|$), and $corr$ again for the linear correlation (please see for example Section 1.3 for details about this function).

Table 1 Comparing similarity estimation results: Basic statistics

$mean(\mathbf{n_e}) = \frac{1}{N_n}\sum_{i\in[1;N_n]}(n_e(i))$	0.3444	$std(\mathbf{n_e} - \mathbf{n_{s1}})$	0.1625
$mean(\mathbf{n_{s1}}) = \frac{1}{N_n}\sum_{i\in[1;N_n]}(n_{s1}(i))$	0.6467	$std(\mathbf{n_e} - \mathbf{n_{s2}})$	0.1500
$mean(\mathbf{n_{s2}}) = \frac{1}{N_n}\sum_{i\in[1;N_n]}(n_{s2}(i))$	0.6061	$std(\mathbf{n_{s1}} - \mathbf{n_{s2}})$	0.0268
$mean(\mathbf{t_e}) = \frac{1}{N_n}\sum_{i\in[1;N_n]}(t_e(i))$	0.4224	$std(\mathbf{t_e} - \mathbf{t_{s1}})$	0.2159
$mean(\mathbf{t_{s1}}) = \frac{1}{N_t}\sum_{i\in[1;N_t]}(t_{s1}(i))$	0.6595	$std(\mathbf{t_e} - \mathbf{t_{s2}})$	0.1992
$mean(\mathbf{t_{s2}}) = \frac{1}{N_t}\sum_{i\in[1;N_t]}(t_{s2}(i))$	0.6327	$std(\mathbf{t_{s1}} - \mathbf{t_{s2}})$	0.0305
$mse(\mathbf{n_e},\mathbf{n_{s1}}) = \frac{1}{N_n}\sum_{i\in[1;N_n]}(n_e(i)-n_{s1}(i))^2$	0.1178	$corr(\mathbf{n_e},\mathbf{n_{s1}})$	0.8179
$mse(\mathbf{n_e},\mathbf{n_{s2}}) = \frac{1}{N_n}\sum_{i\in[1;N_n]}(n_e(i)-n_{s2}(i))^2$	0.0910	$corr(\mathbf{n_e},\mathbf{n_{s2}})$	0.8455
$mse(\mathbf{n_{s1}},\mathbf{n_{s2}}) = \frac{1}{N_n}\sum_{i\in[1;N_n]}(n_{s1}(i)-n_{s2}(i))^2$	0.0024	$corr(\mathbf{n_{s1}},\mathbf{n_{s2}})$	0.9954
$mse(\mathbf{t_e},\mathbf{t_{s1}}) = \frac{1}{N_t}\sum_{i\in[1;N_t]}(t_e(i)-t_{s1}(i))^2$	0.1028	$corr(\mathbf{t_e},\mathbf{t_{s1}})$	0.7634
$mse(\mathbf{t_e},\mathbf{t_{s2}}) = \frac{1}{N_t}\sum_{i\in[1;N_t]}(t_e(i)-t_{s2}(i))^2$	0.0839	$corr(\mathbf{t_e},\mathbf{t_{s2}})$	0.7998
$mse(\mathbf{t_{s1}},\mathbf{t_{s2}}) = \frac{1}{N_n}\sum_{i\in[1;N_n]}(t_{s1}(i)-t_{s2}(i))^2$	0.0016	$corr(\mathbf{t_{s1}},\mathbf{t_{s2}})$	0.9947
Runtime consumption per generation (evaluation based similarity)			2h08'30"
Runtime consumption per generation (structural similarity, per method)			38'02"

Obviously, the structural similarity values tend to be a lot higher than the evaluation based ones – which is not really surprising as even small changes in the formula's structure can affect its evaluation significantly. The mean squared difference between structural and evaluation based similarity values ranges from ∼0.08 to ∼0.12; the respective standard deviations of the similarity differences range from 0.15 to ∼0.216. The much more informative statistic feature is the linear correlation coefficient: Analyzing NO_x tests we see that the correlation between structural and evaluation based similarities is between ∼0.82 (for the additive structural calculation) and ∼0.8455 (for multiplicative structural approach); for the *Thyroid* tests, these are not quite as high, namely ∼0.76 and ∼0.8, respectively.

As we had expected, the correlation between the results calculated using the additive structural model comparison method and the multiplicative one is very high, namely approximately 0.995 for NO_x as well as *Thyroid* tests.

The runtime consumption of the evaluation based similarity estimation method is, of course, a lot higher than the runtime consumption caused by structural population diversity analysis: Although only 400 validation samples are evaluated for evaluation based similarity estimation, structural similarity calculation consumes only approximately a fourth as much runtime.

Even more detailed results discussion becomes possible by partitioning all pairs of corresponding similarity values into five groups with equal range. This means that we collect all structural similarity results in the intervals $[0.0 \ldots 0.2]$, $[0.2 \ldots 0.4]$, $\ldots$, $[0.8 \ldots 1.0]$; of course, we also collect all evaluation based similarity values in the same intervals. Thus, what we get is a number of partitions of data sets which are defined and summarized in Table 2.

Table 2 Partitions formed for detailed comparison of similarity estimation results

Partition Index	Index and Data Set Definitions
a0	$I_{a0} = \{i : (0.0 \le n_e(i) \le 0.2)\}; n_e^{a0} = n_e(I_{a0}), n_{s1}^{a0} = n_{s1}(I_{a0}), n_{s2}^{a0} = n_{s2}(I_{a0})$
a1	$I_{a1} = \{i : (0.2 < n_e(i) \le 0.4)\}; n_e^{a1} = n_e(I_{a1}), n_{s1}^{a1} = n_{s1}(I_{a1}), n_{s2}^{a1} = n_{s2}(I_{a1})$
a2	$I_{a1} = \{i : (0.4 < n_e(i) \le 0.6)\}; n_e^{a2} = n_e(I_{a2}), n_{s1}^{a2} = n_{s1}(I_{a2}), n_{s2}^{a2} = n_{s2}(I_{a2})$
a3	$I_{a1} = \{i : (0.6 < n_e(i) \le 0.8)\}; n_e^{a3} = n_e(I_{a3}), n_{s1}^{a3} = n_{s1}(I_{a3}), n_{s2}^{a3} = n_{s2}(I_{a3})$
a4	$I_{a1} = \{i : (0.8 < n_e(i) \le 1.0)\}; n_e^{a4} = n_e(I_{a4}), n_{s1}^{a4} = n_{s1}(I_{a4}), n_{s2}^{a4} = n_{s2}(I_{a4})$
b0	$I_{b0} = \{i : (0.0 \le n_{s1}(i) \le 0.2)\}; n_e^{b0} = n_e(I_{b0}), n_{s1}^{b0} = n_{s1}(I_{b0}), n_{s2}^{b0} = n_{s2}(I_{b0})$
b1	$I_{b1} = \{i : (0.2 < n_{s1}(i) \le 0.4)\}; n_e^{b1} = n_e(I_{b1}), n_{s1}^{b1} = n_{s1}(I_{b1}), n_{s2}^{b1} = n_{s2}(I_{b1})$
b2	$I_{b2} = \{i : (0.4 < n_{s1}(i) \le 0.6)\}; n_e^{b2} = n_e(I_{b2}), n_{s1}^{b2} = n_{s1}(I_{b2}), n_{s2}^{b2} = n_{s2}(I_{b2})$
b3	$I_{b3} = \{i : (0.6 < n_{s1}(i) \le 0.8)\}; n_e^{b3} = n_e(I_{b3}), n_{s1}^{b3} = n_{s1}(I_{b3}), n_{s2}^{b3} = n_{s2}(I_{b3})$
b4	$I_{b4} = \{i : (0.8 < n_{s1}(i) \le 1.0)\}; n_e^{b4} = n_e(I_{b4}), n_{s1}^{b4} = n_{s1}(I_{b4}), n_{s2}^{b4} = n_{s2}(I_{b4})$
c0	$I_{c0} = \{i : (0.0 \le n_{s2}(i) \le 0.2)\}; n_e^{c0} = n_e(I_{c0}), n_{s1}^{c0} = n_{s1}(I_{c0}), n_{s2}^{c0} = n_{s2}(I_{c0})$
c1	$I_{c1} = \{i : (0.2 < n_{s2}(i) \le 0.4)\}; n_e^{c1} = n_e(I_{c1}), n_{s1}^{c1} = n_{s1}(I_{c1}), n_{s2}^{c1} = n_{s2}(I_{c1})$
c2	$I_{c2} = \{i : (0.4 < n_{s2}(i) \le 0.6)\}; n_e^{c2} = n_e(I_{c2}), n_{s1}^{c2} = n_{s1}(I_{c2}), n_{s2}^{c2} = n_{s2}(I_{c2})$
c3	$I_{c3} = \{i : (0.6 < n_{s2}(i) \le 0.8)\}; n_e^{c3} = n_e(I_{c3}), n_{s1}^{c3} = n_{s1}(I_{c3}), n_{s2}^{c3} = n_{s2}(I_{c3})$
c4	$I_{c4} = \{i : (0.8 < n_{s2}(i) \le 1.0)\}; n_e^{c4} = n_e(I_{c4}), n_{s1}^{c4} = n_{s1}(I_{c4}), n_{s2}^{c4} = n_{s2}(I_{c4})$
d0	$I_{d0} = \{i : (0.0 \le t_e(i) \le 0.2)\}; t_e^{d0} = t_e(I_{d0}), t_{s1}^{d0} = t_{s1}(I_{d0}), t_{s2}^{d0} = t_{s2}(I_{d0})$
d1	$I_{d1} = \{i : (0.2 < t_e(i) \le 0.4)\}; t_e^{d1} = t_e(I_{d1}), t_{s1}^{d1} = t_{s1}(I_{d1}), t_{s2}^{d1} = t_{s2}(I_{d1})$
d2	$I_{d1} = \{i : (0.4 < t_e(i) \le 0.6)\}; t_e^{d2} = t_e(I_{d2}), t_{s1}^{d2} = t_{s1}(I_{d2}), t_{s2}^{d2} = t_{s2}(I_{d2})$
d3	$I_{d1} = \{i : (0.6 < t_e(i) \le 0.8)\}; t_e^{d3} = t_e(I_{d3}), t_{s1}^{d3} = t_{s1}(I_{d3}), t_{s2}^{d3} = t_{s2}(I_{d3})$
d4	$I_{d1} = \{i : (0.8 < t_e(i) \le 1.0)\}; t_e^{d4} = t_e(I_{d4}), t_{s1}^{d4} = t_{s1}(I_{d4}), t_{s2}^{d4} = t_{s2}(I_{d4})$
e0	$I_{e0} = \{i : (0.0 \le t_{s1}(i) \le 0.2)\}; t_e^{e0} = t_e(I_{e0}), t_{s1}^{e0} = t_{s1}(I_{e0}), t_{s2}^{e0} = t_{s2}(I_{e0})$
e1	$I_{e1} = \{i : (0.2 < t_{s1}(i) \le 0.4)\}; t_e^{e1} = t_e(I_{e1}), t_{s1}^{e1} = t_{s1}(I_{e1}), t_{s2}^{e1} = t_{s2}(I_{e1})$
e2	$I_{e2} = \{i : (0.4 < t_{s1}(i) \le 0.6)\}; t_e^{e2} = t_e(I_{e2}), t_{s1}^{e2} = t_{s1}(I_{e2}), t_{s2}^{e2} = t_{s2}(I_{e2})$
e3	$I_{e3} = \{i : (0.6 < t_{s1}(i) \le 0.8)\}; t_e^{e3} = t_e(I_{e3}), t_{s1}^{e3} = t_{s1}(I_{e3}), t_{s2}^{e3} = t_{s2}(I_{e3})$
e4	$I_{e4} = \{i : (0.8 < t_{s1}(i) \le 1.0)\}; t_e^{e4} = t_e(I_{e4}), t_{s1}^{e4} = t_{s1}(I_{e4}), t_{s2}^{e4} = t_{s2}(I_{e4})$
f0	$I_{f0} = \{i : (0.0 \le t_{s2}(i) \le 0.2)\}; t_e^{f0} = t_e(I_{f0}), t_{s1}^{f0} = t_{s1}(I_{f0}), t_{s2}^{f0} = t_{s2}(I_{f0})$
f1	$I_{f1} = \{i : (0.2 < t_{s2}(i) \le 0.4)\}; t_e^{f1} = t_e(I_{f1}), t_{s1}^{f1} = t_{s1}(I_{f1}), t_{s2}^{f1} = t_{s2}(I_{f1})$
f2	$I_{f2} = \{i : (0.4 < t_{s2}(i) \le 0.6)\}; t_e^{f2} = t_e(I_{f2}), t_{s1}^{f2} = t_{s1}(I_{f2}), t_{s2}^{f2} = t_{s2}(I_{f2})$
f3	$I_{f3} = \{i : (0.6 < t_{s2}(i) \le 0.8)\}; t_e^{f3} = t_e(I_{f3}), t_{s1}^{f3} = t_{s1}(I_{f3}), t_{s2}^{f3} = t_{s2}(I_{f3})$
f4	$I_{f4} = \{i : (0.8 < t_{s2}(i) \le 1.0)\}; t_e^{f4} = t_e(I_{f4}), t_{s1}^{f4} = t_{s1}(I_{f4}), t_{s2}^{f4} = t_{s2}(I_{f4})$

Now we can analyze these partitions separately: For each partition we have calculated the linear correlation between evaluation based, additive structural and multiplicative structural similarities as well as the mean squared difference between these respective values; Table 3 summarizes these partition-wise statistics. Additionally, the frequency of each partition is also given: The frequency of a partition is hereby given by the number of pairs of values included divided by the number of all pairs of values available, $frequ(I_{ki}) = \frac{|I_{ki}|}{\sum_{j \in [0;4]} I_{kj}}$ for $k \in \{a,b,c,d,e,f\}$ and $i \in [0;4]$.

Table 3 Comparing similarity estimation results: Detailed partition-wise statistics

$freq(I_{a0})=0.3172$	$corr(n_e^{a0},n_{s1}^{a0})=0.6294$	$corr(n_e^{a0},n_{s2}^{a0})=0.6772$	$freq(I_{a1})=0.2609$	$corr(n_e^{a1},n_{s1}^{a1})=0.8407$	$corr(n_e^{a1},n_{s2}^{a1})=0.8574$
	$mse(n_e^{a0},n_{s1}^{a0})=0.1061$	$mse(n_e^{a0},n_{s2}^{a0})=0.0751$		$mse(n_e^{a1},n_{s1}^{a1})=0.1083$	$mse(n_e^{a1},n_{s2}^{a1})=0.0818$
$freq(I_{a2})=0.2595$	$corr(n_e^{a2},n_{s1}^{a2})=0.7886$	$corr(n_e^{a2},n_{s2}^{a2})=0.8047$	$freq(I_{a3})=0.1272$	$corr(n_e^{a3},n_{s1}^{a3})=0.6963$	$corr(n_e^{a3},n_{s2}^{a3})=0.7376$
	$mse(n_e^{a2},n_{s1}^{a2})=0.1364$	$mse(n_e^{a2},n_{s2}^{a2})=0.1106$		$mse(n_e^{a3},n_{s1}^{a3})=0.1279$	$mse(n_e^{a3},n_{s2}^{a3})=0.1077$
$freq(I_{a4})=0.0352$	$corr(n_e^{a4},n_{s1}^{a4})=0.7174$	$corr(n_e^{a4},n_{s2}^{a4})=0.7559$			
	$mse(n_e^{a4},n_{s1}^{a4})=0.1184$	$mse(n_e^{a4},n_{s2}^{a4})=0.0983$			
$freq(I_{b0})=0.0974$	$corr(n_{s1}^{b0},n_e^{b0})=0.3815$	$corr(n_{s1}^{b0},n_{s2}^{b0})=0.9890$	$freq(I_{b1})=0.1222$	$corr(n_{s1}^{b1},n_e^{b1})=0.6744$	$corr(n_{s1}^{b1},n_{s2}^{b1})=0.9931$
	$mse(n_{s1}^{b0},n_e^{b0})=0.1407$	$mse(n_{s1}^{b0},n_{s2}^{b0})=0.0057$		$mse(n_{s1}^{b1},n_e^{b1})=0.0884$	$mse(n_{s1}^{b1},n_{s2}^{b1})=0.0028$
$freq(I_{b2})=0.1363$	$corr(n_{s1}^{b2},n_e^{b2})=0.7591$	$corr(n_{s1}^{b2},n_{s2}^{b2})=0.9962$	$freq(I_{b3})=0.2451$	$corr(n_{s1}^{b3},n_e^{b3})=0.8350$	$corr(n_{s1}^{b3},n_{s2}^{b3})=0.9963$
	$mse(n_{s1}^{b2},n_e^{b2})=0.0985$	$mse(n_{s1}^{b2},n_{s2}^{b2})=0.0026$		$mse(n_{s1}^{b3},n_e^{b3})=0.1080$	$mse(n_{s1}^{b3},n_{s2}^{b3})=0.0024$
$freq(I_{b4})=0.3990$	$corr(n_{s1}^{b4},n_e^{b4})=0.7677$	$corr(n_{s1}^{b4},n_{s2}^{b4})=0.9975$			
	$mse(n_{s1}^{b4},n_e^{b4})=0.1337$	$mse(n_{s1}^{b4},n_{s2}^{b4})=0.0013$			
$freq(I_{c0})=0.1160$	$corr(n_{s2}^{c0},n_e^{c0})=0.4119$	$corr(n_{s2}^{c0},n_{s1}^{c0})=0.9888$	$freq(I_{c1})=0.1335$	$corr(n_{s2}^{c1},n_e^{c1})=0.7667$	$corr(n_{s2}^{c1},n_{s1}^{c1})=0.9961$
	$mse(n_{s2}^{c0},n_e^{c0})=0.0997$	$mse(n_{s2}^{c0},n_{s1}^{c0})=0.0059$		$mse(n_{s2}^{c1},n_e^{c1})=0.0580$	$mse(n_{s2}^{c1},n_{s1}^{c1})=0.0023$
$freq(I_{c2})=0.1584$	$corr(n_{s2}^{c2},n_e^{c2})=0.8229$	$corr(n_{s2}^{c2},n_{s1}^{c2})=0.9963$	$freq(I_{c3})=0.2728$	$corr(n_{s2}^{c3},n_e^{c3})=0.8764$	$corr(n_{s2}^{c3},n_{s1}^{c3})=0.9967$
	$mse(n_{s2}^{c2},n_e^{c2})=0.0730$	$mse(n_{s2}^{c2},n_{s1}^{c2})=0.0027$		$mse(n_{s2}^{c3},n_e^{c3})=0.0794$	$mse(n_{s2}^{c3},n_{s1}^{c3})=0.0021$
$freq(I_{c4})=0.3193$	$corr(n_{s2}^{c4},n_e^{c4})=0.7528$	$corr(n_{s2}^{c4},n_{s1}^{c4})=0.9969$			
	$mse(n_{s2}^{c4},n_e^{c4})=0.1205$	$mse(n_{s2}^{c4},n_{s1}^{c4})=0.0011$			
$freq(I_{d0})=0.3241$	$corr(t_e^{d0},t_{s1}^{d0})=0.4233$	$corr(t_e^{d0},t_{s2}^{d0})=0.4777$	$freq(I_{d1})=0.1239$	$corr(t_e^{d1},t_{s1}^{d1})=0.8323$	$corr(t_e^{d1},t_{s2}^{d1})=0.8409$
	$mse(t_e^{d0},t_{s1}^{d0})=0.1964$	$mse(t_e^{d0},t_{s2}^{d0})=0.1572$		$mse(t_e^{d1},t_{s1}^{d1})=0.0336$	$mse(t_e^{d1},t_{s2}^{d1})=0.0295$
$freq(I_{d2})=0.2216$	$corr(t_e^{d2},t_{s1}^{d2})=0.8455$	$corr(t_e^{d2},t_{s2}^{d2})=0.8606$	$freq(I_{d3})=0.1919$	$corr(t_e^{d3},t_{s1}^{d3})=0.8471$	$corr(t_e^{d3},t_{s2}^{d3})=0.8607$
	$mse(t_e^{d2},t_{s1}^{d2})=0.0703$	$mse(t_e^{d2},t_{s2}^{d2})=0.0587$		$mse(t_e^{d3},t_{s1}^{d3})=0.0688$	$mse(t_e^{d3},t_{s2}^{d3})=0.0566$
$freq(I_{d4})=0.1385$	$corr(t_e^{d4},t_{s1}^{d4})=0.7956$	$corr(t_e^{d4},t_{s2}^{d4})=0.8109$			
	$mse(t_e^{d4},t_{s1}^{d4})=0.0433$	$mse(t_e^{d4},t_{s2}^{d4})=0.0375$			
$freq(I_{e0})=0.1079$	$corr(t_{s1}^{e0},t_e^{e0})=0.2693$	$corr(t_{s1}^{e0},t_{s2}^{e0})=0.9853$	$freq(I_{e1})=0.1043$	$corr(t_{s1}^{e1},t_e^{e1})=0.3053$	$corr(t_{s1}^{e1},t_{s2}^{e1})=0.9854$
	$mse(t_{s1}^{e0},t_e^{e0})=0.1435$	$mse(t_{s1}^{e0},t_{s2}^{e0})=0.0077$		$mse(t_{s1}^{e1},t_e^{e1})=0.2216$	$mse(t_{s1}^{e1},t_{s2}^{e1})=0.0024$
$freq(I_{e2})=0.1193$	$corr(t_{s1}^{e2},t_e^{e2})=0.4652$	$corr(t_{s1}^{e2},t_{s2}^{e2})=0.9954$	$freq(I_{e3})=0.2412$	$corr(t_{s1}^{e3},t_e^{e3})=0.8559$	$corr(t_{s1}^{e3},t_{s2}^{e3})=0.9986$
	$mse(t_{s1}^{e2},t_e^{e2})=0.2120$	$mse(t_{s1}^{e2},t_{s2}^{e2})=0.0015$		$mse(t_{s1}^{e3},t_e^{e3})=0.0479$	$mse(t_{s1}^{e3},t_{s2}^{e3})=0.0006$
$freq(I_{e4})=0.4274$	$corr(t_{s1}^{e4},t_e^{e4})=0.8376$	$corr(t_{s1}^{e4},t_{s2}^{e4})=0.9985$			
	$mse(t_{s1}^{e4},t_e^{e4})=0.0641$	$mse(t_{s1}^{e4},t_{s2}^{e4})=0.0006$			
$freq(I_{e0})=0.1305$	$corr(t_{s2}^{e0},t_e^{e0})=0.3430$	$corr(t_{s2}^{e0},t_{s1}^{e0})=0.9860$	$freq(I_{e1})=0.0978$	$corr(t_{s2}^{e1},t_e^{e1})=0.3380$	$corr(t_{s2}^{e1},t_{s1}^{e1})=0.9964$
	$mse(t_{s2}^{e0},t_e^{e0})=0.0837$	$mse(t_{s2}^{e0},t_{s1}^{e0})=0.0069$		$mse(t_{s2}^{e1},t_e^{e1})=0.2230$	$mse(t_{s2}^{e1},t_{s1}^{e1})=0.0021$
$freq(I_{e2})=0.1456$	$corr(t_{s2}^{e2},t_e^{e2})=0.6722$	$corr(t_{s2}^{e2},t_{s1}^{e2})=0.9960$	$freq(I_{e3})=0.2435$	$corr(t_{s2}^{e3},t_e^{e3})=0.8513$	$corr(t_{s2}^{e3},t_{s1}^{e3})=0.9986$
	$mse(t_{s2}^{e2},t_e^{e2})=0.1307$	$mse(t_{s2}^{e2},t_{s1}^{e2})=0.0012$		$mse(t_{s2}^{e3},t_e^{e3})=0.0505$	$mse(t_{s2}^{e3},t_{s1}^{e3})=0.0007$
$freq(I_{e4})=0.3826$	$corr(t_{s2}^{e4},t_e^{e4})=0.8501$	$corr(t_{s2}^{e4},t_{s1}^{e4})=0.9985$			
	$mse(t_{s2}^{e4},t_e^{e4})=0.0518$	$mse(t_{s2}^{e4},t_{s1}^{e4})=0.0005$			

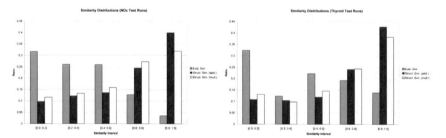

Fig. 2 Distribution of similarity values calculated using structural and evaluation based similarity functions

Figure 2 shows the distributions of structural and evaluation based similarity estimation for the NO_x and *Thyroid* tests separately. As we see in both charts the structural similarity values are significantly higher than the evaluation based ones.

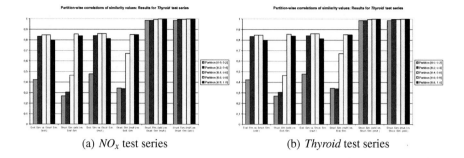

(a) NO_x test series   (b) *Thyroid* test series

Fig. 3 Partition-wise correlations of similarity values for NO_x and *Thyroid* test series

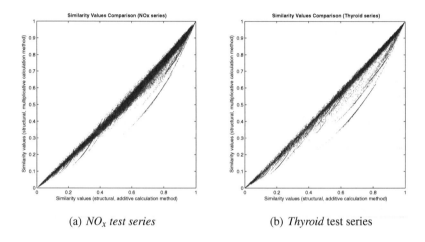

(a) NO_x test series   (b) *Thyroid* test series

Fig. 4 Similarity values comparison: Structural (additive calculation) vs. structural (multiplicative calculation)

Regarding results correlations, the figures documented in Table 3 can be summarized in the following way: The correlations between structural and evaluation based similarity are approximately in the range between 0.3 and 0.85. Especially low correlation coefficients are calculated for the comparison of structural and evaluation based similarities, especially when the structural similarity is considered very low (<0.4). This impression becomes even more clear when we analyze Figures 3(a) and 3(b) which give the partition wise correlations of similarity values. In each of the 6 series shown in each of these figures we show the correlations of similarity values calculated by each possible pair of methods; in each case those partitions of value pairs are selected that correspond to the values calculated by the first method mentioned in the respective label. So, for example, in the first series we see the partition-wise correlations of similarity values calculated by the evaluation based and the additive structural method; the values are classified in partitions with respect to the evaluation specific similarities.

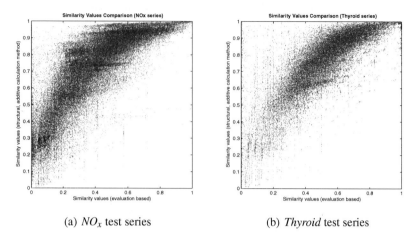

(a) NO_x test series   (b) *Thyroid* test series

Fig. 5 Similarity values comparison: Evaluation based vs. structural (additive calculation)

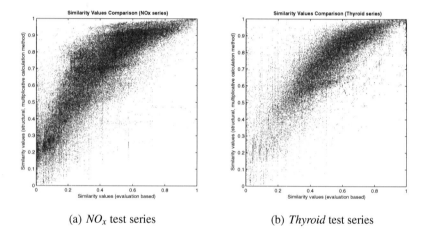

(a) NO_x test series   (b) *Thyroid* test series

Fig. 6 Similarity values comparison: Evaluation based vs. structural (multiplicative calculation)

The Figures 3(a) and 3(b) show clearly that the structural similarity estimation methods calculate very similar values (with high correlations for trees that are very different as well as for those which are considered rather similar). Furthermore, the correlation of structural and evaluation based similarity values is rather low in the case of low structural similarities (<0.4).

Finally, for graphically illustrating the direct comparison of similarity values calculated by the three estimation methods chosen we have randomly chosen 100,000 structure tree comparison cases both from the NO_x and the *Thyroid* tests. The respectively correspondent similarity values are drawn against each other in the

Figures 4 – 6. On the one hand there is no high correlation which can be seen when comparing structural and evaluation based similarity values, but on the other hand the high correlation between the similarities calculated by the structural similarity estimation methods becomes obvious.

4 Conclusion

In this paper we have summarized a series of GP test runs incorporating evaluation based as well as structural similarity estimation for measuring the genetic diversity in GP populations.

In general, evaluation based similarity calculation consumes a lot more runtime than structural comparison, and on average it also tends to produce lower similarity values. The results show that in most cases there is a linear correlation of approximately $0.4 - 0.9$ for the results returned by the evaluation based and structural methods; not very surprisingly, this correlation is positive, but not very high. Especially in some cases showing very low structural similarity there can be significantly different results when using the evaluation based similarity methods.

Furthermore, we have also compared additive and multiplicative structural similarity estimation. These two variants tend to produce rather similar results with high correlations for pairs of structure trees with low as well as rather high similarities; the results retrieved by the multiplicative structural method show a higher correlation with those calculated using the evaluation based similarity function.

Thus, analyzing these correlations, we see that structural and evaluation based similarity measures give non-redundant information about the similarity of structure trees used in GP; both types of similarity measures should therefore be used for analyzing GP populations and algorithms.

Acknowledgments

The work described in this paper was done within the Translational Research Project L284-N04 "GP-Based Techniques for the Design of Virtual Sensors" funded by the Austrian Science Fund (FWF).

References

[1] Affenzeller, M., Wagner, S.: Offspring selection: A new self-adaptive selection scheme for genetic algorithms. In: Adaptive and Natural Computing Algorithms, pp. 218–221. Springer, Springer Computer Science, Heidelberg (2005)

[2] Affenzeller, M., Winkler, S., Wagner, S., Beham, A.: Genetic Algorithms and Genetic Programming - Modern Concepts and Practical Applications. Chapman & Hall / CRC (2009)

[3] Burke, E.K., Gustafson, S., Kendall, G.: Diversity in genetic programming: An analysis of measures and correlation with fitness. IEEE Transactions on Evolutionary Computation 8(1), 47–62 (2004)

[4] Deb, K., Goldberg, D.E.: An investigation of niche and species formation in genetic function optimization. In: Proceedings of the Third International Conference on Genetic Algorithms, pp. 42–50. Morgan Kaufmann, San Francisco (1989)

[5] Ekart, A., Nemeth, S.Z.: A metric for genetic programs and fitness sharing. In: Poli, R., Banzhaf, W., Langdon, W.B., Miller, J.F., Nordin, P., Fogarty, T.C. (eds.) EuroGP 2000. LNCS, vol. 1802, pp. 259–270. Springer, Heidelberg (2000)

[6] Keijzer, M.: Efficiently representing populations in genetic programming. In: Angeline, P.J., Kinnear Jr., K.E. (eds.) Advances in Genetic Programming, vol. 2, ch.13, pp. 259–278. MIT Press, Cambridge (1996)

[7] Koza, J.R.: Genetic Programming: On the Programming of Computers by Means of Natural Selection. The MIT Press, Cambridge (1992)

[8] Langdon, W.B., Poli, R.: Foundations of Genetic Programming. Springer, Heidelberg (2002)

[9] Levenshtein, V.I.: Binary codes capable of correcting deletions, insertions, and reversals. Soviet Physics Doklady 10(8), 707–710 (1966)

[10] McKay, R.I.B.: Fitness sharing in genetic programming. In: Whitley, D., Goldberg, D., Cantu-Paz, E., Spector, L., Parmee, I., Beyer, H.G. (eds.) Proceedings of the Genetic and Evolutionary Computation Conference (GECCO 2000), pp. 435–442. Morgan Kaufmann, Las Vegas (2000)

[11] McPhee, N.F., Hopper, N.J.: Analysis of genetic diversity through population history. In: Banzhaf, W., et al. (eds.) Proceedings of the Genetic and Evolutionary Computation Conference, vol. 2, pp. 1112–1120. Morgan Kaufmann, Orlando (1999)

[12] O'Reilly, U.M.: Using a distance metric on genetic programs to understand genetic operators. In: IEEE International Conference on Systems, Man, and Cybernetics, Computational Cybernetics and Simulation, Orlando, Florida, USA, vol. 5, pp. 4092–4097 (1997)

[13] Rosca, J.P.: Entropy-driven adaptive representation. In: Rosca, J.P. (ed.) Proceedings of the Workshop on Genetic Programming: From Theory to Real-World Applications, Tahoe City, California, USA, pp. 23–32 (1995)

[14] Wagner, S.: Heuristic optimization software systems – modeling of heuristic optimization algorithms in the heuristiclab software environment. PhD thesis, Johannes Kepler University Linz (2009)

[15] Winkler, S.: Evolutionary system identification - modern concepts and practical applications. PhD thesis, Institute for Formal Models and Verification, Johannes Kepler University Linz (2008)

[16] Winkler, S., Affenzeller, M., Wagner, S.: Using enhanced genetic programming techniques for evolving classifiers in the context of medical diagnosis - an empirical study. Genetic Programming and Evolvable Machines 10(2), 111–140 (2009)

Artificial Bee Colony Optimization: A New Selection Scheme and Its Performance

Andrej Aderhold, Konrad Diwold, Alexander Scheidler, and Martin Middendorf

Abstract. The artificial bee colony optimization (ABC) is a population based algorithm for function optimization that is inspired by the foraging behaviour of bees. The population consists of two types of artificial bees: employed bees (EBs) which scout for new good solution in the search space and onlooker bees (OBs) that search in the neighbourhood of solutions found by the EBs. In this paper we study the influence of the populations size on the optimization behaviour of ABC. Moreover, we investigate when it is advantageous to use OBs. We also propose two variants of ABC which use new methods for the position update of the artificial bees. Empirical tests were performed on a set of benchmark functions. Our findings show that the ideal population size and whether it is advantageous to use OBs depends on the hardness of the optimization goal. Additionally the newly proposed variants of the ABC outperform the standard ABC significantly on all test functions. In comparison to several other optimization algorithm the best ABC variant performs better or at least as good as all reference algorithms in most cases.

1 Introduction

Swarm intelligence [6] is a subfield of biological inspired computation that applies concepts found in the collective behaviour of swarms such as social insects to problems in various domains such as robotics or optimization [5]. In recent years a number of bee inspired optimization methods have been proposed (the interested reader can refer to Baykasoglu et al.'s overview of bee inspired optimization methods [3]).

One behaviour of honey bees that has inspired optimization methods is foraging. Although it is a decentralized process that works at the basis of decisions of

Andrej Aderhold · Konrad Diwold · Alexander Scheidler · Martin Middendorf
Department of Computer Science, University of Leipzig, Johannisgasse 26,
04103 Leipzig, Germany
e-mail: aaderhold@gmail.com,
    {kdiwold,scheidler,middendorf}@informatik.uni-leipzig.de

J.R. González et al. (Eds.): NICSO 2010, SCI 284, pp. 283–294, 2010.
springerlink.com   © Springer-Verlag Berlin Heidelberg 2010

individual bees, a colony is still able to maintain a good ratio of exploitation and exploration of food sources and can adapt toward changing needs for food if necessary [4]. The waggle dance has been identified as a communication mechanism that allows scout bees that found a food site to promote this site to other foragers [19]. Besides distance and direction to a site the bee can also encode its quality. Utilizing this mechanism foragers can distribute on the available resources regarding their profitability. A recent study has shown [7] that recruitment strategies as they are used in honeybees are especially beneficial if resources are of poor quality, few in number, and of variable quality.

The artificial bee colony optimization algorithm (ABC) is an algorithm that is inspired by principles of the foraging behaviour of honeybees and was introduced by Karaboga [9] in 2005. The ABC algorithm has been applied to various problem domains including the training of artificial neural networks [11, 16], the design of a digital infinite impulse response (IIR) filters [10], solving constrained optimization problems [13], and the prediction of the tertiary structures of proteins [2]. Its optimization performance has been tested and compared to other optimization methods such as Genetic Algorithms (GA), Particle Swarm Optimization (PSO), Particle Swarm Inspired Evolutionary Algorithm (PS-EA), Differential Evolution (DE), and different evolutionary strategies [1, 12, 14, 15].

The ABC algorithm works with a population of artificial bees. The bees are divided into two groups, one group of bees — called employed bees (EBs) — is responsible for finding new promising solutions and the other group of bees — called onlooker bees (OBs) — for performing local search at these solutions. It should be mentioned that the EB are sometimes divided into two subgroups the EBs that stay at a location and the EBs that search for a new location. The latter one are called scouts.

In this paper we study a central aspect of ABC which has not been studied before. That is the influence of the size of the bee population and of the ratio between the number of employed bees and onlooker bees on the performance of the algorithm. Moreover, we propose two variants of the standard ABC algorithm that use new methods for the selection of new locations. The performance of the new variants of ABC and the standard ABC is tested against several other population based optimization heuristics.

This article is structured as follows. In Section 2 the ABC is described. The new variants of ABC are introduced in Section 3. The experimental setup is given in Section 4 and the experimental results are described in Section 5. Conclusion are given in Section 6.

2 Artificial Bee Colony Optimization (ABC)

The ABC algorithm [9] is a population based algorithm for function optimization that can be seen as a minimal honeybee foraging model. The artificial bee population consists of two types of bees: employed bees (EB) and onlooker bees (OB). In ABC the search space represents an environment and each point in the search space corresponds to a food source (solution) that the artificial bees can exploit. The quality of a food source is given by the value of the function to be optimized at the corresponding location. Initially the EBs scout and each EB decides to exploit a food source it has found. The number of EBs thus corresponds to the number of food sources that are currently exploited in the system. EBs communicate their food sources to the OBs. Based on the quality of a food source the OBs decide whether or not to visit it. Good food sources will attract more OBs. Once an OB has chosen a food source it flies there and tries to find a better location in its neighborhood by using a local search strategy. If the quality of a new location found by the OB is better than the quality of the location originally communicated by of the corresponding EB, the EB will change its location and promote the new food source. Otherwise, the EB remains on its current food source. If the solution of an EB has not been improved for a certain number of steps the EB will abandon the food source and scout for a new one (i.e., it decides for a new food source in search space).

More formally: Given a *dim* dimensional function F and a population of n virtual bees consisting of n_{eb} employed bees and n_{ob} onlooker bees (i.e., $n = n_{eb} + n_{ob}$). Initially and when scouting an EB i ($i \in [1 \ldots n_{eb}]$) is placed on a randomly chosen position $p_i = (x_1^i, \ldots, x_{dim}^i)$ in the search space. At the beginning of an iteration each EB i tries to improve its current position by creating a new candidate position p_i^* using the following local search rule

$$p_i^* = (x_1^i, \ \ldots \ , x_j^i + rand(-1,1)(x_j^k - x_j^i), \ \ldots \ , x_{dim}^i) \tag{1}$$

where $j \leq dim$ is a randomly chosen dimension, $k \neq i$) denotes a randomly chosen EB (called *reference EB*), and $rand(-1,1)$ is a real valued random number drawn from a uniform distribution between -1 and 1. Note, that only one dimension is changed via Equation 1. Based on the follwoing greedy selection mechanism each EB decides whether to discard p_i in favor of p_i^*

$$p_i = \begin{cases} p_i & \text{if } F(p_i) > F(p_i^*) \\ p_i^* & \text{else} \end{cases} \tag{2}$$

where $F(p)$ denotes the fitness at location p.

After each EB has updated its location, every OB chooses one of the current EB locations by using a standard roulette wheel selection so that the probability P_i of choosing the location p_i of EB i is

$$P_i = \frac{F(p_i)}{\sum_{k=1}^{n_{eb}} F(p_k)}. \tag{3}$$

After an OB has chosen the location of an EB i it tries to find a better location using Equation 1. In response, the corresponding EB updates its position as described before in case the OB has found a better location. The algorithm monitors the number of steps an EB remains on the same position. When the number of steps an EB has spent at the same location reaches a limit $l \geq 1$ the EB abandons its position and scouts for a new one. In [15] the impact of l was investigated and as a good value $l = n_e \cdot dim$ was proposed. The algorithm stops when a certain stop criterion (e.g., maximum number of iterations, or a good function value has been found) is met. An outline of ABC is given in Algorithm 6.

Algorithm 6. Artificial Bee Colony

```
 1: place each employed bee on a random position in the search space
 2: while  stop criterion not met  do
 3:    for all employed bees do
 4:       if # steps on same position = l then
 5:          choose random position in search space
 6:       else
 7:          try to find better position (according to equations 1 and 2)
 8:          if better position found then
 9:             move from current position to found position
10:          end if
11:       end if
12:    end for
13:    for all onlooker bees do
14:       choose an employed bee and move to its position (according to Equation 3)
15:       try improve position (according to equations 1 and 2)
16:    end for
17: end while
```

For the standard ABC algorithm it was defined that the number of employer bees equals the number of onlooker bees, i.e., $n_{eb} = n_{ob} = n/2$. Thus algorithm ABC depends only on the parameters n and l. In [15] experiments with different population sizes n were performed with the conclusion that, a population size of 50-100 bees can provide reasonable convergence behaviour. The parameter l determines how fast solutions are abandoned. In [15] it is argued that $l = n_e \cdot dim$ shows better performance than very high or low values of l. In a very recent study on ABC parameter tuning [1] Akay and Karaboga concluded that for small colony sizes $l = n_e \cdot dim$ might not be sufficient, as the algorithm is not able to explore EB solutions enough before they are abandoned. Hence, it is suggested to use higher values of l for small colonies.

3   ABC Variants

The variants of ABC that are proposed in this section concern the selection of reference EBs when OBs and EBs generate candidate solutions according to Equation 1). In the standard ABC algorithm the reference EBs are selected randomly with uniform distribution. A potential disadvantage is that the location of the chosen reference EB might not fit well to the current location of the bee. The two modifications of the reference selection rule that are proposed in the following aim to overcome this problem.

Including global best solution as reference (ABC_{gBest}). In the proposed ABC_{gBest} the global best solution found so far is used in addition to the randomly chosen reference EB in order to generate new candidate solutions. Note, that this has some similarity to the functioning of a Particle Swarm Optimization (PSO) algorithm where the global best particles influence the position update of the particles [17]. To incorporate the global best solution Equation 1 is altered as follows

$$p_i^* = (x_1^i, \ \dots \ , x_j^i + rand(-1,1)(x_j^k - x_j^i) + rand(0,1)(x_j^{best} - x_j^i), \ \dots \ , x_{dim}^i) \quad (4)$$

where p_i denotes the bees current position, k refers to the randomly chosen reference EB p_k, *best* refers to the best position p_{best} found so far, and $j \leq dim$ denotes a random dimension. To make sure that the global best term in Equation 4 always points towards the global best reference $rand(0,1)$ was used (instead of $rand(-1,1)$).

ABC_{gBest} with additional distance based reference selection ($ABC_{gBestDist}$). Besides including the global best reference in the generation of candidate solutions, in this modification the distance between the current location and a potential reference EB influences the selection probability. Therefore, instead of using the same probability for all reference EBs, an EB (or OB) at position p_i chooses the reference EB $k \in \{1, .., n_{eb}\}$ with $k \neq i$ according to the following probability

$$P_k = \frac{\frac{1}{dist(p_i, p_k)}}{\sum\limits_{j=1, j \neq i}^{n_{eb}} \left(\frac{1}{dist(p_i, p_j)} \right)} \quad (5)$$

where $dist(x, y)$ is the euclidean distance between positions x and y.

After a reference EB has been chosen the new candidate solution is created using Equation 4. As can be seen, the further away a potential reference EB is located from the current position, the smaller is the probability to be selected as a reference. The idea of this modification is to prefer near references because for many types of optimization functions it is more reasonable to search between good positions that are close to each other.

4 Experimental Setup

The performance of ABC, the proposed ABC modifications, and other reference algorithms was tested on several standard benchmark problems (see Table 1 for details). The following five algorithms were used as reference algorithms: the Particle Swarm Optimization (PSO) algorithm from [22], two forms of the hierarchical PSO (H-PSO and $\bigvee$H-PSO) from [8], the differential evolution (DE) algorithm from [18, 21], and the Ant Colony Optimization algorithm for continuous functions (ACO_R) from [20]. The parameter values that were used for these algorithms have been adopted from the given references (see Table 2).

Table 1 Test function names and equations (F), domain space range (R), a standard optimization goal (G_{std}) that is often used in the literature and a harder optimization goal (G_{hrd}). The hard goals were chosen in such a way that a standard ABC (with $n = 100$) will need approximately 10^5 function evaluations to reach them. The dimension of the test functions was $dim = 30$ with the exception of Schaffer's F6 were dimension $dim = 2$ was used

F		R	G_{std}	G_{hrd}
Schaffer's F6	$f_{sc}(\mathbf{x}) = 0.5 + \frac{sin^2(\sqrt{x_1^2 + x_2^2}) - 0.5}{(1 + 0.001(x_1^2 + x_2^2))^2}$	$[-100; 100]^2$	10^{-5}	10^{-25}
Sphere	$f_{sp}(\mathbf{x}) = \sum\limits_{i=1}^{n} x_i^2$	$[-100; 100]^n$	0.01	10^{-10}
Griewank	$f_{gr}(\mathbf{x}) = \frac{1}{4000} \left(\sum\limits_{i=1}^{n} x_i^2 \right) - \prod\limits_{i=1}^{n} cos\left(\frac{x_i}{\sqrt{i}} \right) + 1$	$[-600; 600]^n$	0.1	10^{-9}
Rastrigin	$f_{rg}(\mathbf{x}) = \sum\limits_{i=1}^{n} (x_i^2 - 10cos(2\pi x_i) + 10)$	$[-5.12; 5.12]^n$	100	10^{-7}
Rosenbrock	$f_{rn}(\mathbf{x}) = \sum\limits_{i=1}^{n-1} (100(x_{i+1} - x_i^2)^2 + (x_i - 1)^2)$	$[-30; 30]^n$	100	1
Ackley	$f_{ac}(\mathbf{x}) = -20exp\left(-0.2 \sqrt{\frac{1}{n} \sum\limits_{i=1}^{n} x_i^2} \right) - exp\left(\frac{1}{n} \sum\limits_{i=1}^{n} cos(2\pi x_i) \right) + 20 + e$	$[-32; 32]^n$	0.1	10^{-7}

Table 2 Setting of control parameters used in the final experiment: n is the population size, swarm size, or colony size respectively; n_{eb} is the number of employed bees; n_{ob} is the number of onlooker bees; l is the abandon limit; dim is the dimension of problem function; ω is the inertia weight; $c_{(*)}$ is the constriction factors; CR is the crossover rate; F is the scaling factor; k is the archive size; q is the locality of search; ε is the convergence speed

ABC	PSO	H-PSO	$\surd$H-PSO	DE	ACO_R
$n = 30$	$n = 40$	$n = 31$	$n = 31$	$n = 50$	$n = 2$
$n_{eb} = 15$	$\omega = 0.6$	$\omega = 0.6$	$\omega = [0.729; 0.4]$	$CR = 0.8$	$k = 50$
$n_{ob} = 15$	$c_1 = c_2 = 1.7$	$c_1 = c_2 = 1.7$	$c_1 = c_2 = 1.7$	$F = 0.5$	$q = 0.1$
$l = dim * n_e$					$\varepsilon = 0.85$

All test runs were repeated 100 times. The number of function evaluations that each algorithm required to reach the specified goal — the standard optimization goal (G_{std}) and the hard optimization goal (G_{hrd}) as given in Table 1 — was recorded for each run. To evaluate the significance of the observed performance differences the algorithms were tested pairwise against each other by using a one sided Wilcoxon Rank Sum Test with a significance level of $\alpha = 0.05$.

5 Results

5.1 Population Size

As pointed out in Section 2 population size is one of ABCs two control parameters. In a recent study Karaboga and Basturk [15] investigated the influence of the population size on the performance of ABC. Based on a comparison of fitness improvement per algorithm step they argue that an increase of population size up to a certain value increases the algorithms performance. Their suggestion is to use a population size of 50 - 100 as it provides acceptable convergence speed and good solutions.

A problem with a comparison that is based on the number of algorithmic steps is that an algorithm with a larger population sizes requires more function evaluations per step. For example an ABC with a population of size 100 needs 10 times as many function evaluations as one with a population of 10 per step.

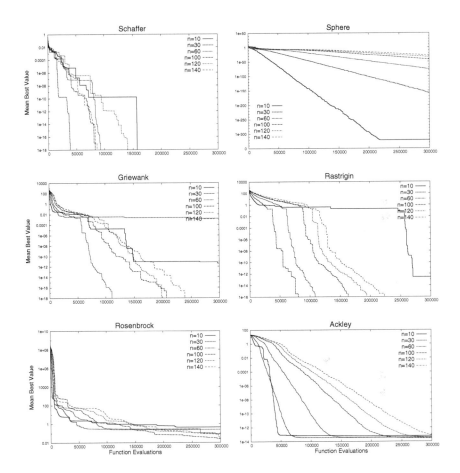

Fig. 1 ABC population size test: Comparing improvement of mean best solution (y-axis) per iteration (x-axis) for different population sizes n over 300000 function evaluations; Standard ABC settings are used except for l. To avoid very small limit values with small population sizes $l = 100$ if $n_e \, dim < 100$

Therefore we have done a comparison of different population sizes with respect to the total number of function evaluations. Figure 1 depicts the average quality of the best found solution over the first $3 \cdot 10^5$ function evaluations for different population sizes $n \in \{10, 30, 60, 100, 120, 140\}$ and all six test functions. As can be seen the relative quality differs in different stages of the optimization process. For most test functions (i.e., Griewank, Rosenbrock, Rastrigin, Ackley) very small populations (i.e., $n = 10$) show a fast convergence at the beginning of the optimization process (i.e., in the first 20 000 evaluation steps). However larger populations perform better in later stages of the optimization process. Only for the Sphere function very small populations perform best throughout the whole optimization process. But this is a very simple optimization function. For the more complex functions as Schaffer Griewank, Rastrigin, and Ackley population size 30-60 performs best for more than 700000 evaluations. Only for the Ackely function the population size 100 is best for a higher number of function evaluations. Thus, our results suggest that a population size of $30 - 60$ seems good for many test functions. This is a slightly smaller populations size than recommended in [12, 15].

5.2 Number of Onlooker Bees

The influence of the number of onlooker bees is studied in this subsection. Table 3 presents the mean number of function evaluations that are necessary to reach the optimization goals for populations containing 15 and 50 EBs. For each number of EBs the ABC with a standard number of onlooker bees (i.e., the number of OBs equals the number of EBs) is compared to a modified ABC were no OBs are used. Each version has been tested on the standard and hard optimization goal (see Table 1).

Table 3 ABC with different number of employed bees n_{eb} and with or without and onlooker bees n_{ob} for the standard optimization goal G_{std} and the hard optimization goal G_{hrd}. Mean number of function evaluations (mean) to reach the goal for the six test functions and significance (sig) comparing the ABC with and without onlooker bees and with the same number of n_{eb}; 'X' denotes significantly better; '-' denotes not significantly better

ABC (n_{eb}, n_{ob})	Schaffer mean/sig	Sphere mean/sig	Griewank mean/sig	Rastrigin mean/sig	Rosenbrock mean/sig	Ackley mean/sig
Standard G_{std}						
(15,0)	17525/–	13915/–	14085/–	3578/–	16073/–	18162/–
(15,15)	14820/–	**7810/X**	**10404/X**	**3162/X**	**10017/X**	**15969/X**
(50,0)	**21644/X**	43345/–	42739/–	10580/–	48751/–	54750/–
(50,50)	23599/–	**15632/X**	**22326/X**	**8909/X**	**18055/X**	**45794/X**
Hard G_{hrd}						
(15,0)	31419/–	**28318/X**	**53337/X**	45041/–	68847/–	**39292/X**
(15,15)	31412/–	32970/–	60383/–	**38361/X**	82921/–	39947/–
(50,0)	68079/–	**89694/X**	**102639/X**	127124/–	118932/–	**124140/X**
(50,50)	74350/–	104216/–	107023/–	**93845/X**	**92714/X**	125344/–

As can be seen the performance regarding the number of OBs differs. Using OBs increases the performance of the algorithm significantly for the standard optimization goal G_{std} in 5 of 6 test unctions for both numbers of EBs. But this is not the case for the hard optimization goal G_{hrd}. For the case of 15 EBs the algorithm containing no OBs performs significantly better for three of the six test functions. Only for the Rastrigin function the algorithm with OBs is able to perform significantly better. For two test functions no significant difference can be constituted. When 50 EBs are used the algorithm with no OBs performs significantly better in 3 of the 6 test cases, whereas the algorithm with OBs performs significantly better for only two test function (Rastrigin and Rosenbrock). For one function no statistic difference can be constituted.

These results suggests that the advantage of using OBs in the ABC algorithm is not so clear for the hard optimization goal G_{hrd} while OBs are advantageous for most cases when only the standard optimization goal G_{std} is given. This questions the standard rule to set the ratio between the number of OB and EBs to 1/2. A more detailed analysis is necessary to fully understand the impact of OBs on the algorithms performance and under what conditions they are useful and when not.

5.3   Comparison of ABC and Other Algorithms

In this section the performance of the standard ABC and the suggested modifications are compared with other optimization algorithms. As the standard optimization

Table 4 Mean number of function evaluations to reach the standard goal G_{std} for ABC, ABC_{gBest}, and $ABC_{gBestDist}$; Population size $n = 30$, number of EBs $n_{eb} = 15$, number of OBs $n_{ob} = 15$; For each test function the significance between each pair of algorithms is shown, 'X' denotes that the algorithm in the corresponding line is significantly better than the algorithm in the corresponding row, '-' denotes no significance

Function	Standard Test Method	Mean	Significance		
			ABC	ABC_{gBest}	$ABC_{gBestDist}$
Schaffer	ABC	19038		–	–
	ABC_{gBest}	6680	X		–
	$ABC_{gBestDist}$	6377	X	–	
Sphere	ABC	7773		–	–
	ABC_{gBest}	6509	X		–
	$ABC_{gBestDist}$	6245	X	X	
Griewank	ABC	10160		–	–
	ABC_{gBest}	9020	X		–
	$ABC_{gBestDist}$	8680	X	X	
Rastrigin	ABC	3218		–	–
	ABC_{gBest}	2506	X		–
	$ABC_{gBestDist}$	2466	X	–	
Rosenbrock	ABC	9800		–	–
	ABC_{gBest}	6682	X		–
	$ABC_{gBestDist}$	7049	X	–	
Ackley	ABC	15759		–	–
	ABC_{gBest}	10118	X		–
	$ABC_{gBestDist}$	10038	X	–	

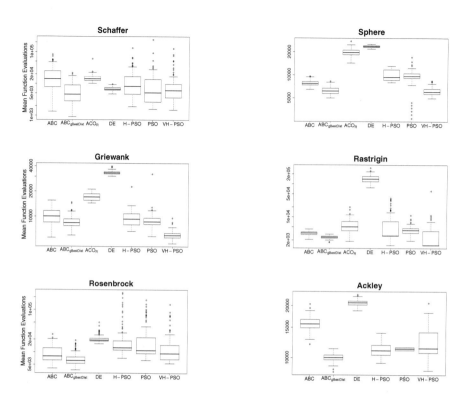

Fig. 2 Boxplots of the number of function evaluations needed to reach the standard optimization goal G_{std} for ABC, $ABC_{gBestDist}$, PSO, $\sqrt{}$H-PSO, H-PSO, DE and ACO_R. Results for ACO_R are omitted when it was not able to reach the optimization goal in 500000 function evaluations

goals will be used in order to compare the algorithms OBs have been used in the tests for ABC and the proposed variants. Table 4 shows the mean number of function evaluations and pairwise significance tests for the standard ABC and the proposed modifications ABC_{gBest} and $ABC_{gBestDist}$ on the test functions.

The results show that the proposed variants of ABC — ABC_{gBest} and $ABC_{gBestDist}$ — improve the performance of ABC on all test functions significantly. $ABC_{gBestDist}$ is able to enhance the performance of ABC_{gBest} on two test functions (i.e., Sphere and Griewank). In the other cases no significant difference between the two ABC variants could be observed.

The standard ABC algorithm and $ABC_{gBestDist}$, the best performing ABC variant, were tested against 5 reference algorithms for the standard optimization goal G_{std}. Figure 2 depicts boxplots of the number of function evaluation for each algorithm on each test function.

In terms of the necessary number of function evaluations the proposed ABC variant $ABC_{gBestDist}$ performs significantly better than all the reference algorithms for two test functions (i.e., Ackley, Rosenbrock). For two test functions (i.e., Sphere and

Schaffer) its performance is on par with the performance of the PSO respectively the $\bigvee$H-PSO algorithm. For the remaining two test functions (i.e., Griewank and Rastrigin) the hierarchical PSO variant $\bigvee$H-PSO outperforms all other algorithms significantly, $ABC_{gBestDist}$ is the second best algorithm for this test functions.

6 Conclusion

In this paper we have proposed two variants — called ABC_{gBest} and $ABC_{gBestDist}$ — of the artificial bee colony optimization (ABC) algorithm. Both variants concern the selection of the reference locations that influence the position update for the artificial bees. Moreover, we investigated the influence of the colony size and the relative number of so called onlooker bees in the artificial bee population on the optimization performance. Experimental results for six standard benchmark test functions suggest that ABC performs better with a smaller population size than used in a standard ABC setup. However, it was also shown that the ideal population size depends on the optimization goal. For harder optimization goals larger populations seem to be advantageous. Whether it is advantageous to use onlooker bees depends also on the optimization goals. For weaker optimization goals using OBs was advantageous for all test functions. But for the harder optimization goals it was in most cases better not to use OBs. This questions the standard division of the population of ABC into an equal number of EBs and OBs. The proposed ABC variants ABC_{gBest} and $ABC_{gBestDist}$ performed better than the standard ABC on all test functions. $ABC_{gBestDist}$ performed slightly better than ABC_{gBest}. In comparison to other optimization algorithms $ABC_{gBestDist}$ was better or at least as good as all tested algorithm on all test functions. Only for two test functions $\bigvee$H-PSO performed better.

Acknowledgments

This work was supported by the Human Frontier Science Program Research Grant "Optimization in natural systems: ants, bees and slime moulds".

References

[1] Akay, B., Karaboga, D.: Parameter tuning for the artificial bee colony algorithm. In: Nguyen, N.T., Kowalczyk, R., Chen, S.-M. (eds.) ICCCI 2009. LNCS, vol. 5796, pp. 608–619. Springer, Heidelberg (2009)
[2] Bahamish, H.A.A., Abdullah, R., Salam, R.A.: Protein tertiary structure prediction using artificial bee colony algorithm. In: Asia International Conference on Modelling & Simulation, pp. 258–263 (2009)
[3] Baykasoglu, A., Oezbakir, L., Tapkan, P.: Artificial Bee Colony Algorithm and Its Application to Generalized Assignment Problem. In: Swarm Intelligence: Focus on Ant and Particle Swarm Optimization, pp. 113–144. Itech Education and Publishing (2007)

[4] Biesmeijer, J.C., de Vries, H.: Exploration and exploitation of food sources by social insect colonies: a revision of the scout-recruit concept. Behavioral Ecology and Sociobiology 49, 89–99 (2001)

[5] Blum, C., Merkle, D. (eds.): Swarm Intelligence: Introduction and Applications. Springer, Heidelberg (2008)

[6] Bonabeau, E., Dorigo, M., Theraulaz, G.: Swarm intelligence: from natural to artificial systems. Oxford University Press, Oxford (1999)

[7] Dornhaus, A., Kluegl, F., Oechslein, C., Puppe, F., Chittka, L.: Benefits of recruitment in honey bees: effects of ecology and colony size in an individual-based model. Behavioral Ecology (2006)

[8] Janson, S., Middendorf, M.: A hierarchical particle swarm optimizer and its adaptive variant. IEEE Transactions on Systems, Man, and Cybernetics – Part B: Cybernetics 35, 1272–1283 (2005)

[9] Karaboga, D.: An idea based on honey bee swarm for numerical optimization. Tech. rep., Erciyes University, Engineering Faculty (2005)

[10] Karaboga, D.: A new design method based on artificial bee colony algorithm for digital IIR filters. Journal of the Franklin Institute 346(4), 328–348 (2009)

[11] Karaboga, D., Akay, B.: Artificial bee colony (abc) algorithm on training artificial neural networks. In: IEEE 15th Signal Processing and Communications Applications, pp. 1–4 (2007)

[12] Karaboga, D., Akay, B.: A comparative study of artificial bee colony algorithm. Applied Mathematics and Computation 214(1), 108–132 (2009)

[13] Karaboga, D., Basturk, B.: Artificial bee colony (ABC) optimization algorithm for solving constrained optimization problems. In: Melin, P., Castillo, O., Aguilar, L.T., Kacprzyk, J., Pedrycz, W. (eds.) IFSA 2007. LNCS (LNAI), vol. 4529, p. 789. Springer, Heidelberg (2007)

[14] Karaboga, D., Basturk, B.: A powerful and efficient algorithm for numerical function optimization: artificial bee colony (ABC) algorithm. Journal of Global Optimization 39(3), 459–471 (2007)

[15] Karaboga, D., Basturk, B.: On the performance of artificial bee colony (ABC) algorithm. Applied Soft Computing 8(1), 687–697 (2008)

[16] Karaboga, D., Akay, B., Ozturk, C.: Artificial bee colony (ABC) optimization algorithm for training feed-forward neural networks. In: Torra, V., Narukawa, Y., Yoshida, Y. (eds.) MDAI 2007. LNCS (LNAI), vol. 4617, p. 318. Springer, Heidelberg (2007)

[17] Kennedy, J., Eberhart, R.: Particle swarm optimization. In: Proc. IEEE International Conference on Neural Networks, vol. 4, pp. 1942–1948 (1995)

[18] Krink, T., Filipic, B., Fogel, G., Thomsen, R.: Noisy optimization problems - a particular challenge for differential evolution? In: Proc. Congress on Evolutionary Computation. IEEE Press, Los Alamitos (2004)

[19] Seeley, T.D.: The wisdom of the hive. Harvard University Press, Cambridge (1995)

[20] Socha, K., Dorigo, M.: Ant colony optimization for continuous domains. European Journal of Operational Research 185(3), 1155–1173 (2008)

[21] Storn, R., Price, K.: Differential evolution - a simple and efficient heuristic for global optimization over continuous spaces. Journal of Global Optimization 11, 341–359 (1997)

[22] Trelea, I.C.: The particle swarm optimization algorithm: convergence analysis and parameter selection. IPL: Information Processing Letters 85, 317–325 (2003)

A Heuristic-Based Bee Colony Algorithm for the Multiprocessor Scheduling Problem

Pascal Rebreyend, Cedric Clugery, and Emmanuel Hily

Abstract. The multiprocessor scheduling is one of the NP-complete scheduling problems. This problem comes when a known parallel program must be executed on a parallel computer. Different methods and algorithms have been tested for this scheduling problem. This paper presents and tests a hybrid bee algorithm. In this approach, the bee algorithm is combined with a heuristic in order to produce quickly good solutions. The choosen heuristic is a greedy approach and hybridization is done using the indirect representation. The heuristic is a list heuristic and the bee algorithm has to find the best order for the ordered list of tasks used by the heuristic. Experimental results on different benchmarks will be presented and analized, as well as a comparison with other hybrid approaches.

1 Introduction

Combinatorial optimization problems are an important research field in computer science because most of these problems are really important in business and daily life. Most of the real optimization problems belong to the NP-class such as Traveling Salesman problem, Vehicle routing problems,... Among these problems, there is an important interest for scheduling problem. Most of them are NP-complete but come from different areas. Aside timetabling which is difficult to handle since it deal with human wishes and considerations, most of scheduling problems are

Pascal Rebreyend
School of Technology and Business Studies, Högskolan Dalarna
e-mail: prb@du.se

Cedric Clugery
Université de Bretagne Occidentale
e-mail: cedric.clugery@gmail.com

Emmanuel Hily
Université de Bretagne Occidentale
e-mail: emmanuel.hily@gmail.com

J.R. González et al. (Eds.): NICSO 2010, SCI 284, pp. 295–304, 2010.
springerlink.com   © Springer-Verlag Berlin Heidelberg 2010

NP-complete optimization problems of processes. In general, people focus on processes inside factories. In our case, we will focus on scheduling tasks on an homogenous multiprocessor computer. Nowadays, parallelism is widely used as a programming scheme for complex computations such as simulations, physics computations, solvers,... Aside its importance as optimization problem, this scheduling problem has the advantage to be both simple and representing well scheduling problems in general. Therefore it's a good choice to investigate the efficiency and behaviour of an hybrid bee-based algorithm.

This multiprocessor scheduling has been studied with lot of different approaches which represent well the main possibilities to find good approximate solutions for an NP-complete problem. Among these approaches, the most interesting one seems to be hybrid evolutionary methods such as genetic algorithms combined with heuristics [5]. Two different approaches exist to combine a heuristic and an evolutionary algorithm: the *direct* and *indirect* representation. Even if they differ a lot, both methods give simular results [1]. But new evolutionary algorithms such as bee colonies haven't been tested yet on this problem. The scope of this work is to investigate the efficiency of bee colonies for this problem and how they can be combined with heuristics.

2   Problem Description

The multiprocessor scheduling is a scheduling problem which belongs to the class of NP-complete problems ([5]). Given a parallel program decomposed into different tasks and a multiprocessor computer , the goal is to find the optimal (or sub-optimal) schedule of tasks. In the case we are dealing with, the number of processor is fixed, all processors are the same. To assume that all processors are identical is motivated by the fact that this reflects the architecture of parallel computers. If this is not the case, algorithms and approaches explain in this paper can be easily deal with this fact and efficiency can be still achieved since the choice of cpu is already part of the solving process. Obviously, we will take into account communication time to reflect the latency of the communication network.

Formelly, we can describe the multiprocessor scheduling as an acyclic digraph $D = (V, A)$ where:

- The set of vertices $V = t_1, \ldots, t_n$ represents the set of the n tasks which compose the program. Each vertex is labeled with a value representing the size of the task.
- Each arc (t_{i_1}, t_{i_2}) of A represents that, at the end of its execution, the task t_{i_1} sends a message which is required by t_{i_2} in order to start its execution. Each arc is labeled with a value representing the size of the corresponding message.
- The goal is to find the optimal schedule s. A schedule s can be represented by a vector $s_1, \ldots, s_n$ where s_j is the ordered list of tasks on the processor j. Each S_j is therefore a vector too. It has been shown and proven that this is way of describing the solution (and to built the real schedule by a greedy scheme) reduce the search space but do not exclude all optimal solution [3]. Therefore in this article we will look to find this set of vectors.

The communication time between two tasks is null if they are schedule on the same processor. Otherwise, the formula used is $\alpha + \lambda * s$ where α is the time needed to start the communication, s the amount of data to transmit and α the time needed to transmit one data. The amount of data to transmit between the two tasks is labeled on each arc of teh graph.

3 Artificial Bee Colony

Studies and modelizations of insects colonies such as ants and bees have shown that swarm intelligence can solve complex problems ([6]). Bee colonies is one of the latest development in this area. Like other methods (ant colonies,...), the aim is to stimulate their behaviour to solve our complex problems. Bee lives in "hives" and like ants, they have to explore the surroundings to find food sources. But the comparison between bees and ants stops at this point: the communication between bees is completly different from communication between ants. Communication between bees is done by a dance. Before to go out and bring back some food, bees are watching other bees dancing on the "dance floor" and the dance is their way of describing where the food is. This scheme has lot of differences with the usage of pheromons by ants. With pheromon, the "talk" of different ants is naturally agregated as the quantity of pheromon increases on the path. At the same time, the description of one solution is spread out on its set of paths. On the other hand, on the dance floor it's only a set of differents solutions which are presented at the same time.

In the litterature, descriptions of bee behaviour and bee algorithms differ but all of them have a common part: a dance floor where bees express their solution an dthey work on a complete solution.

Few researchers have worked with bee colonies on scheduling problems and have shown interesting results ([2, 4, 12]). Chong and al ([2]) are focused on the job shop scheduling. This problem shows some similarities with the multiprocessor scheduling problem such as minimizing the makespan. Authors define the *profitability rating* of a bee as the inverse of the makespan of the solution. This profitability rating is used to comute the duration of the dance of the corresponding bee. A lookup-table is used to adjust the probability for a bee to follow a particular path. Experiments are done on problems up to 50 jobs and with 5 to 20 machines. Results shows that the bee approach performs slightly better than an ant colony system but cannot match a tabu search. Wong and al ([12]) are working on the job schop sscheduling too. Their work is based on the big valley landscape structure ([11]). In their work, authors starts to generate feasible solutions using different dispatching rules or heuristics. Another particularity of thei bee colony is to add to each some memory using a tabu list. Authors also introduce a measure of the distance between two solutions to replace a solution by solutions which are not too "far" from it. INstead of using a lookup table as in [2], authors have tried different strategies for a bee to pick and follow a dance and after initial experiment decided to use the round robin method. Experiments are done on small-size problem as Chang and al ([2]). Results obtained are in the line with top of the art methods for this problem but their bee approach seems to perform well in the long term. Davidović and al ([4]) work on the multiprocessor scheduling problem but they do not take into account communication times. Their algorithm consists of two passes: a forward pass where bee constructs solutions and a forward pass where communication between bees is done. In the forward pass, the processing time of a task is used to biased the probability for a task to be choosen. Bee are producing partial solutions and durng the backward pass, bee decides to continue to complete the partial solution or to explore the search space. Their algorithm is test with graphs up to 100 tasks scheduled on 4 processors and in few second the optimal solution is found.

Anyway, a real comparison with other existing methods is interesting to investigate. An interesting point is that descriptions of algorithms based on bee colonies differ a lot between authors. Especially, often the application of biological facts to the specific problem is confusing.

4   Proposed Method

As explained in the previous section, bee colonies work on a complete solution. Like others evolutionary methods, they suffer from a slow convergence to interesting solutions. It's why in general they are mixed with other methods such as heuristic in order to reach quickly good solutions. This meta-method is called hybrid algorithm.

4.1   Background on Hybrid Methods

Research on this problem has shown the efficiency of hybrid methods. Best results known so far comes mainly from hybrid algorithms which mixed a heuristic and an evolutionary algorithm such as genetic algorithms as shown in [9]. Hybridization between an evolutionary algorithm and a heuristic can be done in two different ways ([1]):

The *indirect* approach is based on a list heuristic. A list heuristic is a heuristic which take as input a total order among elements. For scheduling problems, elements are often the different tasks. Based on this order, the heuristic will produce a final solution using the knowledge embedded in the algorithm. The goal of this evolutionary method is to find an order among elements as best as possible. Often, the list heuristic used is a greedy approach which schedules as soon as possible each task. One main advantage of this approach is in the ease of building an hybrid algorithm. The drawback is that the heuristic is a gateway to the solution and therefore its choice is very sensitive and this kind of algorithm can perform well or badly depending of the instance and the poor flexibilty of this approach. This approach is often used with a simulated annealing and a list algorithm

The *direct* approach as been designed to overcome this problem. In this approach, the heuristic knowledge is integraded directly into the evolutionary algorithm. In most cases, the evolutionary part is a genetic algorithm. Instead of being a TSP-like representation (total order) like in the indirect method, the genetic algorithm works on a complete solution of the problem. The heuristic is introduced into the genetic algorithm via the crossover and/or the mutation operators. This approach is more robust since by design the complete solution is not built only through the heuristic but also by the evolutionary algorithm. This approach is more flexible: the heuristic is just a redesign of an operator and more than one heuristic can be used at the same time. The main drawback is that this approach needs more work to design the hybrid algorithms and the complexity of the new operator can be high ([10]). And for each evolutionary algorithm some work is needed to built the complete method.

Practical results on this scheduling problems show similar results. We have decided to follow the first approach, the indirect method to test bee colonies on our scheduling problem.

4.2   The Greedy Heuristic

The chosen heuristic is a simple greedy approach. This choice is guided by the need to have a heuristic which does not excludes too many potential solutions. Therefore we have decided to use a randomized heuristic. By randomized, we mean that we use the greedy scheme to filter the different choices but in case of equality regarding the starting time between different tasks to schedule, we will pick at random a task. The choice of processor will be done ramdomly too among all processor minimizing the starting time of the corresponding task. This can be seems as a weak heuristic but this weakness is in fact some flexibility and this is a wish in order to be able to guide the heuristic by the evolutionary method. The heuristic picks at random a free task (i.e a task for which all predecessors are already scheduled) and assigns to one processor at random. The starting time is the earliest one. This is repeated until the complete schedule is built.

4.3   The Dance Floor

In our case, we will work with a fix number of bees and like other methods, we will simulate the evolution by iterations of a main loop. For a shake of simplicity, we will assume that a bee has two possible states: either the bee is dancing on the floor, either the bee is looking for some food source.

At each generation, bees who are looking on the dance floor will either, with a probability 0.1, generates a random solution or look at random a bee dancing on the floor (probability 0.9) and picks its solution except for the first iteration where all bees generate random solutions. When a bee observes the dance of another bees, this communication is prone to error or inacurracy. For us, this means that the bee will look for a solution in the neighborhood of the solution of the dancing bee. This idea is close to the mutation of the Genetic Algorithm or on how simulated annealing and tabu search work. In our case, each bee represents the solution by an ordered list of tasks. We have choosen to swap some pairs of tasks. after some experiments, a number of pairs has been fixed at 5 % of the tasks.

Once a bee has generated a new solution, the bee may stay on the dance floor. The general idea is that, the better is the solution, the longer should dance the bee and expose its solution. This is simply done by, at the end of each generation, keeping the bees representing the best 5% of solutions. Other possibilities exist in the litterature but this is one is simple and has two advantages: the number of bee dancing is kept constant and we don't compute in advance how long a bee will dance, regardless of the quality of solutions which will enter the dance floor later.

We have tested two different list heuristics to build the complete schedule from the total order amon the tasks. Both heuristics are based on the same engine: We built the schedule step by step. At each step, a task is scheduled on one processor. In order to do that, the list of free tasks (i.e tasks for which all predecessors in the graph are already schedule) is built and the one with the higher order of priority according the list build by the bee part of the algorithm is choosen. For both heuristics, the processor is chosen at random between all processor on which the starting time of the task is the earliest one.

4.4 Heuristic 1

In this heuristic, we filter the list of free tasks such as to keep only tasks with the earliest starting time. Using this method, we use the knowledge represented by the greedy approach. Since this approach is not design to make distinctions between tasks with the same starting time, the bee colony will be used to choose the task to schedule among all tasks with the minimum starting time.

4.5 Heuristic 2

In the second heuristic, we release the constraint about the starting time of free tasks. Therefore we choose at random among all free tasks. This lead to give relatively more power to the evolutionary algorithmand therefore may extend the search space and avoid particular local minimas. The greedy idea is still used in the schedule since we are scheduling each as soon as possible on only on a processor where the starting time is the minimum for this task.

4.6 Parameters

The number of bees is equal to the number of tasks. The number of bees is one of the parameters of bee colony algorithms. This number is connected to the number of parallel explorations of solutions by bees and to the number of bees dancing on the dance floor. Obviously, the bigger the problem (or the higher is its dimension), the more bees should be in the system since the search space is bigger and has more dimensions. It's why we have adapted this rule and in practice, like hybrid genetic algorithm, this parameter has not strong effect on the efficiency as far as its value is average. Some initial experiments with a fixed number of bees has confirmed us in this choice.

Other parameters have been choosen after some experiments: The probability for a bee to watch a given dance is 0.9 (and 0.1 to explore at ranom). Like the number of bees in the systems, results show that this hybrid system is not very sensitive to this parameter, like most of hybrid systems. An interesting point is that, since the number of bees is fixed and a bee is either dancing, either watching the dancefloor and exploring the surroundings, we keep 5 % of the bees on the dance floor at the end of each generation and we don't decide in advance how long the bee will dance.

Table 1 Summary of results, 2 hours run

Graph	size	Best known results before	Heuri Random 2 hours	bee-1 2 hours	bee-2 2 hours
Bellford	m	**71 936 050**	81 088 350	73 254 250	73 254 250
	l	193 794 250	210 392 700	**185 960 150**	**185 960 150**
Diamond-1	m	**131 940 750**	134 422 800	**131 940 750**	**131 940 750**
	l	**276 758 950**	301 345 850	295 760 200	303 439 950
Diamond-2	m	**127 224 450**	154 219 000	133 327 550	137 920 400
	l	**218 954 400**	286 231 500	240 218 850	278 276 200
Diamond-3	m	**176 982 100**	191 046 900	179 024 600	179 024 600
	l	**228 590 500**	245 601 300	229 520 750	228 642 100
Diamond-4	m	**132 875 550**	161 903 850	160 468 850	159 578 500
	l	**164 264 000**	193 110 750	186 117 300	186 117 300
Divconq	m	**97 307 880**	101 788 840	99 460 640	98 524 780
	l	**169 043 350**	175 227 610	172 662 500	171 417 990
fft	m	**29 888 250**	32 085 550	30 986 900	30 986 900
	l	102 647 300	105 023 800	**100 875 950**	**100 875 950**
Gauss	m	**226 882 900**	249 342 850	241 115 100	234 992 850
	l	**307 395 800**	353 121 550	330 620 250	315 658 650
Iteratif	m	**16 733 350**	22 099 550	21 382 050	21 808 700
	l	**47 936 700**	54 820 850	52 661 200	54 542 100
ms-gauss	m	**4 598 559 550**	4 605 112 750	4 605 112 750	4 605 112 750
	l	**2 559 388 750**	2 710 832 350	2 561 870 800	2 559 444 800
prolog	m	**60 499 350**	63 292 900	60 575 800	60 575 800
	l	**258 529 350**	261 322 900	261 322 900	258 605 800
qcd	m	**1 173 100 600**	1 176 047 050	1 176 047 050	1 176 047 050
	l	**2 038 576 050**	2 041 522 500	2 041 522 500	2 041 522 500
elbow		**6630**	**6630**	**6630**	**6630**
stanford		**627**	647	629	629
ssc	5	**74**	106	104	107
	6	**77**	127	124	128
	7	**82**	150	145	150
	8	**86**	177	173	176
	9	**89**	187	184	181
Average distance from the best in %		0.193	21.183	16.349	20.307

5 Experiments

Experiments are done using different benchmarks. Some benchmarks has been generated by the tool ANDES-SYNTH ([8]). These synthetic graphs represent the most common parallel programs and values represents an IBM-SP1. Graphs called ssc come from a controller and elbow and stanford graphs represent a robot shoulder and arm ([7]). Results are presented in the table (1). Best known results (oublished in [9]) before this project are presented in the first column. The computations were performed on UPPMAX resources under Project p2009019. All other columns represent results achieved after 2 hours of computation on a single nod of an X3455 IBM computer (AMD opteron) of the Uppmax cluster Isis.

From the results, we clearly see that the bee algorithm is more efficient than the random heuristic used by it but at the same time, it suffer from this approach by giving poor results on some graphs (like the ssc family). By comparison to detailed results of other methods ([9], results are mixed. Obviously, the heuristic used can be improved by using other criterias such as the level of the task in the graph.

The figure 1 shows for 4 different graphs results obtained by the different methods proposed here with the respect of their processing time. They represent well the

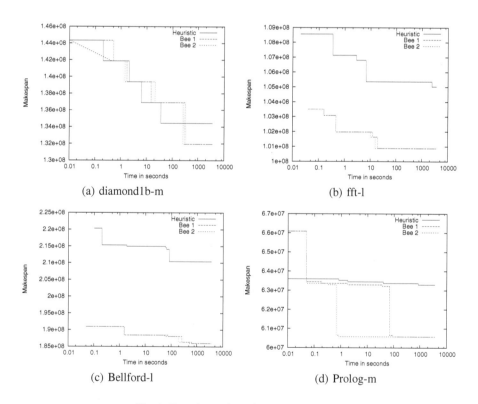

(a) diamond1b-m

(b) fft-1

(c) Bellford-1

(d) Prolog-m

Fig. 1 Experimental results on some graphs

general behaviour of other graphs of the benchmark. The two bee algorithms have similar results and outperformed the heuristice on which they are based after some time.

6 Future Work

Results leads to think about the connection between the evolutionary method and the heuristic. To develop a hybrid method using the direct representation can be an idea. Another idea can be to improve the bee colony algorithm: we may add extra information to the solution to guide more the heuristic (and therefore use a less strong greedy method), or modify the bee algorithm in the way that a bee is looking at several dances and mix them. But at this stage the question is what will be then the difference with a Genetic Algorithm in practice?

7 Conclusion

Bee algorithms have been recently tested on different problems. Aside their novelty, methods based on bees colonies are often simple to develop. Experiments show that bee methods give some improvments in comparison of the heuristic used in this approach, especially for huge problems. But, on the other hand, like other indirect approaches ([9]), the choice heuristic is very important and results are very sensitive to the input data. Experiments confirm previous conclusions about the indirect hybrid approach used b other evolutionnary algorithms: easy to implement, fast results but strongly connected of the heuristic used and find the compromise between the power of each method is rather difficult and unflexible. Anyway, bee colonies are a good solution to such problems.

References

[1] Chamaret, B., Rebreyend, P., Sandnes, F.E.: Scheduling problems: A comparison of hybrid genetic algorithms. In: Proceedings of the 2nd IASTED International Conference on Parallel and Distributed Computing and Networks, pp. 210–213. ACTA Press, Brisbane (1998) ISBN 0-88986-237-0, ISSN 1027-2658

[2] Chong, C.S., Sivakumar, A.I., Low, M.Y.H., Gay, K.L.: A bee colony optimization algorithm to job shop scheduling. In: Perrone, L.F., Lawson, B., Liu, J., Wieland, F.P. (eds.) Winter Simulation Conference, WSC, pp. 1954–1961 (2006)

[3] Corrêa, R., Ferreira, A., Rebreyend, P.: Integrating list heuristic into genetic algorithms for multiprocessor scheduling. In: Eighth IEEE Symposium on Parallel and Distributed Processing, pp. 462–469. IEEE Computer Society, New-Orleans (1996) ISSN-ISBN 0-8186-7683-3

[4] Davidovic, T., Selmic, M., Teodorovic, D.: Scheduling independent tasks: Bee colony optimization approach. In: Mediterranean Conference on Control and Automation, pp. 1020–1025 (2009),
http://doi.ieeecomputersociety.org/10.1109/
MED.2009.5164680

[5] Hou, E.S., Ansari, N., Ren, H.: A genetic algorithm for multiprocessor scheduling. IEEE Transactions on Parallel and Distributed Systems 5(2), 113–120 (1994), http://doi.ieeecomputersociety.org/10.1109/71.265940

[6] Karaboga, D., Basturk, B.: On the performance of artificial bee colony (abc) algorithm. Applied Soft Computing 8(1), 687–697 (2008),
http://www.sciencedirect.com/science/article/
B6W86-4NWCGRR-G/2/422ccff5df9d32a5bf8517068ca2a094

[7] Kasahara, H., Narita, S.: Practical multiprocessor scheduling algorithms for efficient parallel processing. IEEE Transactions on computers C-33(11), 1023–1029 (1984)

[8] Kitajima, J.: Modèles quantitatifs d'algorithmes parallèles. PhD thesis, LMC-IMAG (1994)

[9] Rebreyend, P.: Algorithmes génétiques hybrides en optimisation combinatoires. PhD thesis, Lip, ENS-Lyon, France (1999),
http://pascal.rebreyend.free.fr/Fichiers/these.pdf

[10] Rebreyend, P., Sandnes, F., Megson, G.: Static multiprocessor task graph scheduling in the genetic paradigm: A comparison of genotype representations. Research Report RR1998-25, LIP-ENS-Lyon, 46 allée d'Italie, F-69364 Lyon Cedex 07, France (1998),
ftp://ftp.ens-lyon.fr/pub/LIP/Rapports/RR/RR1998/
RR1998-25.ps.Z

[11] Reeves, C.: Landscapes, operators and heuristic search. Annals of Operations Research 86(0), 473–490 (1986),
http://dx.doi.org/10.1023/A:1018983524911

[12] Wong, L.P., Puan, C.Y., Low, M.Y.H., Chong, C.S.: Bee colony optimization algorithm with big valley landscape exploitation for job shop scheduling problems. In: Mason, S.J., Hill, R.R., Mönch, L., Rose, O., Jefferson, T., Fowler, J.W. (eds.) Winter Simulation Conference, WSC, pp. 2050–2058 (2008)

A Bumble Bees Mating Optimization Algorithm for Global Unconstrained Optimization Problems

Yannis Marinakis, Magdalene Marinaki, and Nikolaos Matsatsinis

Abstract. A new nature inspired algorithm, that simulates the mating behavior of the bumble bees, the Bumble Bees Mating Optimization (BBMO) algorithm, is presented in this paper for solving global unconstrained optimization problems. The performance of the algorithm is compared with other popular metaheuristic and nature inspired methods when applied to the most classic global unconstrained optimization problems. The methods used for comparisons are Genetic Algorithms, Island Genetic Algorithms, Differential Evolution, Particle Swarm Optimization, and the Honey Bees Mating Optimization algorithm. A high performance of the proposed algorithm is achieved based on the results obtained.

1 Introduction

In the last years, several biological and natural processes have been influencing the methodologies in science and technology in an increasing manner. Among the most popular nature inspired approaches, when the task is optimization within complex domains of data or information, are those methods representing successful animal and micro-organism team behaviour, such as the Particle Swarm Optimization [14], the artificial immune systems [6], the Ant Colony Optimization [7], etc. Also, a number of swarm intelligence algorithms, based on the behaviour of the bees have

Yannis Marinakis · Nikolaos Matsatsinis
Decision Support Systems Laboratory,
Department of Production Engineering and Management,
Technical University of Crete, 73100 Chania, Crete, Greece
e-mail: {marinakis,nikos}@ergasya.tuc.gr

Magdalene Marinaki
Industrial Systems Control Laboratory,
Department of Production Engineering and Management,
Technical University of Crete, 73100 Chania, Crete, Greece
e-mail: magda@dssl.tuc.gr

J.R. González et al. (Eds.): NICSO 2010, SCI 284, pp. 305–318, 2010.
springerlink.com   © Springer-Verlag Berlin Heidelberg 2010

been presented [4]. These algorithms are divided, mainly, in two categories according to their behaviour in the nature, the foraging behaviour and the mating behaviour. The most important approaches that simulate the foraging behaviour of the bees are the Artificial Bee Colony (ABC) Algorithm proposed by [13], the Virtual Bee Algorithm proposed by [27], the Bee Colony Optimization Algorithm proposed by [24], the BeeHive algorithm proposed by [26], the Bee Swarm Optimization Algorithm proposed by [8] and the Bees Algorithm proposed by [22]. Contrary to the fact that there are many algorithms that are based on the foraging behaviour of the bees, the main algorithm proposed based on the marriage behaviour is the Honey Bees Mating Optimization Algorithm (HBMO), that was presented in ([1, 2]) and simulates the mating process of the queen of the hive. Since then, it has been used on a number of different applications [3, 10, 11, 15–20, 23].

In this paper, a new algorithm that simulates the mating behavior of the Bumble bees, the **Bumble Bees Mating Optimization (BBMO)** algorithm is presented, analysed and used for solving unconstrained optimization problems. This algorithm is a population-based swarm intelligence algorithm that simulates the mating behavior that a swarm of bumble bees perform. An hybridized initial version of the algorithm was presented in [21] for clustering. The other methods used for the comparisons are the Genetic Algorithms [12] and their variants, the Differential Evolution [25], the Particle Swarm Optimization [14] and its variants and the Honey Bees Mating Optimization algorithm [1, 3]. The test functions used are the Rosenbrock, the Sphere, the Rastrigin and the Griewank.

The rest of the chapter is organized as follows. In the next section an analytical description of the proposed algorithm is given. In the third section, the test functions used are given while in the fourth section computational results are presented and analyzed. The last section gives the conclusions and future research.

2 The Proposed Bumble Bees Mating Optimization Algorithm

In this section, initially, the bumble bees behavior is presented, while in the following the proposed algorithm based on this behavior is presented and analyzed in detail.

2.1 Bumble Bees Behavior

Bumble bees are social insects that form colonies consisting of the queen, many workers (females) and the drones (males). Queens are the only members of the nest to survive from one season to the next, as they spend the winter months hibernating in a protected underground overwintering chamber. Upon emerging from hibernation, a queen collects pollen and nectar from flowers and searches for a suitable nest site and when she finds such a place, she prepares wax pots to store food and wax cells into which eggs are laid ([28–31]).

The bumble bee queen can lay fertilized or unfertilized eggs. The fertilized eggs have chromosomes from the queen and a male or males she mated with the previous year and they develop into workers while the unfertilized eggs contain chromo-somes from the queen alone and they develop into males. After the emergence of the first workers, the queen no longer forages as the workers take over the respon-sibilities of collecting food (foragers) and the queen remains in the nest laying eggs and tending to her young. Some workers, also, remain in the nest and help raise the brood (household workers). Males do not contribute in collecting food or helping rear young as the sole purpose of the males are to mate with the queens. Bumble bee workers are able to lay haploid eggs when the queen's ability to suppress the workers' reproduction diminishes. These eggs are developed into viable male bum-ble bees ([28–31]).

A few days after the males leave the nest, new queens will emerge. After new queens and males have gone, the colony begins to deteriorate. The founder queen stops laying eggs and grows weak from old age while the remaining workers con-tinue to forage for food but only for themselves. Away from the colony, the new queens and males live off nectar and pollen and spend the night on flowers or in holes. The queens are eventually mated (often more than once), the sperm from the mating is stored in spermatheca and she searches for a suitable location for diapause. Three different mating behaviors exist in bumble bees. The first mating behavior is where a male perches on a tall structure and waits for queens to fly by and he will pursue them for mating once one queen is spotted. The second mating behavior is when males create a scent trail, marking their flight path with pheromones and, thus, queens of the same species will be attracted to the pheromones and follow the scent trail. The third mating behavior is where males wait at the entrance of a bumble bee nest for queens to leave ([28–31]).

2.2 BBMO for Global Unconstrained Optimization

In the BBMO algorithm, there are three kind of bumble bees in the colony, the queen, the workers and the drones (males). Initially, a number of bees are selected randomly. Each bee (a bee corresponds to an individual in the population) represents a candidate solution of the problem. Let n be the total number of variables. The bees are represented by vectors of dimension n. We use a real valued representation where initially the values of each of the bees are random numbers between 0 and 1. Afterwards, the fitness of each bee is calculated using each one of the test functions depending of the problem (see section 3) and the best bee is selected as the queen. All the other bees in the initialization phase of the algorithm are the drones.

The queen selects the drones that are used for mating by using the second mating behavior where it is assumed in the algorithm that the fittest males let larger amount of pheromone in their flight paths and, thus, the queen selects the most promis-ing paths. This procedure is realized by sorting of all drones based on their fitness function. Each time the queen successfully mates with a drone, the genotype of the drone is stored in her spermatheca until the maximum number of matings has been reached.

After the mating, the queen finds a place to hibernate and in the next year (a year corresponds to an iteration) finds a place to create the hive and to begin to lay eggs. There are three kinds of bees that a queen lays: new queens, workers and drones. The first two kinds of bees are created by crossover of the genotype of the queen and the genotype of the drones using a specific crossover operator. In this crossover operator, the points are selected randomly from the selected drones and from the queen. Thus, initially a crossover operator number is selected (Cr_1) that controls the fraction of the parameters that are selected for the drones and the queen. The Cr_1 value is compared with the output of a random number generator, $rand_i(0,1)$. If the random number is less or equal to the Cr_1 the corresponding value is inherited from the queen, otherwise it is selected, randomly, from the solutions of one of the drones' genotypes that are stored in spermatheca. Thus, if the solution of the brood i is denoted by $b_{ij}(t)$ (t is the iteration number and j is the dimension of the problem ($j = 1, \cdots, n$)), the solution of the queen is denoted by $q_j(t)$ and the solution of the drone k is denoted by $d_{kj}(t)$:

$$b_{ij}(t) = \begin{cases} q_j(t), & \text{if } rand_i(0,1) \leq Cr_1 \\ d_{kj}(t), & \text{otherwise.} \end{cases} \quad (1)$$

The fittest of the broods are selected as new queens while the rest are the workers. The new queens are selected to be equal to the maximum number of the queens. Initially, the new queens are fed from the old queen (or queens) and, afterwards, from the workers and the old queen (or queens). The reason that we use this procedure is to improve the genotype (solution) of each new queen. This is achieved by using a local search phase where each new queen selects which of the workers and the old queen (or queens) are going to feed her by using the following equation:

$$nq_{ij} = nq_{ij} + (b_{max} - \frac{(b_{max} - b_{min}) * lsi}{lsi_{max}}) * (nq_{ij} - q_j) +$$
$$\frac{1}{M} * \sum_{k=1}^{M} (b_{min} - \frac{(b_{min} - b_{max}) * lsi}{lsi_{max}}) * (nq_{ij} - w_{kj}) \quad (2)$$

where nq_{ij} is the solution of the new queen i, q_j is the the solution of the old queen (or queens), w_{kj} is the solution of the worker, M is the number of the workers that each queen selects for feeding her and it is different for each queen, b_{max}, b_{min} are two parameters with values in the interval $(0,1)$ that control if the new queen is fed from the old queen (or queens), from the workers or from both of them, lsi is the current local search iteration and lsi_{max} is the maximum number of local search iterations. Initially, the new queens are fed more from the old queen (or queens) and as the local search iterations increase, then only the workers feed the new queen. The appropriate choice of the values of b_{max} and b_{min} controls the feeding process, i.e. in order to have the feeding process described previously, a large value for b_{max} and a value almost equal to zero for b_{min} are necessary. Afterwards, the new queens leave from the hive.

The drones are produced by mutate the old queen's (or old queens') genotype or by mutate the fittest workers' genotype using a random mutation operator. In this mutation operator, the changes in the genotype of the old queens or the workers are performed randomly.

The drones, then, leave from the hive and they are looking for new queens for mating. As the drones leave from the hive they are moving in a swarm in order to find the best places to wait for the new queens to find them by their marked flight paths. The movement of the drones away from the hive is calculated from the following equation:

$$d_{ij} = d_{ij} + \alpha * (d_{kj} - d_{lj}) \tag{3}$$

where d_{ij}, d_{kj} and d_{lj} are the solutions of the drones i, k, l respectively and α is a parameter that determines the percentage that the drone i is affected by the two other drones k and l. The new queen select the drones that are used for mating by the procedure described previously. In the next generation, the best fertilized queens survive and all the other members of the population die. A pseudocode of the proposed algorithm is presented in Table 1.

It should be noted that the proposed Bumble Bees Mating Optimization (BBMO) algorithm that is inspired from the mating behavior of the bumble bees, it has a number of differences compared to another nature inspired algorithm that is based on the mating behavior of honey bees, the Honey Bees Mating Optimization (HBMO) algorithm [1, 15, 20]. The *Honey Bees Mating Optimization* algorithm simulates the mating process of the queen of the hive, where there are three kinds of bees, the queen, the drones and the workers. The mating process of the queen begins when the queen flights away from the nest performing the mating flight during which the drones follow the queen and mate with her in the air. The main differences of the two algorithms are:

- In the BBMO the workers are different solutions while in the HBMO they are local search phases. This helps the exploration abilities of the population by searching in different places in the solution space.
- In the BBMO after the mating of the queen three kinds of bumble bees are produced, the new queens and the workers (by using a crossover operator) and the drones (by using a mutation operator). On the other hand, in the HBMO after the mating of the queen two kinds of honey bees are produced, the queen and the drones (both of them by using a crossover operator). By using in the proposed algorithm a mutation operator to produce new solutions we have the possibility to obtain completely different solutions.
- In the BBMO the fittest of the broods produced by the crossover operator are the new queens and all the others are the workers while in the HBMO the fittest of the broods is the new queen and all the others are the drones.
- In the BBMO the drones are produced by mutation of the queen or by mutation of the fittest workers. In the HBMO the drones are all the bees produced by the crossover operator except of the queen. By using in the proposed algorithm

Table 1 Bumble Bees Mating Optimization Algorithm

Algorithm Bumble Bees Mating Optimization Algorithm
Definition of parameters for the main phase of the algorithm
    Definition of the maximum number of iterations
    Definition of the maximum number of matings
    Definition of the maximum number of queens
Initialization Phase
    Generate the initial population of the bumble bees
    Calculation of the fitness function of each bumble bee
    Selection of the bee with the best fitness function as the queen
    Selection of the rest bees as the drones
    Sorting the drones according to their fitness' functions
    Selection of the drones for mating by the queen
    Storing the drones' genotype to queen's spermatheca
Main Phase
do while the maximum number of iterations has not been reached
    Creation of the broods by using a crossover operator
    Calculation of the fitness function of each brood
    Sorting the broods according to their fitness' functions
    Selection of the best broods as the new queens
    Selection of the rest broods as the workers
    Feeding of the new queens by the old queens and the workers
    Creation of a percentage of the drones by mutating of the old queens' genotypes
    Creation of the rest of the drones by mutating of the workers' genotypes
    Calculation of the fitness function of each drone
    Calculation of the moving direction of the drones away from the hive
    Sorting the drones according to their fitness' functions
    do while the maximum number of matings for each new queen has not been reached
        Selection of the drones for mating by each new queen
        Storing the drones' genotypes to each new queen's spermatheca
    enddo
    Survival of the new queens for the next iteration
    Dying of all the other members (workers and drones) of the population
enddo
return The best queen (best solution found)

a mutation operator to produce new solutions we have the possibility to obtain completely different solutions.

- In the BBMO the drones are moving away of the hive and this affects their solutions.
- The feeding procedure in the BBMO is as described previously using the Equation (2) while in the HBMO the feeding procedure is local search phases that are applied independently in each brood.

3   Test Functions

In this paper, four functions are used in order to show the effectiveness of the proposed BBMO algorithm when used for global unconstrained optimization problems. The test functions used are the Rosenbrock, the Sphere, the Rastrigin and the Griewank. The Rosenbrock function is given by:

$$f(x) = \sum_{i=1}^{n-1} (100(x_{i+1} - x_i^2)^2 + (x_i - 1)^2) \tag{4}$$

The Sphere function is given by:

$$f(x) = \sum_{i=1}^{n} x_i^2 \tag{5}$$

The Rastrigin function is given by:

$$f(x) = \sum_{i=1}^{n} (x_i^2 - 10cos(2\pi x_i) + 10) \tag{6}$$

The Griewank function is given by:

$$f(x) = \sum_{i=1}^{n} \frac{(x_i - 100)^2}{4000} - \prod_{i=1}^{n} cos(\frac{x_i - 100}{\sqrt{(i)}}) + 1 \tag{7}$$

In functions Sphere, Rastrigin and Griewank the global minima is $f(x^*) = 0$ with $x^* = (0, \cdots, 0)$, and in Rosenbrock the global minima is $f(x^*) = 0$ with $x^* = (1, \cdots, 1)$.

4   Results

The algorithm was implemented in Fortran 90 and was compiled using the Lahey f95 compiler on a Centrino Mobile Intel Pentium M750 at 1.86GHz, running Suse Linux 9.1. The parameter settings for the Bumble Bees Mating Optimization algorithm were selected after thorough empirical testing. A number of different alternative values were tested and the ones selected are those that gave the best computational results. Thus, the selected parameters are: The number of the total bees (workers - males - queens) is set equal to 100, the number of generations is set equal to 10000. Usually the total number of bees is divided in 5 queens, 45 workers and 50 males but as it is presented in Table 2 the algorithm is tested and for different number of queens. The lsi_{max} is set equal to 100, the b_{max} is set equal to 0.99, the b_{min} is set equal to 0.001 and the α is set equal to 0.8. All the algorithms used in the comparisons are population based algorithms and, thus, in order to have fair comparisons we test the algorithms using the same number of individuals (or particles for the PSO or bees for the Honey Bees Mating Optimization) and generations (or iterations). Thus, we have the same function evaluations.

In Table 2, the performance of the proposed Bumble Bees Mating Optimization algorithm in Rosenbrock function is presented. In this Table, the final cost of the Rosenbrock for six different variables ($n = 2, 4, 8, 10, 20, 50$) is presented. The effectiveness of the proposed algorithm is given using different number of queens, namely, $q = 1, 2, 5, 10, 20$. As it can be observed in all cases the proposed algorithm finds the optimum when the number of variables is less or equal to 10. When the number of variables becomes equal to 20, the optimum is found with the use of 5 or 10 queens. The algorithm did not find a solution near to the optimum only in the case when we use 50 variables. The combination of parameters that gave the best results is when 5 queens are used and, thus, in all other Tables that are presented in this section, the algorithm uses 5 queens. After the selection of the final parameters, 50 different runs with the selected parameters were performed for each of the problems. The results presented in Tables are the best results found for each problem.

Table 2 Results of Bumble Bees Mating Optimization algorithm for the Rosenbrock

Queens						
	2	4	8	10	20	50
1	0.00	0.00	0.00	0.00	1.28E-09	31.69
2	0.00	0.00	0.00	0.00	1.93E-09	75.85
5	0.00	0.00	0.00	0.00	0.00	24.37
10	0.00	0.00	0.00	0.00	0.00	76.63
20	0.00	0.00	0.00	0.00	2.31E-09	77.94

We, also, tested the algorithm using 100000 iterations. In Table 3, the results of all the functions used are presented. As it can be observed for all test functions when the number of variables is less or equal to 20, the proposed algorithm finds the optimum. When the number of variables is equal to 50 the proposed algorithm finds the optimum in Sphere and Griewank test functions. For the Rastrigin function if the number of iterations is equal to 10000 the solution is close to the optimum and is equal to 1.59E-08, while when the number of iterations becomes equal to 100000 the optimum is found. Only in the case of the Rosenbrock function for $n = 50$ the optimum was not found but the increase of iterations to 100000 leads the algorithm to find a better solution near to the optimum.

A comparison with other population based metaheuristic approaches for the solution of the same test functions is presented in Tables 4 and 5. In these Tables, besides the proposed algorithm, five other algorithms are used for the solution of the four test functions. The algorithms are a Honey Bees Mating Optimization algorithm, a Genetic Algorithm, an Island Genetic Algorithm [9], a Differential Evolution algorithm and a Particle Swarm Optimization algorithm. In all algorithms, we used

Table 3 Results of Bumble Bees Mating Optimization algorithm for the four functions

Function	Iterations	n					
		2	4	8	10	20	50
Rosenbrock	10000	0.00	0.00	0.00	0.00	0.00	24.37
Sphere		0.00	0.00	0.00	0.00	0.00	0.00
Rastrigin		0.00	0.00	0.00	0.00	0.00	1.59E-08
Griewank		0.00	0.00	0.00	0.00	0.00	0.00
Rosenbrock	100000	0.00	0.00	0.00	0.00	0.00	1.62E-02
Sphere		0.00	0.00	0.00	0.00	0.00	0.00
Rastrigin		0.00	0.00	0.00	0.00	0.00	0.00
Griewank		0.00	0.00	0.00	0.00	0.00	0.00

the same parameters as in the previous comparisons, the same number of individuals (or particles for the PSO or bees for the HBMO) and two different number of generations (or iterations in PSO), namely 10000 (Table 4) and 100000 (Table 5). As all these algorithms have a number of different variants we use in the comparisons the variant that worked better for global unconstrained optimization problems. Thus, the selected variants for the final comparisons are, for the Differential Evolution the rand/1/bin, where "rand" corresponds to the target vector (a random target vector), "1" corresponds to the number of different vectors and "bin" corresponds to the crossover operator (for more details for the notation in differential evolution algorithms please see [25]), for the PSO the Constriction PSO [5], for the Genetic Algorithms the combination with Linear Crossover and Roulette Wheel Selection and for the Island Genetic Algorithms, the combination with Linear Crossover, Tournament selection, ten different islands and migration of the best individuals after 100 generations.

As it can be observed the proposed BBMO algorithm performs better compared to the other population based metaheuristic algorithms used in the comparisons. The BBMO algorithm performed better than the HBMO as the performance of HBMO algorithm was, in general, very good but HBMO found the optimum in less cases than the BBMO. For $n=2$, independently of the number of iterations, the HBMO algorithm found the optimum, for $n=4$, 8, 10, the optimum was not found in all cases but values near to the optimum were found (these values were even closer to the optimum or became equal to the optimum when the number of iterations was equal to 100000), for $n=20$, 50 the results of HBMO were less efficient than the cases where a smaller number of variables was used for all test functions but also in these cases an increase in the performance of HBMO was observed when 100000 iterations were used. It should be noted that for the case of $n=50$, the results of the BBMO are much better than the ones of the HBMO. The BBMO algorithm performed better than the GA as the GA found the optimum in less cases than the BBMO. In the cases where the optimum was not found by the GA, values near to the optimum were found and these values were improved when the number of iterations was equal to 100000. When the number of variables was equal to 50,

Table 4 Comparisons of the BBMO algorithm with other metaheuristics for the four functions (10000 iterations)

	Function	Iterations	2	4	8	10	20	50
						n		
BBMO	Rosenbrock	10000	0.00	0.00	0.00	0.00	0.00	24.37
	Sphere		0.00	0.00	0.00	0.00	0.00	0.00
	Rastrigin		0.00	0.00	0.00	0.00	0.00	1.59E-08
	Griewank		0.00	0.00	0.00	0.00	0.00	0.00
HBMO	Rosenbrock		0.00	1.32E-05	8.35E-03	5.90E-02	6.38	46.07
	Sphere		0.00	0.00	0.00	0.00	1.39E-07	0.67
	Rastrigin		0.00	0.00	0.00	1.58E-09	3.93E-05	4.03
	Griewank		0.00	0.00	0.00	0.00	0.00	1.44E-02
GA	Rosenbrock		0.00	0.00	1.30	0.90	5.28	26.85
	Sphere		0.00	0.00	6.75E-08	7.31E-08	1.10E-06	9.83E-06
	Rastrigin		0.00	2.32E-07	1.33E-05	2.05E-05	2.31E-04	2.53E-03
	Griewank		0.00	0.00	0.00	0.00	4.77E-07	5.90E-06
IGA	Rosenbrock		3.21E-08	8.67E-02	4.52	16.87	74.79	461.52
	Sphere		0.00	1.02E-03	0.12	0.29	0.81	7.51
	Rastrigin		1.50E-08	0.26	7.43	14.19	58.76	287.68
	Griewank		0.00	5.13E-04	6.06E-02	0.13	0.36	0.98
DE	Rosenbrock		0.00	0.00	0.00	0.00	0.00	0.00
	Sphere		0.00	0.00	0.00	0.00	0.00	0.00
	Rastrigin		0.00	0.00	0.00	0.00	1.99	18.22
	Griewank		0.00	0.00	0.00	0.00	0.00	0.00
PSO	Rosenbrock		0.00	0.00	4.97E-06	3.92E-09	5.62E-08	71.31
	Sphere		0.00	0.00	0.00	0.00	0.00	0.00
	Rastrigin		0.00	0.00	1.99	1.99	6.97	34.85
	Griewank		0.00	0.00	0.00	0.00	0.00	0.00

the GA's results were inferior than the ones obtained by the GA for all the other cases. The improvement achieved to these results when the number of iterations was increased was not so significant as the one performed by the BBMO for the corresponding case. The BBMO algorithm performed better than the IGA as the IGA found the optimum in less cases than the BBMO. The IGA did not find the optimum for all test functions even when the number of variables was equal to 2. The values found by IGA were in some cases far from the optimum. However, a small improvement in the results of IGA was performed when the number of iterations was equal to 100000 but still in some cases the values found were far from the optimum. The BBMO algorithm performed slightly better than the DE. The DE gave the optimum in most of the cases, only for Rastrigin test function and for $n=20$ and $n=50$ the optimum was not found. However, when the number of iterations was increased, these values were not improved, contrary to BBMO algorithm where an increase to the number of iterations always led to an improvement of the solution. The BBMO algorithm performed better than the PSO algorithm. The PSO algorithm

Table 5 Comparisons of the BBMO algorithm with other metaheuristics for the four functions (100000 iterations)

	Function	Iterations	2	4	8	10	20	50
						n		
BBMO	Rosenbrock	100000	0.00	0.00	0.00	0.00	0.00	1.62E-02
	Sphere		0.00	0.00	0.00	0.00	0.00	0.00
	Rastrigin		0.00	0.00	0.00	0.00	0.00	0.00
	Griewank		0.00	0.00	0.00	0.00	0.00	0.00
HBMO	Rosenbrock		0.00	5.34E-07	7.56E-03	4.80E-02	6.38	41.00
	Sphere		0.00	0.00	0.00	0.00	0.00	5.59E-02
	Rastrigin		0.00	0.00	0.00	0.00	2.02E-06	3.00
	Griewank		0.00	0.00	0.00	0.00	0.00	4.66E-03
GA	Rosenbrock		0.00	0.00	6.69E-02	8.50E-02	0.24	18.35
	Sphere		0.00	0.00	0.00	0.00	4.14E-09	7.98E-08
	Rastrigin		0.00	6.87E-09	4.47E-08	4.11E-08	7.76E-07	1.65E-05
	Griewank		0.00	0.00	0.00	0.00	0.00	0.00
IGA	Rosenbrock		0.00	1.47E-02	3.27	9.12	70.95	418.67
	Sphere		0.00	2.61E-04	5.53E-02	0.16	0.81	7.07
	Rastrigin		0.00	7.20E-02	3.38	9.04	49.70	260.71
	Griewank		0.00	1.06E-04	2.74E-02	8.01E-02	0.34	0.97
DE	Rosenbrock		0.00	0.00	0.00	0.00	0.00	0.00
	Sphere		0.00	0.00	0.00	0.00	0.00	0.00
	Rastrigin		0.00	0.00	0.00	0.00	1.99	18.22
	Griewank		0.00	0.00	0.00	0.00	0.00	0.00
PSO	Rosenbrock		0.00	0.00	0.00	0.00	0.00	0.11
	Sphere		0.00	0.00	0.00	0.00	0.00	0.00
	Rastrigin		0.00	0.00	1.99	1.99	6.97	34.85
	Griewank		0.00	0.00	0.00	0.00	0.00	0.00

performed efficiently but in some cases did not find the optimum and the increase in the number of iterations did not manage to improve the results of PSO in all cases, as for example for the Rastrigin test function for $n=8, 10, 20, 50$ the results were not improved at all.

A statistical analysis based on the Mann-Whitney U-test is presented in Table 6. In this Table, a value equal to 1 indicates a rejection of the null hypothesis at the 5% significance level, which means that the proposed method is statistically significant different from the other methods. On the other hand, a value equal to 0 indicates a failure to reject the null hypothesis at the 5% significance level, meaning that no statistical significant difference exists between the two methods. As it can be seen from this Table, the proposed method is statistically significant different from HBMO, GA, IGA and PSO in 10000 iterations, while in 100000 iterations the proposed method is statistically significant different from HBMO, GA and IGA.

Table 6 Results of Mann - Whitney test

| | \multicolumn{6}{c}{10000 iterations} | | | | | |
	BBMO	HBMO	GA	IGA	DE	PSO
BBMO	-	1	1	1	0	1
HBMO	1	-	0	1	1	0
GA	1	0	-	1	1	0
IGA	1	1	1	-	1	1
DE	0	1	1	1	-	1
PSO	1	0	0	1	1	-
	\multicolumn{6}{c}{100000 iterations}					
	BBMO	HBMO	GA	IGA	DE	PSO
BBMO	-	1	1	1	0	0
HBMO	1	-	0	1	1	0
GA	1	0	-	1	1	0
IGA	1	1	1	-	1	1
DE	0	1	1	1	-	0
PSO	0	0	0	1	0	-

5 Conclusions

In this paper, an algorithm based on the mating behavior of the bumble bees, the Bumble Bees Mating Optimization algorithm, was proposed for the solution of global unconstrained optimization problems. This algorithm was analytically presented and tested using four test functions, the Rosenbrock, the Sphere, the Rastrigin and the Griewank. The results of the algorithm were compared with the results of other popular metaheuristic and nature inspired methods, like Genetic Algorithms, Island Genetic Algorithms, Differential Evolution, Particle Swarm Optimization and the Honey Bees Mating Optimization algorithm. The results obtained showed the efficiency of the proposed algorithm and its high performance compared to the other metaheuristic algorithms.

References

[1] Abbass, H.A.: A monogenous MBO approach to satisfiability. In: International Conference on Computational Intelligence for Modelling, Control and Automation, CIMCA 2001, Las Vegas, NV, USA (2001)

[2] Abbass, H.A.: Marriage in honey-bee optimization (MBO): a haplometrosis polygynous swarming approach. In: The Congress on Evolutionary Computation (CEC 2001), Seoul, Korea, pp. 207–214 (May 2001)

[3] Afshar, A., Haddad, O.B., Marino, M.A., Adams, B.J.: Honey-bee mating optimization (HBMO) algorithm for optimal reservoir operation. J. Franklin. Inst. 344, 452–462 (2007)

[4] Baykasoglu, A., Ozbakir, L., Tapkan, P.: Artificial bee colony algorithm and its application to generalized assignment problem. In: Chan, F.T.S., Tiwari, M.K. (eds.) Swarm Intelligence, Focus on Ant and Particle Swarm Optimization, pp. 113–144. I-Tech Education and Publishing (2007)

[5] Clerc, M., Kennedy, J.: The particle swarm: explosion, stability and convergence in a multi-dimensional complex space. IEEE T. Evolut. Comput. 6, 58–73 (2002)

[6] Dasgupta, D. (ed.): Artificial immune systems and their application. Springer, Heidelberg (1998)

[7] Dorigo, M., Stützle, T.: Ant colony optimization. A Bradford Book. The MIT Press, Cambridge (2004)

[8] Drias, H., Sadeg, S., Yahi, S.: Cooperative bees swarm for solving the maximum weighted satisfiability problem. In: Cabestany, J., Prieto, A.G., Sandoval, F. (eds.) IWANN 2005. LNCS, vol. 3512, pp. 318–325. Springer, Heidelberg (2005)

[9] Engelbrecht, A.P.: Computational intelligence: An introduction, 2nd edn. John Wiley and Sons, England (2007)

[10] Fathian, M., Amiri, B., Maroosi, A.: Application of honey bee mating optimization algorithm on clustering. Appl. Math. Comput. 190, 1502–1513 (2007)

[11] Haddad, O.B., Afshar, A., Marino, M.A.: Honey-bees mating optimization (HBMO) algorithm: A new heuristic approach for water resources optimization. Water Resour. Manag. 20, 661–680 (2006)

[12] Holland, J.H.: Adaptation in natural and artificial systems. University of Michigan Press, Ann Arbor (1975)

[13] Karaboga, D., Basturk, B.: On the performance of artificial bee colony (ABC) algorithm. Appl. Soft Comput. 8, 687–697 (2008)

[14] Kennedy, J., Eberhart, R.: Particle swarm optimization. IEEE International Conference on Neural Networks 4, 1942–1948 (1995)

[15] Marinaki, M., Marinakis, Y., Zopounidis, C.: Honey bees mating optimization algorithm for financial classification problems. Appl. Soft Comput. (2009), doi: 10.1016/j.asoc.2009.09.010

[16] Marinakis, Y., Marinaki, M.: ŞA hybrid honey bees mating optimization algorithm for the probabilistic traveling salesman problem. In: IEEE Congress on Evolutionary Computation (CEC 2009), Trondheim, Norway (2009)

[17] Marinakis, Y., Marinaki, M., Dounias, G.: Honey bees mating optimization algorithm for the vehicle routing problem. In: Krasnogor, N., Nicosia, G., Pavone, M., Pelta, D. (eds.) Nature inspired cooperative strategies for optimization - NICSO 2007. Studies in Computational Intelligence, vol. 129, pp. 139–148. Springer, Berlin (2008)

[18] Marinakis, Y., Marinaki, M., Matsatsinis, N.: A hybrid clustering algorithm based on Honey Bees Mating Optimization and Greedy Randomized Adaptive Search Procedure. In: Maniezzo, V., Battiti, R., Watson, J.-P. (eds.) LION 2007 II. LNCS, vol. 5313, pp. 138–152. Springer, Heidelberg (2008)

[19] Marinakis, Y., Marinaki, M., Matsatsinis, N.: Honey bees mating optimization for the location routing problem. In: IEEE International Engineering Management Conference (IEMC – Europe 2008), Estoril, Portugal (2008)

[20] Marinakis, Y., Marinaki, M., Dounias, G.: Honey bees mating optimization algorithm for large scale vehicle routing problems. Nat. Comput. (2009), doi: 10.1007/s11047-009-9136-x

[21] Marinakis, Y., Marinaki, M., Matsatsinis, N.: A hybrid bumble bees mating optimiza-
     tion – GRASP algorithm for clusterin. In: Corchado, E., Wu, X., Oja, E., Herrero, Á.,
     Baruque, B. (eds.) HAIS 2009. LNCS, vol. 5572, pp. 549–556. Springer, Heidelberg
     (2009)
[22] Pham, D.T., Ghanbarzadeh, A., Koc, E., Otri, S., Rahim, S., Zaidi, M.: The bees algo-
     rithm - A novel tool for complex optimization problems. In: IPROMS 2006 Proceeding
     2nd International Virtual Conference on Intelligent Production Machines and Systems,
     Oxford. Elsevier, Amsterdam (2006)
[23] Teo, J., Abbass, H.A.: A true annealing approach to the marriage in honey bees opti-
     mization algorithm. Int. J. Comput. Intell. Appl. 3(2), 199–211 (2003)
[24] Teodorovic, D., Dell'Orco, M.: Bee colony optimization - A cooperative learning ap-
     proach to complex transportation problems. Advanced OR and AI Methods in Trans-
     portation. In: Proceedings of the 16th Mini - EURO Conference and 10th Meeting of
     EWGT, pp. 51–60 (2005)
[25] Storn, R., Price, K.: Differential evolution - A simple and efficient heuristic for global
     optimization over continuous spaces. J. Global Optim. 11(4), 341–359 (1997)
[26] Wedde, H.F., Farooq, M., Zhang, Y.: BeeHive: An efficient fault-tolerant routing al-
     gorithm inspired by honey bee behavior. In: Dorigo, M., Birattari, M., Blum, C., Gam-
     bardella, L.M., Mondada, F., Stützle, T. (eds.) ANTS 2004. LNCS, vol. 3172, pp. 83–94.
     Springer, Heidelberg (2004)
[27] Yang, X.S.: Engineering optimizations via nature-inspired virtual bee algorithms. In:
     Mira, J., Álvarez, J.R. (eds.) IWINAC 2005. LNCS, vol. 3562, pp. 317–323. Springer,
     Heidelberg (2005)
[28] http://www.bumblebee.org
[29] http://www.everythingabout.net/articles/biology/animals/
     arthropods/insects/bees/bumble_bee
[30] http://bumbleboosters.unl.edu/biology.shtml
[31] http://www.colostate.edu/Depts/Entomology/courses/en570/
     papers_1998/walter.htm

A Neural-Endocrine Architecture for Foraging in Swarm Robotic Systems

Jon Timmis, Lachlan Murray, and Mark Neal

Abstract. This paper presents the novel use of the Neural-endocrine architecture for swarm robotic systems. We make use of a number of behaviours to give rise to emergent swarm behaviour to allow a swarm of robots to collaborate in the task of foraging. Results show that the architecture is amenable to such a task, with the swarm being able to successfully complete the required task.

1 Introduction

Swarm robotic systems have many potential uses, ranging from the cleanup of hazardous waste or search and rescue operations at disaster sites that are often too dangerous for humans to respond effectively to or areas that need large coverage for monitoring (such as the ocean) and are simply too large a task for a single robot to cope. Good reviews of swarm robotics and associated issues can be found in [9] and [5]. However, in order to develop such systems, the task of *foraging* is used as a standard test arena for new approaches. Foraging is a popular task for mobile autonomous robots, both individual robots and swarms have been shown to successfully complete various types of foraging problem. The basic principles of foraging involve an agent collecting objects that are spread throughout the environment and returning them to some specified location. The task is completed once all of the

Jon Timmis
Department of Electronics and Department of Computer Science, University of York, Heslington, York. UK
e-mail: jtimmis@cs.york.ac.uk

Lachlan Murray
Department of Electronics, University of York, Heslington, York. UK
e-mail: ljm505@ohm.york.ac.uk

Mark Neal
Department of Computer Science, Aberystwyth University, Aberystwyth, Wales. UK
e-mail: mjn@aber.ac.uk

J.R. González et al. (Eds.): NICSO 2010, SCI 284, pp. 319–330, 2010.
springerlink.com   © Springer-Verlag Berlin Heidelberg 2010

objects in the environment have been collected. Part of our on-going work is the development of a neural-endocrine architecture for deployment in ocean going robotic systems, and the eventual construction of a swarm of ocean going vessels that would be able to operate for prolonged periods of time.

This paper investigates and extends our previous work on a neural endocrine control architecture developed in [2, 3, 6, 7]. Until now its effectiveness at controlling a collection of robots has not been investigated, though work on using two robots has been undertaken in the context of task switching [8]. The addition of more robots brings added complexity to the system, it is necessary that a multi-robot control system not only encompasses the ability to control individual robots, but is also capable of appropriately handling the interactions with other robots. If we are to work towards developing an ocean going version of such a system then the understanding of the ability of our architecture to operate in a swarm of robots is essential. In order to assess the effectiveness of the system it was necessary to design a task for the robots performance to be measured on, the task chosen was a variant of foraging and was one of the most complicated tasks that the neural endocrine control architecture has been applied to. Specifically, in this paper we: investigate whether the neural endocrine control architecture is capable of controlling a multi-robot system; investigate how effective the architecture is at controlling a multi-robot system and investigate the capabilities of the architecture at a new and complex task.

2   Neural Endocrine Control Architecture

The neural endocrine control architecture of [2] is a combination of standard perceptron artificial neural networks, with a novel artificial endocrine system that has the ability to affect the weights of the neural networks, depending on external and internal factors. Here we review the basic neural endocrine architecture, for a more detailed description the reader is directed to [3, 6].

2.1   Artificial Endocrine Systems

The Artificial Endocrine System (AES) described here is based on the original design proposed by [2, 3] as well as subsequent modifications made by [6].

As is the case in the biological endocrine system, the two main components of an AES are glands and hormones. Artificial glands (g) release artificial hormones when they are stimulated. Stimulation can be caused by both the internal state of the system and external stimuli. In [6] *signal values* A_i were obtained by summing sensor inputs and similar gland *activation values* were calculated from the combination of sensor values and the internal state of the robot. The stimulation of a gland (R_g) as given by [6] is shown in equation 1 where α_g is the *stimulation rate*, that is the rate at which a hormone is released from a gland g.

$$R_g(t) = \alpha_g \sum_i A_i(t) \tag{1}$$

Our previous work, unpublished, investigated a second method of stimulation that also takes into account the current *concentration* of hormone $c_g(t)$ this is given by equation 2. As can be seen in equation 2 the amount of hormone released in this method is subject to a negative feedback mechanism, the reason for including this is to prevent the system from becoming over saturated with a particular type of hormone.

$$R_g(t) = \frac{\alpha_g}{1 + c_g(t-1)} \sum_i A_i(t) \tag{2}$$

Every hormone has an associated decay rate (β_g) which takes a value from $[0,1]$, this means that without stimulation the concentration of a hormone will eventually be reduced to an insignificant amount. The concentration of a particular hormone c_g at time $t+1$ is given by equation 3.

$$c_g(t+1) = \beta_g c_g(t) + R_g(t+1) \tag{3}$$

2.2 Neural Endocrine Systems

Artificial hormones can only affect artificial neurons. In line with the biological endocrine system not all of the neurons in a system will be sensitive to all hormones, the sensitivity of a neuron i to the hormone released by a particular gland g is given by s_{ig}. The effect that hormones have on neurons can be calculated by equation 4 which takes into account the sensitivity of inputs to particular hormones and the concentration of those hormones using an artificial endocrine system with ng glands.

$$u = \sum_{i=0}^{nx} x_i \cdot w_i \sum_{g=0}^{ng} c_g \cdot s_{ig} \tag{4}$$

The most common form of coordination between networks is a cooperative approach whereby the outputs of each network are simply summed together. The resulting behaviour of a multi-network neural endocrine system is dependent on the current hormone levels of the system. High levels of a particular hormone will affect some networks more than others, giving these networks more or less influence over the global result when the network outputs are summed together.

3 System Design

3.1 Behaviours

In this work, we make use of eleven different behaviours, the majority of which can be categorised into the three different groups: taxes, reflexes and fixed-action patterns (FAP). One of the behaviours, *wander*, can not easily be classified by type. We also observe resultant emergent behaviours not programmed into the system.

Wander: A wander behaviour is necessary to ensure that robots keep exploring the environment even if none of their other behaviours are currently being stimulated, without a wander behaviour an unstimulated robot would just remain stationary. To implement a wander behaviour we take into account the current hormone levels of the system.

3.1.1 Reflexes

Reflexes are involuntary, spontaneous responses to stimuli, which last only as long as the stimulus that initiates them. The foraging task of this work requires only a single reflex behaviour. Because of their spontaneous and sporadic nature reflex behaviours do not require a neural endocrine control network, their response is simply tied directly to their stimulus.

Signal bin: As robots will have no awareness of the location of the bin, in order to improve their chances of finding it a signal bin behaviour is required, allowing robots to communicate the approximate location of the bin to others. In this case, robots signal that the bin is in their vicinity by the use of a light or beacon. The strength of the response should always be the same, i.e. the brightness of the light should not be effected by the closeness of the bin, it should either be on if the bin is in-sight, or off otherwise.

3.1.2 Taxes

Taxes are behavioural responses that cause agents to move towards, or away from certain stimuli. This work involves six taxes behaviours, two of which are repellent and four of which are attractive. Taxes behaviours are well suited to control using neural endocrine networks because both their inputs and outputs are continuous and should vary according to the current state of the system, i.e. the hormone levels. Robots have the capability of signalling via an LED, and observing that signal on other robots.

Obstacle avoid: Prevents robots from crashing into the walls of the environment, or obstacles within the environment. The response of an obstacle avoid behaviour should be proportional to the distance between a robot and its nearest obstacle, such that a robot responds more urgently to obstacles that are nearer. The inputs to the network of an obstacle avoid behaviour come from a range finding sensor, for example a sonar.

Separation: Prevents robots from crashing into each other. The stimuli of a separation behaviour, also the inputs to the behaviour's network, are the locations of other robots, these can be determine using a camera device. In a similar manner to obstacle avoidance, the strength of a response should be proportional to the distance between a robot and its neighbours, such that the closer a fellow robot is, the faster the robot should retreat.

Cohesion: Attracts a robot its neighbours. As with separation, a cohesion behaviour is useful in the development of emergent global behaviours. The strength of the stimulus should have an effect on the strength of the response so that robots are less attracted to neighbours that are closer, reducing the chance of collisions. The inputs to a cohesion behaviour's network are similar to those of a separation behaviour and come from the positions of their neighbours via a camera device.

Seek rubbish: Robots should be stimulated by the presence of a piece of rubbish, which can be detected using a camera. Robots should be attracted to the location of the rubbish with a strength of response that is relative to the how far away the rubbish is, the further away, the stronger the attraction.

Seek power: Robots should be attracted to charging stations. Inputs are provided in the same manner as the seek rubbish behaviour, using a camera device, and the strength of the response is once again relative to the distance of the stimulus.

Seek bin: A seek bin behaviour is very similar to both the seek power and seek rubbish behaviours, however in this case robots should be attracted to the bin. The stimulus is the presence of the bin, and the strength of response is relative to the distance between the robot and the bin.

3.1.3 Fixed-Action-Patterns

Fixed Action Patterns (FAP) are behaviours that continue even if the stimulus that triggered them is not present, usually they run uninterrupted until completion. Their response is always identical and so they are not suitable for control using neural endocrine networks, like reflexes they can be implemented by directly tying stimulus to response.

Pickup rubbish: A pickup rubbish behaviour should be stimulated when a robot is close enough to a piece of rubbish and is not already carrying some. The behaviour should involve the robot moving towards the piece of rubbish and either successfully or unsuccessfully picking it up, both of which should result in the end of the pattern, however if the pickup is unsuccessful it is possible that the behaviour will be re-stimulated immediately.

Drop Rubbish: If a robot is carrying a piece of rubbish and is close enough to the bin, the drop rubbish pattern should be stimulated. The pattern starts with the robot approaching the bin and continues until the robot has either successfully or unsuccessfully dropped the rubbish into the bin.

Recharge: A recharge behaviour should be stimulated when a robot is close enough to a charging station and its internal state dictates that it needs to recharge. The behaviour should begin with the robot moving towards the charging station and attempting to dock with it, if the robot fails to dock, the pattern should end, if the robot successfully docks the pattern should continue until the robot is fully charged.

3.2 Neural-Endocrine Design

In section 2, it was noted that not every hormone in a system must affect every neuron. In all previous work the approach has been to make all the neurons of a single network sensitive to the same hormones, for example in a system with two hormones h_a and h_b and two networks N_a and N_b, a possible configuration would be that all the neurons of N_a are sensitive to h_a and all the neurons of N_b are sensitive to h_b, this is shown in figure 1. The alternative is to make different neurons of the same network sensitive to different hormones, for example in a system with two hormones h_a and h_b and a single network of seven nodes $\{n_1, n_2, ..., n_7\}$, nodes $n_1 -$ n_4 might be sensitive to h_a and nodes $n_5 - n_7$ might be sensitive to h_b, this is shown in figure 2. Since each network in a system corresponds to a single behaviour, it seems sensible that, as is the case in the previous approach, each network should be affected by the same hormones. For simplicity, here each network is only associated with a single gland-hormone pair. The sensitivity of a neuron i to a particular gland g is denoted s_{ig}, in theory s_{ig} can take any value, however in this work s_{ig} only takes the value 1 or 0, representing full or no sensitivity of i to g.

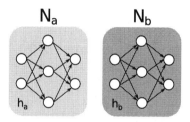

Fig. 1 Two networks, the neurons of which are all sensitive to the same hormone

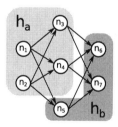

Fig. 2 A single network with different neurons sensitive to different hormones

Having decided that all neurons in a network will be sensitive to the same hormones, that each network will only respond to one hormone and that sensitivity is only ever 1 or 0 it is possible to refine equation 4 from section 2. Removing the sensitivity and multiplicity of 4 leaves 5 where c_g is the concentration of the network's only associated hormone. For simplicity, here each network is only associated with a single gland-hormone pair.

$$u = \sum_{i=0}^{nx} x_i \cdot w_i \cdot c_g \qquad (5)$$

The activation of a gland can be calculated from a combination of both internal and external properties of the system. Each gland is associated with a single activation parameter which changes over time according to a dedicated function and is represented here as a_g. The stimulation of a gland, as was seen in section 2 can be calculated in one of two ways. Early implementations had shown success with using a negative feedback mechanism, we therefore adopted that approach in this work. The stimulation of a gland is calculated using 6, which is a slightly adjusted version of 2 in order to take into account the new representation of activation.

$$R_g(t) = \frac{\alpha_g \cdot a_g(t)}{1 + c_g(t-1)} \qquad (6)$$

The final consideration with ANN-AES integration, is what values of stimulation (α_g) and decay (β_g) rate are used by the networks. The stimulation rate helps determine the amount of hormone released by a gland at a particular time-step and the decay rate determines how long the hormone remains in the system, hence they both have a big influence on the behavioural response. Values of α_g and β_g can vary widely between different networks. These values were chosen experimentally, however an automated learning process could be adopted.

3.2.1 Network Size and Weights

ANNs can be defined by four properties: the number of hidden layers, the number of nodes in each of the hidden layers, the number of nodes in the input layer and the number of nodes in the output layer. For more information on neural networks, the reader is directed to [1]. The number of nodes in the input layer of a network are determined by the number of sensor values needed to define the stimulus of that behaviour, for example in the case of obstacle avoidance which is stimulated by the presence of nearby objects, the number of sonar devices (two in this piece of work) determines the number of input nodes. The number of output nodes is determined by the actuator that the response affects, in most cases, where the response affects the locomotion of the robot, it is the number of inputs to the motors that decides the number of output nodes (which again in this study is two).

The number of hidden layers and the number of hidden layer nodes is less dependent on the behaviour, and puts more pressure on the designer to choose sensible values. It was known from our previous work that the networks required would be relatively simple, consequently we only include one hidden layer.

With regards to setting the weights, we used a combination of determining the weights by hand and back-propogation [1].

3.2.2   Coordination of Different Behaviour Types

Behaviours that are encapsulated as neural endocrine networks are coordinated by summing their outputs. We have not discussed how these behaviours are coordinated with the other types of behaviour, such as the fixed-action-patterns and reflexes. The signal bin behaviour is a reflex, it does not affect anything other than the state of the robot's beacon and so it does not need to be coordinated with the other behaviours. In terms of the FAPs, when stimulated, these will always take complete control of the robot's motors, inhibiting any of the suggested commands from the other behaviours. It is very rare for conflicts to arise between different FAPs since it is never the case that a robot will want to both drop and pickup rubbish at the same time and because the bin and charging posts are positioned far apart (in the experiments carried out in this work) there will never be a conflict between wanting to charge and wanting to drop rubbish. However, we recognise that this is an avenue for further exploration.

3.2.3   Environments

In order to test the adaptability of the system it was necessary to test the performance of the robots in two different environments. Both of the environments were designed with the capabilities of the robots in mind, for example, it was known that because the robots had only two sonar sensors, both of which were located at the front, they would struggle to find their way out of concave obstacles with small internal angles. When faced with concave obstacles robots can be indecisive about which way to turn and in the end may either end up stalling or crashing into the obstacle. Another deficiency caused by the poor sonar coverage is that if an obstacle is too small (smaller than the width of the robot) when a robot approaches it head on, its sonar devices will not recognise it and the robot will crash. Due to these problems, both of the environments were designed to contain no concave obstacles (with small internal angles) and no obstacles smaller than the width of a robot.

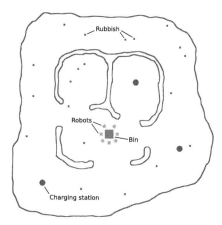

Fig. 3 Environment used for experiments

The first environment, referred to as *world 1*, can be seen in figure 3, it contains a single bin, shown by the large square; three charging stations, represented by the circles; and twenty pieces of rubbish, depicted as very small squares, the robots are the squares located by the bin. The world was made deliberately challenging by placing the bin in the centre of the environment and surrounding it with obstacles. The reason for placing the bin in a difficult position, was the expectation that to reach it, robots would fair better if they cooperated with each other, for example by signalling and flocking. A second world was used, but space restricts the inclusion of those results.

4 Experiments

All experiments were carried out in the Player/Stage environment [4], running on Linux. We simulated Pioneer mobile robots, containing sonars, a camera, a gripper and a beacon. All code is available on request. The variant of foraging that was chosen for this project is known as rubbish or garbage collection. The task of rubbish collection used here involves a group of robots collecting pieces of rubbish that are randomly distributed throughout the environment and returning them to a bin. In order to make the task slightly more complex and to model the real world closer robots are required to monitor their power levels and when they are running low find a charging station at which to recharge.

4.1 Results for Neuro-endocrine Swarms

The success of the system is measured in terms of the *amount* of rubbish that was collected. Graphs are presented to show how the success of the system changed as the number of robots was varied. The total amount of rubbish collected by the group as a whole, as well as the number of pieces collected per robot are analysed.

4.2 Results

Figures 4(a) and 4(b) show the success of the robots after periods of 300 and 1200 seconds respectively. Each boxplot shows the results of ten different runs with ten different starting positions for the rubbish. Both the graphs show a strong positive correlation between the number of robots and the number of pieces of rubbish collected, until the case where five robots were used, at which the performance starts to level out and even drops in figure 4(b). The levelling out is expected in 4(b) since the maximum number of pieces that can be collected is twenty, but the fact that it is observed in 4(a) and that the performance drops in 4(b) indicates that interference starts to have an effect after five robots. The case with five robots also had the smallest interquartile range showing that five robots not only performed the best, but did so consistently.

The first outlier in 4(b), where the number of robots was three and the number of pieces picked up was six, was caused by one robot crashing, and the other robots crashing into the obstruction formed by the other robots, which emphasises the importance of redundancy in multi-robot systems. The outlier where the number of robots was five and the number of pieces collected was sixteen can be attributed, at least partly, to the simulator and the way the bin is represented. Since robots cannot see the inside of the bin from the outside, there is always the danger that collisions can occur as one robot travels into and one robot travels out of the bin, this is what happened in case of this outlier, two robots crashed whilst entering and leaving the bin which meant that when other robots came to drop rubbish there was a pileup effect. Only one other crash at the bin was observed in the seventy experiments of world 1, again for an experiment involving five robots however in this case it did not involve all of the robots and two were able to continue functioning, resulting in nineteen pieces being collected.

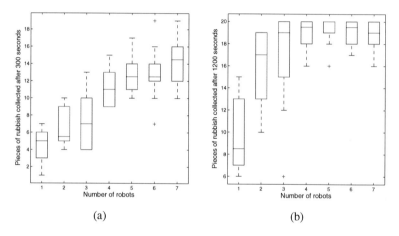

(a)   (b)

Fig. 4 Graphs showing the number of pieces of rubbish collected over periods of 300 (a) and 1200 (b) seconds, with varying numbers of robots between one and seven: World 1

Figure 5 shows the number of pieces of rubbish collected per robot after 300 and 1200 seconds, as to be expected, in both graphs the number of pieces drops as more robots are added. What is interesting about figure 5(b) is that the smallest interquartile range is observed for the case where there were five robots, showing that a group of five robots is most consistent on an individual level as well as a group level as indicated by figure 4(b). The outliers in figure 5(b) relate to the same runs as in figure 4(b).

What is interesting to note from the observation of the experimental runs are the *emergence* of certain types of behaviour: specifically flocking of robots and dispersion of robots. Flocking emerges from the combination of obstacle avoidance, seek bin, signal bin, separation and cohesion and dispersion emerges from the combination of obstacle avoidance and separation, simply stated it is the spreading out of

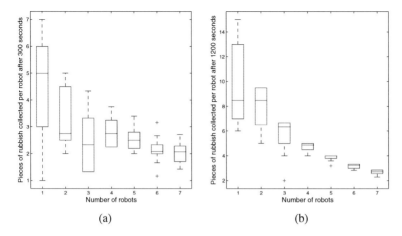

Fig. 5 Graphs showing the number of pieces of rubbish collected per robot over periods of 300 (a) and 1200 (b) seconds, with varying numbers of robots between one and seven: World 1

robots over the environment to ensure the greatest amount of coverage. Robots recharge when necessary, and collaborate together, through flocking etc. to remove as much garbage as possible from the environment. We have not undertaken a comparison between other approaches as yet, this would be outside the scope of a conference paper. However, we have investigated the efficiency and the effect of speed up on the swarm (how does adding more swarm members effect the overall performance), but have not room to report those results here. In summary, however, we have been able to show that there is an optimal number of robots for each world to achieve the best performance in garbage collection.

5 Conclusions

This work has adapted the neural-endocrine architecture for the development of swarm robotic systems. An architecture has been proposed for the task of foraging and has been showed to allow for good collection of garbage over two basic environments. The work has also shown us that the simple neural-endocrine approach can easily be used for the development of such swarm systems. We observe that too many robots in the environment causes a potential problem (to be expected) for the optimal collection of garbage. The work presented in this paper is also the most complex task that the neural-endocrine approach has been used for to date. This gives us confidence in our approach and further work will investigate the actual role of each behaviour, and its importance to the overall performance of the system, and developing neural-endocrine systems on an ocean-going platform.

Acknowledgements. This work is funded by EOARD, grant number FA-8655-07-3061.

References

[1] Haykin, S.: Neural Networks - A Comprehensive Foundation. Prentice-Hall, Englewood Cliffs (1999)
[2] Neal, M., Timmis, J.: Timidity: A useful emotional mechanism for robot control? Informatica 27(2), 197–204 (2003)
[3] Neal, M., Timmis, J.: Once more unto the breach: Towards artificial homeostasis? In: Recent Developments in Biologically Inspired Computing, pp. 340–365. Idea Group, USA (2005), http://www.cs.kent.ac.uk/pubs/2005/1948
[4] Player: The Player Project (2009) http://playerstage.sourceforge.net (accessed: April 23, 2009)
[5] Şahin, E., Winfield, A.: Special issues on swarm robotics. Swarm Intelligence 2(2-4), 69–72 (2008)
[6] Timmis, J., Neal, M., Thorniley, J.: An adaptive neuro-endocrine system for robotic systems. In: IEEE Workshop on Robotic Intelligence in Informationally Structured Space. Part of IEEE Workshops on Computational Intelligence, pp. 129–136 (2009)
[7] Vargas, P., Moioli, R., de Castro, L.N., Timmis, J., Neal, M., Von Zuben, F.: Artificial homeostatic system: A novel approach. In: Capcarrère, M.S., Freitas, A.A., Bentley, P.J., Johnson, C.G., Timmis, J. (eds.) ECAL 2005. LNCS (LNAI), vol. 3630, pp. 293–306. Springer, Heidelberg (2005)
[8] Walker, J., Wilson, M.: A performance sensitive hormone-inspired system for task distribution amongst evolving robots. In: Proceedings of IEEE/RSJ 2008 International Conference on Intelligent Robots and Systems (2008)
[9] Winfield, A.F., Nembrini, J.: Safety in numbers: Fault tolerance in robot swarms. International Journal of Modelling, Identification and Control 1(1), 3–37 (2006)

Using Entropy for Evaluating Swarm Intelligence Algorithms

Gianluigi Folino and Agostino Forestiero

Abstract. In the last few years, the bio-inspired community has experienced a growing interest in the field of Swarm Intelligence algorithms applied to real world problems. In spite of the large number of algorithms using this approach, a few methodologies exist for evaluating the properties of self-organizing and the effectiveness in using these kinds of algorithm. This paper presents an entropy-based model that can be used to evaluate self-organizing properties of Swarm Intelligence algorithms and its application to SPARROW-SNN, an adaptive flocking algorithm used for performing approximate clustering. Preliminary experiments, performed on a synthetic and a real-world data set confirm the presence of self-organizing characteristics differently from the classical flocking algorithm.

1 Introduction

Swarm Intelligence (SI) [1] is an innovative computational method for solving problems that originally took its inspiration from the biological examples provided by social insects such as ants, termites, bees, etc. These systems are typically made up of a population of simple agents interacting directly or indirectly (by acting on their local environment) with each other. Indirect interaction, i.e. when an individual modify the environment and the other responds to this change, is named *stigmergy* [5]. This mechanism permits to reduce direct communication among agents, and must be taken into account when designing artificial systems. In practice, an agent deposits something in the environment that makes no direct contribution to the task being undertaken, but is used to influence the subsequent behavior that is task related. Although there is normally no centralized control structure dictating how individual agents should behave, local interactions between such agents often lead to the emergence of global behavior. Examples of systems like these can be found in nature, including ant colonies, bird flocking, animal herding, bacteria molding

Gianluigi Folino · Agostino Forestiero
Institute for High Performance Computing and Networking (ICAR-CNR)
e-mail: {folino,forestiero}@icar.cnr.it

J.R. González et al. (Eds.): NICSO 2010, SCI 284, pp. 331–343, 2010.

and fish schooling. The advantages of SI are twofold. Firstly, it offers intrinsically distributed algorithms that can use parallel computation quite easily. Secondly, the use of multiple agents supplies a high level of robustness, as the failure of a few individuals does not alter too much the behavior of the overall system.

Clustering is the act of partitioning an unlabeled dataset into groups of similar objects. Each group, called a cluster, consists of objects that are similar between themselves and dissimilar to objects of other groups.

The SPARROW-SNN (Shared Nearest-Neighbor similarity), better described in [4], couples an adaptive flocking algorithm with a shared nearest neighbor (SNN) [3] cluster algorithm to discover clusters with differing sizes, shapes in noise and high dimensional data.

In the last few years, innovative algorithms based on SI models [7] [8][10][2] have been introduced to solve real world problems in a decentralized fashion (i.e. the clustering problem, correlated to the SPARROW-SNN algorithm).

In spite of the large number of algorithms using this approach, a few methodologies exist for evaluating the properties of self-organizing and the effectiveness of using these kinds of algorithm. In this work, a methodology based on the concept of entropy inspired by the paper [11], is illustrated. The main principle stated in the paper is that the key to reduce disorder in a multi-agent system and to achieve a coherent global behavior is coupling that system to another in which disorder increases. This corresponds to a macro-level where the order increases, i.e. a coherent behavior arises, and a micro-level where an increase in disorder is the cause for this coherent behavior at the macro-level.

Using this approach, the self organizing properties of SI algorithms can be experimentally evaluated, considering macro and micro levels of entropy. The method is applied to evaluate self-organizing properties of the SPARROW-SNN, the adaptive flocking algorithm cited above. However, it can be easily applied to any type of SI algorithm.

The rest of this paper is organized as follows: Section 2 presents the multi agent adaptive flocking algorithm for searching interesting objects and shows how this algorithm can be used as a basis for clustering spatial data, combining it with a local merging strategy based on the SNN algorithm; Section 3 introduces a methodology based on the concept of entropy useful to evaluate self-organizing properties of SI based algorithms. Section 4 shows how entropy can be experimentally evaluated and used to assess the goodness of the flocking algorithm.

Finally, section 5 draws some conclusions.

2 An Adaptive Flocking Algorithm

In this section, a multi-agent adaptive flocking algorithm is presented, which has the advantage of being easily implementable on parallel and distributed machines and is robust compared to the failure of individual agents. First, the rules governing the flock model originally introduced by Reynolds [13] are explained; then, the modified behavioral rules of the swarm agents are illustrated. They add an adaptive behavior to the flock and make it more effective in searching points, which have some desired properties in the space.

The flocking algorithm was proposed by Reynolds as a method for simulating the flocking behavior of birds on a computer both for animation and as a way to study emergent behavior. Flocking is an example of emergent collective behavior: there is no leader, i.e., no global control. Flocking behavior emerges from the local interactions. In the flock algorithm each agent has direct access to the geometric description of the whole scene, but reacts only to flock mates within a certain small radius. The basic flocking model consists of three kind of simple steering behavior:

Separation gives an agent the ability to maintain a certain distance from others nearby. This prevents agents from crowding too closely together, allowing them to scan a wider area.

Cohesion supplies an agent with the ability to cohere (approach and form a group) with other nearby agents. Steering for cohesion can be computed by finding all agents in the local neighborhood and computing the average position of the nearby agents. The steering force is then applied in the direction of that average position.

Alignment gives an agent the ability to align with other nearby characters. Steering for alignment can be computed by finding all agents in the local neighborhood and averaging together the 'heading' vectors of the nearby agents.

Our flocking algorithm extends Reynolds's rules and is inspired by a work presented by Macgill [9], first introducing colored agents.

The algorithm starts with a fixed number of agents that occupy a randomly generated position in the search space. Each agent moves around the spatial data, testing the neighborhood of each location in order to verify whether a point can have some desired properties. Each agent follows the rules of movement described in Reynolds' model. In addition, this model considers four different kinds of agents, classified on the basis of some properties of data in their neighborhood. Different agents are characterized by a different color: red, revealing interesting patterns in the data, green, a medium one, yellow, a low one, and white, indicating a total absence of patterns.

The main idea behind this approach is to take advantage of the colored agent in order to explore more accurately the most interesting regions (signaled by the red agents) and avoid the ones without interesting properties (signaled by the white agents). Red and white agents stop moving in order to signal this type of region to the others, while green and yellow ones fly to find denser zones. Indeed, each flying agent computes its heading by taking the weighted average of alignment, separation and cohesion (as illustrated in figure 1).

The following are the main features which make this model different from Reynolds' model:

- *Alignment* and *cohesion* do not consider yellow agents, since they move in a not very attractive zone.
- *Cohesion* is the resultant of heading towards the average position of the green flockmates (centroid), of the attraction towards reds, and of the repulsion from whites, as illustrated in figure 1.
- A *separation* distance is maintained from all the agents, apart from their color.

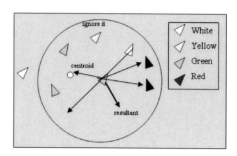

Fig. 1 Computing the direction of a green agent

2.1 Formal Description of the Flock

Consider the search space, in which the swarm moves, having dimension d. Let
N be the number of birds and B be the set of all the birds $\{B_1, B_2, \ldots, B_N\}$. Each
bird B_k can be represented by three d-dimensional vectors: its position in this space
$Pos_k : (x_k^1, x_k^2, \ldots, x_k^d)$, its direction $Dir_k : (dir_k^1, dir_k^2, \ldots, dir_k^d)$, where dir_k^i represents the component along the axis i of the direction of the bird and the color
$Col_k \in \{white, yellow, green, red\}$, indicating the type of bird. We used as distance
between two birds B_a and B_b, the euclidean distance between their respective positions: $dist(B_a, B_b) = \sqrt{\sum_{i=1}^{d}(x_a^i - x_b^i)^2}$.

We define as $dist\_max$ and $dist\_min$ respectively the radius indicating the
limited sight of the birds and the minimum distance that must be maintained
among them. $Neigh(B_k)$ denotes the neighborhood of a bird B_k, i.e. the set $\{B_a \in
B \mid dist(B_k, B_a) \leq dist_max\}$, that is the set of the birds visible from the bird
B_k. Furthermore, we define as $Neigh(col, B_k)$ the set $\{B_\alpha \in B \mid dist(B_k, B_\alpha) \leq
dist_max, Col_\alpha = col\}$, that is the set of the birds, having color col, visible from
the bird B_k.

Each bird moves with speed v, depending on the color of the agents (green agents'
speed is slower, because they are exploring interesting zones). Then, for each iteration t, the new position of a bird B_k can be computed as:

$$\forall i = 1 \ldots d \qquad x_k^i(t+1) = x_k^i(t) + v \times dir_k^i \qquad (1)$$

Note also that for each iteration the new direction of the agent k is obtained summing
the three components of alignment, separation and cohesion:

$$dir_k^i = dir\_al_k^i - dir\_sep_k^i + dir\_co_k^i. \qquad (2)$$

Considering as $dir(B_a, B_b)$ the normalized direction of the vector between a bird B_a
and a bird B_b, these components can be computed using the following formulas (3,
4 and 5):

$$dir\_al_k^i = \frac{1}{|Neigh(green,B_k)|} \cdot \sum_{B_\alpha \in Neigh(green,B_k)} dir_\alpha^i \qquad (3)$$

and considering $centr(green, B_k)$ as the position of the centroid of the green agents in the neighborhood of k with generic coordinate i:
$\frac{1}{|Neigh(green,B_k)|} \cdot \sum_{B_\alpha \in Neigh(green,B_k)} x_\alpha^i$, then:

$$dir\_co_k^i = dir(centr(green,B_k),B_k)^i + attr\_red - rep\_white \qquad (4)$$

where $attr\_red$ is equals to:

$$\sum_{B_\alpha \in Neigh(red,B_k)} dir(B_\alpha,B_k)^i$$

and rep_white is equals to:

$$\sum_{B_\alpha \in Neigh(white,B_k)} dir(B_\alpha,B_k)^i$$

i.e the sum of the attraction towards the centroid, of the attraction towards the red birds and of the repulsion from the white birds;

$$dir\_sep_k^i = \sum_{B_\alpha \in Neigh(B_k),dist(B_\alpha,B_k)<dist\_min} dir(B_\alpha,B_k)^i \qquad (5)$$

2.2 Using the Flocking Algorithm for Clustering Spatial Data

SNN is a clustering algorithm developed by Ertöz, Steinbach and Kumar [3] to discover clusters with differing sizes, shapes and densities in noise and high dimensional data. The algorithm extends the nearest-neighbor non-hierarchical clustering technique by Jarvis-Patrick [6] redefining the similarity between pairs of points in terms of how many nearest neighbors the two points share. Using this new definition of similarity, the algorithm eliminates noise and outliers, identifies representative points, and then builds clusters around the representative points. These clusters do not contain all the points, but rather represent relatively uniform group of points.

SPARROW-SNN combine the strategy of search of the previously described clustering algorithm with the SNN algorithm main principles for discovering clusters of arbitrary form and density. In practice, the flocking algorithm performs a biased sampling of the points of the dataset, as it focuses the search on interesting parts of the search space. Thus, the SNN algorithm was applied to merge the clusters and to eliminate the noise points. A more complete description of the algorithm can be found in [4].

To better understand as SPARROW-SNN works, the pseudocode was shown in figure 2. The algorithm starts with a fixed number of agents placed in a randomly generated position. From their initial position, each agent moves around the spatial

```
for i=1 ... MaxIterations
        foreach agent (yellow, green)
                age=age+1;
                if (age > Max_Life)
                        generate_new_agent();die();
                endif
                if (not visited (current_point))
                        property = compute_local_property(current_point);
                        mycolor= color_agent(property);
                endif
        end foreach
        foreach agent (yellow, green)
                dir= compute_dir();
        end foreach
        foreach agent (all)
                switch (mycolor){
                        case yellow, green: move(dir, speed(mycolor)); break;
                        case white: stop(); generate_new_agent(); break;
                        case red: stop(); generate_new_close_agent(); break; }
        end foreach
end for
```

Fig. 2 The pseudo-code of the adaptive flocking algorithm

data testing the neighborhood of each location in order to verify whether the point can be identified as a *representative* (or core) point.

The *compute_property* function represents the connectivity of the point as defined in the SNN algorithm. In practice, when an agent falls on a data point A, not yet visited, it computes the connectivity, *conn(A)*, of the point, i.e. computes the total number of strong links the points has, according to the rules of the SNN algorithm. Points having connectivity smaller than a fixed threshold (*noise_threshold*) are classified as noise and are considered for removal from clustering. Then a color is assigned to each agent, on the basis of the value of the connectivity computed in the visited point, using the following procedure (called *color_agent()* in the pseudocode):

$$conn > core\_threshold \Rightarrow mycolor = red \quad (speed = 0)$$
$$noise\_threshold < conn \leq core\_threshold \Rightarrow mycolor = green \quad (speed = 1)$$
$$0 < conn < noise\_threshold \Rightarrow mycolor = yellow \quad (speed = 2)$$
$$conn = 0 \Rightarrow mycolor = white \quad (speed = 0)$$

The colors assigned to the agents are: red, revealing representative points, green, border points, yellow, noise points, and white, indicating an obstacle (uninteresting region). After the coloration step, the green and yellow agents compute their movement observing the positions of all other agents that are at most at some fixed distance (*dist_max*) from them and applying the rules described in the previous subsection. In any case, each new red agent (placed on a representative point) will run the merge procedure, so that it will include, in the final cluster, the representative point discovered, together to the points that share with them a significant (greater that P_min) number of neighbors and that are not noise points. The merging phase considers two different cases: when points in the neighborhood have never been visited and when there are points belonging to different clusters. In the former, the same temporary label will be assigned and a new cluster will be constituted; in the latter, all the points will be merged into the same cluster, i.e. they will get the label corresponding to the smallest label. Thus clusters will be built incrementally.

3   An Entropy-Based Model

This section describes the application of a new methodology for understanding and evaluating self-organizing properties in bio-inspired systems. The approach is experimentally evaluated on the flocking system of the previous subsection, but it could be easily applied to any bio-inspired systems.

To this aim, we used a model based on the entropy introduced in [11] by Parunak and Brueckner. The authors adopted a measure of entropy to analyze emergence in multi-agent systems. Their fundamental claim is that the relation between self-organization based on emergence in multi-agent systems and concepts as entropy is not just a loose metaphor, but it can provide quantitative and analytical guidelines for designing and operating agent systems. These concepts can be applied in measuring the behavior of multi-agent systems. The main result, that the above cited paper suggests, concerns the principle that the key to reduce disorder in a multi-agent system and to achieve a coherent global behavior is coupling that system to another in which disorder increases. This corresponds to a macro-level where the order increases, i.e. a coherent behavior arises, and a micro-level where an increase in disorder is the cause for this coherent behavior at the macro-level.

A multi-agent system follows the second law of thermodynamics "Energy spontaneously disperses from being localized to becoming spread out if it is not hindered", if agents move without any constriction. However, if we add information in an intelligent way, the agents' natural tendency to maximum entropy will be contrasted and the system will go towards self-organization. For the sake of simplicity, in the case of the flocking, the attractive behavior of the red birds and the repulsive effect of the white agents add self-organization to the system, while in the case of ant systems, this is typically originated from the pheromone.

Really, as stated in [11], we can observe two levels of entropy: a macro level in which organization takes place, balanced by a micro level in that we have an increase of entropy. For the sake of clarity, in the flocking algorithm, micro-level is represented by red and white agents' positions, signaling respectively interesting and desert zones, and the macro level is computed considering all the agents' positions. So, we expect to observe an increase in micro entropy due to the birth of new red and white agents and, on the contrary, a decrease in macro entropy indicating organization in the coordination model of the agents.

[12] defines autocatalytic property for agent systems as follows: "A set of agents has autocatalytic potential if, in some regions of their joint state space, their interaction causes system entropy to decrease (and thus leads to increased organization). In that region of state space, they are autocatalytic". As for our algorithm, at the beginning, the agents move and spread out randomly. Afterward, the red agents act as catalyzers towards the most interesting zones, organization increases and entropy should decrease. Note that in the case of ants, attraction is produced by the effect of pheromone, while for our flock, it is caused by the attractive power of the red birds (and by the repulsion of the white birds).

Now, a more formal description of the entropy-based model is described. In information theory, entropy can be defined as:

$$S = -\sum_i p_i \log p_i \qquad (6)$$

Now, to adapt this formula to our aims, a location-based (*locational*) entropy is introduced. Consider an agent moving in a space of data divided in a grid $N \times M = K$, where all the cells have the same dimensions. So, if N and M are quite large and each agent is placed in a randomly chosen cell of this grid (as in the first iteration of the flocking algorithm), then the probability that the agent is in one of the K cells of the grid is equal for all the agents.

The entropy can be measured experimentally running the flocking algorithm for T tries and counting how many times an agent falls in the same cell i for each time-step. Dividing this number by T we obtain the probability p_i that the agent be in this cell.

Then, the locational entropy will be:

$$S = -\frac{\sum_{i=1}^{k} p_i \log p_i}{\log k} \qquad (7)$$

In the case of a random distribution of the agents, every state has probability $\frac{1}{k}$, so the overall entropy will be $\frac{\log k}{\log k} = 1$; this explains the factor of normalization log k in the formula. Obviously, in the case of the flocking algorithm, clustering zones will be visited more frequently and the probability will be higher in this zones and lower outside them. Consequentially, the entropy will be lower than 1. This situation can be verified for the Cure dataset (figure 3 a), observing the probability distribution (figure 4) in the grid.

The above equation can be generalized for P agents, summing over all the agents and averaging dividing by P. Equation (7) represents the *macro-entropy*; if we consider only red and white agents, it represents the *micro entropy*.

4   Experimental Results

Using the approach described in the previous section, the micro and macro entropy has been evaluated experimentally. All the experiment have been conducted using the real world North-East dataset, showed in figure 3 b, containing 123,593 points representing postal addresses of three metropolitan areas (New York, Boston and Philadelphia) in the North East States, It comprises a lot of noise represented from distributed rural areas and smaller towns. The artificial Cure dataset (figure 3 a) is also used, as it presents a cluster distribution quite regular and this permits a better understanding of the catalytical properties; in fact, the dataset contains 100,000 points distributed in three circles and two ellipsoids and connected by a chain of outliers and random noise scattered in the entire space.

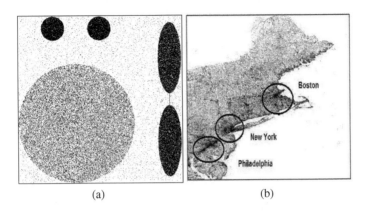

(a)   (b)

Fig. 3 a) CURE dataset. b) North-East dataset

We run the adaptive flocking algorithm (averaged over 100 tries) for 2000 time-steps using 100 agents and the same standard parameters for the flocking algorithm of the work in [4] and computed the probability an agent falls in every cell of the grid (as shown in figure 4 for the CURE dataset). Using these data and settings, we computed the micro and macro locational entropy both for SPARROW-SNN and, for the sake of comparison, for the random search algorithm and for the classical Reynolds model (without the adaptive behavior of the colored birds).

The result of these experiments, for the North-East dataset, is reported in figure 5. As expected, we can observe an increase in micro entropy (figure 5 b) and a decrease in macro entropy (figure 5 a) due to the organization introduced in the coordination model of the agents by the attraction towards red agents and the repulsion of white agents. On the contrary, in random search and standard flock model, the curve of macro entropy is almost constant, confirming the absence of organization.

A similar trend was observed for the Cure dataset (here not reported for the lack of space).

In addition, we conducted simulations in order to verify the property of autocatalysm of our system and to better understand the behavior of our algorithm specifically for the CURE dataset.

In figures 6 a and b, respectively the entropy in the cluster zones and outsides the clusters is reported. Entropy decreases both in cluster zones and outsides the

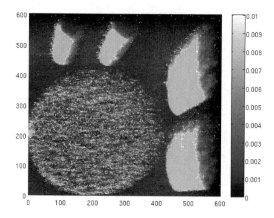

Fig. 4 Probability that an agent falls in a cell of the Grid for the Cure dataset (probability greater than 0.01 is set to 0.01)

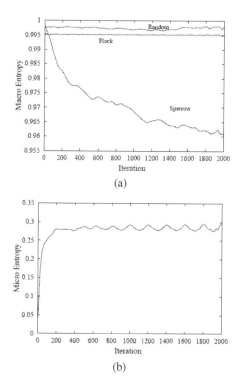

(a)

(b)

Fig. 5 North-East dataset: a) Macro Entropy (all the agents) using SPARROW-SNN, random search and standard flock b) Micro Entropy (red and white agents) using SPARROW-SNN

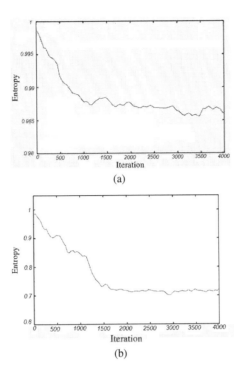

(a)

(b)

Fig. 6 Macro Entropy a) inside the cluster zones and b) outside the cluster zones for the Cure dataset using SPARROW-SNN, random and standard flock

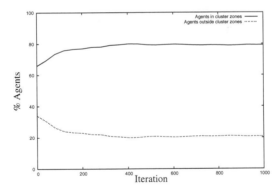

Fig. 7 Percentage of agents exploring cluster and non cluster zones for the Cure dataset using SPARROW-SNN

clusters zones, as the flock visits more frequently cluster zones and keeps away from the other zones (this behavior also causes a decrease in the entropy).

However these curves are not sufficient to verify the effectiveness of the algorithm as organization alone is not sufficient to solve problems, but it must bring the search in the appropriate zones. In fact, the main idea behind our algorithm is to let the flock explore the search space and, when the birds reach a desirable region (zone dense of clusters), an autocatalytic force is applied to the system (red birds) to keep searching in these zones.

Thus, we analyzed the average percentage of birds present in these two different zones (figure 7). In cluster zones we have about the 80% of the entire flock (while the space occupied by the clusters is about 65%) and this confirms the goodness of the algorithm, as in the interesting zones of the clusters, not only there is organization but there also a larger presence of searching agents.

5 Conclusions

This paper shows how an entropy-based model can be used to evaluate self-organizing properties of SI algorithms. Preliminary experiments, conducted using a flocking algorithm successfully employed for performing approximate clustering, demonstrate the presence of self-organizing characteristics differently from random search and classical flocking algorithm. However, entropy alone is not sufficient to assess the goodness of the algorithm in searching the space (i.e. performing clustering) and other measures are needed in order to verify the search is concentrated in interesting zones. Anyway, we believe that this model could be useful to better understand and control the behavior of multi-agent systems and to drive the user for choosing the appropriate parameters. Future works aim to evaluate and compare self-organization properties of other SI models, as Ants Colony Optimization, Particle Swarm Optimization, etc..

References

[1] Bonabeau, E., Dorigo, M., Theraulaz, G.: Swarm intelligence: from natural to artificial systems. Oxford University Press, New York (1999)
[2] Ellabib, I., Calamai, P., Basir, O.: Exchange strategies for multiple ant colony system. Inf. Sci. 177(5), 1248–1264 (2007), http://dx.doi.org/10.1016/j.ins.2006.09.016
[3] Ertoz, L., Steinbach, M., Kumar, V.: A new shared nearest neighbor clustering algorithm and its applications. In: Workshop on Clustering High Dimensional Data and its Applications at 2nd SIAM International Conference on Data Mining, pp. 105–115 (2002)
[4] Folino, G., Forestiero, A., Spezzano, G.: An adaptive flocking algorithm for performing approximate clustering. Inf. Sci. 179(18), 3059–3078 (2009)
[5] Grassé, P.: La Reconstruction du nid et les Coordinations Inter-Individuelles chez Beellicositermes Natalensis et Cubitermes sp. La Théorie de la Stigmergie: Essai d'interprétation du Comportement des Termites Constructeurs in Insect. Soc. 6. Morgan Kaufmann, San Francisco (1959)
[6] Jarvis, R.A., Patrick, E.A.: Clustering using a similarity measure based on shared nearest neighbors. IEEE Transactions on Computers C 22(11) (1973)

[7] Kuntz, P., Snyers, D.: Emergent colonization and graph partitioning. In: SAB 1994: Proceedings of the third international conference on Simulation of adaptive behavior: from animals to animats, vol. 3, pp. 494–500. MIT Press, Cambridge (1994)

[8] Lumer, E.D., Faieta, B.: Diversity and adaptation in populations of clustering ants. In: From Animals to Animats: Proc. of the Third Int. Conf. on Simulation of Adaptive Behaviour, pp. 501–508. MIT Press, Cambridge (1994)

[9] Macgill, J.: Using flocks to drive a geographical analysis engine. In: Bedau, M.A., McCaskill, J.S., Packard, N.H., Rasmussen, S. (eds.) Artificial Life VII: Proceedings of the Seventh International Conference on Artificial Life, pp. 446–453. The MIT Press, Cambridge (2000)

[10] Monmarché, N., Slimane, M., Venturini, G.: On improving clustering in numerical databases with artificial ants. In: Floreano, D., Mondada, F. (eds.) ECAL 1999. LNCS, vol. 1674, pp. 626–635. Springer, Heidelberg (1999)

[11] Parunak, H.V.D., Brueckner, S.: Entropy and self-organization in multi-agent systems. In: AGENTS 2001: Proceedings of the fifth international conference on Autonomous agents, pp. 124–130. ACM Press, New York (2001),
http://doi.acm.org/10.1145/375735.376024

[12] Parunak, H.V.D., Brueckner, S.: Engineering swarming systems. In: Methodologies and Software Engineering for Agent Systems, pp. 341–376. Kluwer, Dordrecht (2004)

[13] Reynolds, C.W.: Flocks, herds and schools: A distributed behavioral model. In: SIGGRAPH 1987: Proceedings of the 14th annual conference on Computer graphics and interactive techniques, pp. 25–34. ACM Press, New York (1987),
http://doi.acm.org/10.1145/37401.37406

Empirical Study of Performance of Particle Swarm Optimization Algorithms Using Grid Computing

Miguel Cárdenas-Montes, Miguel A. Vega-Rodríguez, Antonio Gómez-Iglesias, and Enrique Morales-Ramos

Abstract. This article presents an empirical study of the performance of the Particle Swarm Optimization algorithms catalog. The original Particle Swarm Optimizer has proved to be a very efficient algorithm, being applied in a wide portfolio of optimization problems. Spite of their capacities to find optimal solutions, some drawbacks, such as: the clustering of the particles with the consequent losing of genetic diversity, and the stagnation of the fitness amelioration, are inherent to the nature of the algorithm. Diverse enhancements to avoid these pernicious effects have been proposed during the last two decades. In order to test the improvements proposed, some benchmarks are executed. However, these tests are based on different configurations and benchmark functions, impeding the comparison of the performances. The importance of this study lies in the frequent use of Particle Swarm Optimizer to seek solutions in complex problems in the industry and science. In this work, several improvements of the standard Particle Swarm Optimization algorithm are compared using a identical and extensive catalog of benchmarks functions and configurations, allowing to create a ranking of the performance of the algorithms. A platform of Grid Computing has been used to support the huge computational effort.

Miguel Cárdenas-Montes · Antonio Gómez-Iglesias
CIEMAT, Centro de Investigaciones Energéticas Medioambientales y Tecnológicas,
Avda. Complutense 22, 28040, Madrid, Spain
e-mail: {miguel.cardenas,antonio.gomez}@ciemat.es

Miguel A. Vega-Rodriguez · Enrique Morales-Ramos
ARCO Research Group, Dept. Technologies of Computers and Communications,
University of Extremadura, Escuela Politécnica, Campus Universitario s/n,
10071, Cáceres, Spain
e-mail: mavega@unex.es, enmorales@alumnos.unex.es

J.R. González et al. (Eds.): NICSO 2010, SCI 284, pp. 345–357, 2010.
springerlink.com

1 Introduction

Particle Swarm Optimization (PSO) is an evolutionary computation technique introduced by Kennedy and Eberhart in 1995 [6], and later [4]. The Particle Swarm Optimization concept originated as a simulation of a social system. Initial simulations were modified to incorporate nearest-neighbour velocity matching and multidimensional search and acceleration by distance (Eberhart and Kennedy 1995 [3]; Kennedy and Eberhart 1995 [6]). At some point in the evolution of the algorithm, it was realised that the conceptual model was, in fact, an optimiser. The result was a very simple implementation of a powerful new optimizer.

The Standard Particle Swarm Optimization has been demonstrated to be an efficient and fast optimizer, with a wide applicability to very diverse scientific and technical problems. In spite of the efficiency demonstrated by the SPSO, also some disadvantages have appeared. Mainly the premature convergence which prevents the finding of optimal solutions.

When enhancements are proposed, the original authors execute some benchmarks. These benchmarks includes the most profitable configuration and function for the enhancements proposed, being, in general, a reduce number of them. The benchmarks have been executed with different set of functions and configuration, impeding the matching of the results. In any case, these tests verify the new algorithm proposed versus the standard algorithm, impeding that the diverse enhancements can be compared between them.

The grid computing paradigm [5] [7] has made proof of being able to cover the requirements of a lot of scientific communities, such as, high energy physics, fusion, astrophysics and astronomy, chemistry, biomedicine, etc. The computing capabilities delivered by this paradigm have increased the generation of new science inside of these communities. Moreover, some challenges tackled by them do not have been faced without the strategic support of the grid computing. For these reasons, a platform of grid computing has been selected for the present work. Grid computing has emerged as a powerful paradigm in E-Science, providing to the researchers an immense volume of computational resources distributed along diverse institutions.

As consequent of the set of benchmarks functions selected, taking several configurations for each algorithm, a huge volume of executions appear. In order to cover this volume of executions, a sub-set of the Spanish National Grid Initiative computing platform based on the middleware gLite and the metascheduler GridWay was chosen, providing the necessary computational resources to execute the work proposed.

This paper is organized as follows: in section 2, a resume of the Particle Swarm Optimization algorithms family is introduced, as well as, some reflexions about the weaknesses of the original algorithm. In section 3, the details of the implementation and the production setup are shown. The results are displayed in section 4. And finally, the conclusions and the future work are presented in section 5.

2 Particle Swarm Algorithms Family

In the PSO technique, each particle is represented as a point inside of a N-dimensional space. The dimension (N) of the problem is the number of variables of the problem to be evaluated.

Initially, a set of particles are created randomly. During the process, each particle keeps track of its coordinates in the problem space that are associated with the best solution it has achieved so far. Not only the best historical position of each particle is kept, also the associated fitness is stored. This value is called *localbest*.

Another "best" value that is tracked and stored by the global version of the particle swarm optimizer is the overall best value, and its location, obtained so far by any particle in the population. This location is called *globalbest*.

The PSO concept consists in, at each time step, changing the velocity (accelerating) each particle toward its *localbest* and the *globalbest* locations (in the global version of PSO). Acceleration is weighted by a random term, with separate random numbers being generated for acceleration toward *locallbest* and *globalbest* locations.

2.1 Standard Particle Swarm Optimization

The process for implementing the global version of PSO is as follows:

1. Creation of a random initial population of particles. Each particle has a position vector and a velocity vector on N dimensions in the problem space.
2. Evaluation of the desired (benchmark function) fitness in N variables for each particle.
3. Comparative of the each particle fitness with its *localbest*. If the current value is better than the recorded *localbest*, it is replaced. Additionally, if replacement occurs, the current position is recorded as *localbest position*.
4. For each particle, comparison of the present fitness with the global best fitness, *global best*. If the current fitness improves the *globalbest* fitness, it is replaced, and the current position is recorded as *globalbest position*.
5. Updating the velocity and the position of the particle according to eq. 1 and eq. 2:

$$v_{id}(t + \delta t) \leftarrow v_{id}(t) + c_1 \cdot Rand() \cdot (x_{id}^{localbest} - x_{id}) +$$
$$c_2 \cdot Rand() \cdot (x_{id}^{globalbest} - x_{id}) \qquad (1)$$

$$x_{id}(t + \delta t) \leftarrow x_{id}(t) + v_{id} \qquad (2)$$

6. If the end execution criterion – fitness threshold or number of generations– is not met, back to the step 2.

Apparently, in eq. 1 a velocity is added to a position. However, this addition occurs over a single time increment (iteration), so the equation keeps its coherency.

2.2 Weaknesses of Standard Particle Swarm Optimization

Diverse authors ([8], [1]) have demonstrated that the particles in SPSO oscillate in damped sinusoidal waves until they converge to new positions. These new positions are between the global best position and their previous best position. During this oscillation, a position visited can have better fitness than its previous *local best*, reactivating the oscillation. This movement is continuously repeated by all particles until the convergence is reached or any end execution criteria is met.

However, in some cases, where the global optimum has not a direct path between current position and the local minimum already reached, the convergence is prevented. In this case, the efficiency of the algorithm diminishes. From the computational point of view, a lot of CPU-time is wasted exploring the area of suboptimal solution already discovered.

In order to avoid this pernicious effect, diverse alternatives to SPSO formulation have been proposed. Frequently, these enhancements are based on effects present in the nature, enforcing the image of the PSO algorithm as a mechanism presents in the nature.

2.3 Inertial Weight

Historically, this modification was the first enhancement proposed for the SPSO [4]. It consists in a progressive reduction of the importance of the previous velocity by a factor, called Inertial Weight, and being the algorithm modified, IWPSO, shown in eq. 3.

$$v_{id}(t + \delta t) \leftarrow \mu \cdot v_{id}(t) + c_1 \cdot Rand() \cdot (x_{id}^{localbest} - x_{id}) +$$
$$c_2 \cdot Rand() \cdot (x_{id}^{globalbest} - x_{id}) \qquad (3)$$

In our implementation, the Inertial Weight, μ, diminishes linearly from 0.9 to 0.4 throughout the number of generations.

2.4 Particle Swarm Optimization with Massive Extinction

The fossil record shows the existence of massive extinction throughout the history of the Earth. After these massive extinctions, the remove of the stagnant groups from the niches creates opportunities for new species. The importance of this mechanism is based on the fact that allows to flourish new species genetically very different of the former ones.

The massive extinction (ME) can be adapted to the SPSO in several ways. The simplest one is to re-initialise the position and velocity of the particles with a fitness below a threshold predefined and after a number of generation established [11]. The remove of particles stagnated allows to create new particles able to explore new areas of the search space. In this way, the new algorithm is called Particle Swarm Optimizer with Massive Extinction [11] (PSOME).

In the PSOME, two new parameters appear. The first one is the threshold under which the particles are reinitialized. And, the second one is the number of generation after which the ME mechanism is activated. These two parameters have to be carefully selected in order to maximize the contribution of the ME mechanism to reinitialize the population only when the stagnation of the amelioration of the fitness appears. Otherwise, pernicious effect will be introduced in the population, such as: premature reinitialization whereas particles are approaching to optimal solutions.

In the our implementation, the value of the period to reinitialize the particles is each 10% of the number of generations, and the threshold of fitness above the particles are reinitialized is the 10% of the *global best fitness*.

2.5 *Fitness Distance Ratio Based Particle Swarm Optimization*

The SPSO has foundations in a learning process for all particles. This process is based on the capacity to learn from the particle's own experience and from the experience of the most successful particle.

The Fitness-Distance-Ratio modification for the PSO algorithm (FDRPSO) proposes that particles are also able to learn from the experience of the neighboring particles having a better fitness that itself spite of it is not the global best [9]. For each particle, the FDRPSO algorithm selects only one other particle at a time when modify the velocity. This particle is chosen satisfying two criteria:

- The particle chosen must be near of the particle being updated.
- The particle chosen must have visited a position of better fitness.

One of the simplest way to satisfy these two criteria is to maximize the ratio of the fitness difference to one dimensional distance. In other words, the *dth* dimension of the *ith* particle's velocity is updated using a particle called the *nbest*, with prior best position P_j, chosen to maximize the expression 4.

$$\frac{Fitness(P_j) - Fitness(X_i)}{P_{jd} - X_{id}} \tag{4}$$

The expression 4 is called Fitness-Distance-Ratio, suggesting the name of the algorithm.

The FDRPSO algorithm modifies the original velocity eq. 1 adding a new term based on the best experience of the best near neighbor (*nbest*). Thus eq. 1 results in eq. 5.

$$v_{id}(t + \delta t) \leftarrow v_{id}(t) + c_1 \cdot Rand() \cdot (x_{id}^{localbest} - x_{id}) +$$
$$c_2 \cdot Rand() \cdot (x_{id}^{globalbest} - x_{id}) +$$
$$c_3 \cdot Rand() \cdot (x_{id}^{nbest} - x_{id}) \tag{5}$$

In this work, the value chosen for the new parameter was $c_3 = 2$, as proposed the original authors in order to maximize the performance. The values of the other parameters are identical for the rest of the survey, $c_1 = c_2 = 1$.

2.6 Dissipative Particle Swarm Optimization

As it has been exposed, as much the swarm evolves, going to equilibrium, the evolution process falls in the stagnation. To prevent this trend, a dissipative PSO (DPSO) is constructed introducing a negative entropy through additional chaos for the particles [12].

The simplest way to implement this approach is by the reset of the velocities and positions by new ones randomly generated, eq. 6.

$$IF(rand() < c_v)THEN v_{id} = rand() \cdot v_{max}$$
$$IF(rand() < c_l)THEN x_{id} = RANDOM(lowerlimit, upperlimit) \tag{6}$$

In an open system, the individuals in the social swarm are not only governed by the historical experiences, *global best* and *local best*, but also, they are affected by the environment. Due to the changes in the environment, the best historical positions may be not longer compatible. This alternate environment can have a stronger influence that the social learning, driving the individual to move toward directions in principle incompatibles with the present *global best* and *local best*.

The new positions and velocities assigned to the particle will allow to explore a different area in the search space, escaping from the local optima.

2.7 A Diversity-Guide Particle Swarm Optimizer

In the SPSO, the fast information flowing between particles seems to be the reason for clustering of particles. Diversity declines rapidly, leaving the SPSO algorithms with great difficulties of escaping local optima. Consequently, the clustering leads to low diversity with a fitness stagnation as final result.

An accepted hypothesis explains that maintenance of high diversity is crucial for preventing premature convergence in SPSO. The introduction of a repulsive phase in the PSO, as well as, the attractive already presents, tries to overcome the problem of premature convergence. In order to control the algorithm, a diversity measure is introduced. The results is an algorithm that alternates between phases of attraction and repulsion, giving the name to the algorithm, ARPSO [10].

During the repulsion phase, the particle is no longer attracted to, but instead repelled by the *global best* and *local best*. In this case, the equation governing the movement inverts the sign of *global best* and *local best* terms, eq. 7. In the other hand, the attractive phase is still governed by the same eq. 1 that SPSO.

$$v_{id}(t + \delta t) \leftarrow v_{id}(t) - c_1 \cdot Rand() \cdot (x_{id}^{localbest} - x_{id}) - $$
$$c_2 \cdot Rand() \cdot (x_{id}^{globalbest} - x_{id}) \tag{7}$$

The swarm contracts during the attraction phase, consequently the diversity decreases. In ARPSO, when diversity measure drops below a lower bound, d_{low}, it switches to the repulsion phase. Similarly, when the diversity reaches a upper bound, d_{high}, it switches back to the attraction phase. The final mechanism is an algorithm that alternates between phases of attraction and repulsion, or low diversity and high diversity.

The diversity parameter is defined by eq. 8. The inputs for this diversity measure are: $|S|$ is the swarm size, $|L|$ is the length of the longest diagonal in the search space, N is the dimensionality of the problem, P_{ij} is the j'th value of the i'th particle and p_j is the j'th value of the average point, p.

$$diversity(S) = \frac{1}{|S| \cdot |L|} \cdot \sum_{i=1}^{|S|} \sqrt{\sum_{j=1}^{N} (p_{ij} - p_j)^2} \qquad (8)$$

2.8 Mean Particle Swarm Optimization

The Mean Particle Swarm Optimization [2] proposes an alternative equation to calculate the velocity of the particles. Instead of comparing the particle's current position with *global best* and *local best*, it is compared with a linear combination of them. Thus, the equation resulting for the velocity is the eq. 9. Clearly, MeanPSO seems to be a suitable name for this modified PSO, see Fig. 1.

$$v_{id}(t + \delta t) \leftarrow v_{id}(t) + c_1 \cdot Rand() \cdot \left(\frac{x_{id}^{localbest} + x_{id}^{globalbest}}{2} - x_{id} \right) +$$

$$c_2 \cdot Rand() \cdot \left(\frac{x_{id}^{localbest} - x_{id}^{globalbest}}{2} - x_{id} \right) \qquad (9)$$

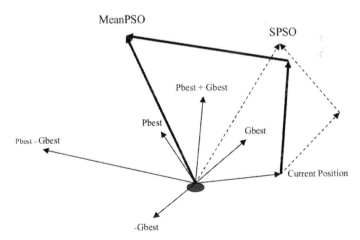

Fig. 1 Comparative movement of a particle in SPSO and MeanPSO

3 Production Setup

The empirical study was conducted using a set of benchmarks, where diverse functions widely used in these studies were selected. These functions were selected in order that the set has a mixture of multimodal (functions: f_1, f_2, f_6 and f_8) and monomodal functions (functions: f_3, f_4, f_5, f_7, f_9, f_{10} and f_{11}). For each benchmark function and algorithm, a set of identical configurations was executed. These configurations show the most characteristic values of dimensionality, population size and number of generations. The benchmark functions selected are presented in the table 1.

Table 1 Benchmark functions used in the survey

Expression	Search Space	Optimum
$f_1 = \sum_{i=1}^{D}[sin(x_i) + sin(\frac{2 \cdot x_i}{3})]$	$[3, 13]$	$\approx -1.21598 \cdot D$
$f_2 = \sum_{i=1}^{D-1}[sin(x_i \cdot x_{i+1}) + sin(\frac{2 \cdot x_i \cdot x_{i+1}}{3})]$	$[3, 13]$	$-2D + 2$
$f_3 = \sum_{i=1}^{D}[(x_i + 0.5)^2]$	$[-100, 100]$	0
$f_4 = \sum_{i=1}^{D}[(x_i)^2 - 10 \cdot cos(2\pi x_i) + 10]$	$[-5.12, 5.12]$	0
$f_5 = \sum_{i=1}^{D}[(x_i)^2]$	$[-5.12, 5.12]$	0
$f_6 = \sum_{i=1}^{D}[x_i \cdot sin(10 \cdot \pi \cdot x_i)]$	$[-1, 2]$	$\approx -1.95 \cdot D$
$f_7 = 20 + 20 \cdot \exp(-20 \cdot \exp(-0.2\sqrt{\frac{\sum_{i=1}^{D} x_i^2}{D}})) - \exp(\sum_{i=1}^{D}\frac{cos(2\pi x_i)}{D})$	$[-30, 30]$	0
$f_8 = 418.9828 \cdot D - \sum_{i=1}^{D}[x_i \cdot sin(\sqrt{|x_i|})]$	$[-500, 500]$	0
$f_9 = \sum_{i=1}^{D-1}[100 \cdot (x_{i+1} - x_i^2)^2 + (x_i - 1)^2]$	$[-5.12, 5.12]$	0
$f_{10} = \sum_{i=1}^{D}[i \cdot (x_i)^2]$	$[-5.12, 5.12]$	0
$f_{11} = \sum_{i=1}^{D}[(x_i)^2] + [\sum_{i=1}^{D}(\frac{i}{2} \cdot x_i)]^2 + [\sum_{i=1}^{D}(\frac{i}{2} \cdot x_i)]^4$	$[-5.12, 5.12]$	0

In order to avoid statistical fluctuations, a total of 400 tries of each configuration, each algorithm and each benchmark function have been executed. In these tries, the powerful machinery of the grid was used to support the computational activity.

To manage the complexity of the problem, involving several algorithms and benchmark functions, and the set of configuration of each of them; the grid jobs were created with 50 tries of each configuration of one specific algorithm and function. This structure assures the optimization of the time execution for the grid environment. A total of 8 runs were executed to reach the statistical level desired.

Each job was composed with a shellscript that handled the execution, and a tarball containing the source code (C++) of the program and the configuration files. When the job arrives to the Worker Node, it executes the instructions of the shellcript: rollout the tarball, compile the source code, execute the 50 tries of each configuration for a particular algorithm and benchmark function, and finally resume in a tarball all the output files. When the job finishes, the middleware recuperates the output tarball containing the the output files.

All PSO algorithms shares some common parameters, such as, $c_1 = c_2 = 1$ in eq. 1, and the maximum velocity, $V_{max} = 5$. Furthermore, the configuration values of dimensionality (20, 100), population size (10, 100) and number of generations (100, 1000, 10000) were established.

This methodology has demonstrated to be quite flexible face to Middleware and Operating System updates. During the production period, the infrastructure suffers some major upgrades, both gLite and Scientific Linux Operating System, being the production methodology completely transparent to these changes. Spite of this flexibility to surf over computational resources being updated, the heterogeneity of the grid made that the efficiency of job success was only 40%. In the most of the cases, the jobs aborted or be lost due to major failures, such as: unexpected stops or network connectivity.

Finally, the production was composed by a total of 616 jobs, resulting from the 11 functions and 7 types of PSO algorithms tested; and 8 runs per function and algorithm established. As consequence that each job has 400 tries, the number of total tries executed was 246,400. On the other hand, the mean CPU-time employed for run was 148.7 hours, then the total CPU-time for the eight runs was 1,189.6 hours.

4   Results

In the tables 2 and 3 a resume of the best results obtained for each fitness function, configuration and PSO algorithm are presented. For each function and configuration the function that obtains the best result is presented. In the case of several functions obtaining equal best results, all them are presented.

Table 2 Results of benchmarks for the functions f_1, f_2, f_3, f_4, f_5 and f_6 after 400 tries

Dim.	Pop.	Gen.	f_1	f_2	f_3	f_4	f_5	f_6
100	10	100	DPSO	DPSO	MeanPSO	MeanPSO	MeanPSO	MeanPSO
		1000	DPSO	PSOME	MeanPSO	MeanPSO	MeanPSO	MeanPSO
		10000	DPSO SPSO PSOME	DPSO	MeanPSO	MeanPSO	MeanPSO	MeanPSO
	100	100	DPSO	SPSO	MeanPSO	MeanPSO	MeanPSO	MeanPSO
		1000	SPSO	SPSO	MeanPSO	MeanPSO	MeanPSO	MeanPSO
		10000	DPSO SPSO PSOME	DPSO	MeanPSO	MeanPSO	MeanPSO	MeanPSO
20	10	100	SPSO	IWPSO	MeanPSO	MeanPSO	MeanPSO	MeanPSO
		1000	SPSO	DPSO	MeanPSO	MeanPSO	MeanPSO	MeanPSO
		10000	DPSO SPSO PSOME	DPSO	MeanPSO	MeanPSO	MeanPSO	MeanPSO
	100	100	DPSO	IWPSO	MeanPSO	MeanPSO	MeanPSO	MeanPSO
		1000	DPSO SPSO IWPSO	PSOME	MeanPSO	MeanPSO	MeanPSO	MeanPSO
		10000	DPSO SPSO PSOME IWPSO	DPSO	MeanPSO	MeanPSO	IWPSO	MeanPSO

Table 3 Results of benchmarks for the functions f_7, f_8, f_9, f_{10} and f_{11} after 400 tries

Dim.	Pop.	Gen.	f_7	f_8	f_9	f_{10}	f_{11}
100	10	100	MeanPSO FDRPSO	PSOME	MeanPSO	MeanPSO	MeanPSO
		1000	MeanPSO FDRPSO	FDRPSO	MeanPSO	MeanPSO	MeanPSO
		10000	MeanPSO FDRPSO	FDRPSO	SPSO	MeanPSO	MeanPSO
	100	100	MeanPSO FDRPSO	FDRPSO	MeanPSO	MeanPSO	MeanPSO
		1000	MeanPSO FDRPSO	FDRPSO	MeanPSO	MeanPSO	MeanPSO
		10000	MeanPSO FDRPSO	MeanPSO	SPSO	MeanPSO	MeanPSO
20	10	100	MeanPSO FDRPSO	FDRPSO	MeanPSO	MeanPSO	MeanPSO
		1000	MeanPSO FDRPSO	DPSO	SPSO	MeanPSO	MeanPSO
		10000	MeanPSO FDRPSO	FDRPSO	SPSO	MeanPSO	MeanPSO
	100	100	MeanPSO FDRPSO	FDRPSO	MeanPSO	MeanPSO	MeanPSO
		1000	MeanPSO FDRPSO	DPSO	MeanPSO	MeanPSO	MeanPSO
		10000	MeanPSO FDRPSO DPSO	DPSO	SPSO	MeanPSO	MeanPSO

5 Analysis and Conclusions

As consequence of the results obtained, the following conclusions can be extracted:

- Thanks to the tests executed in this work, a ranking of the most efficient Particle Swarm Algorithm can be created.
- The MeanPSO is the most powerful algorithm, obtaining the best result in 91 from the total 132 tests, the 69% of the tests, being the algorithm dominant for the functions f_3, f_4, f_6, f_{10} and f_{11}. Moreover, the MeanPSO produces the best results in 11 of the 12 configurations for f_5, and for f_7 shares the best results with FDRPSO for all configurations.
- However, in the functions f_1 and f_2 MeanPSO does not obtain any best result independently of the configuration. Furthermore, for the function f_8 it obtains only one best result. Consequently, the marriage between the function and the algorithm is critical in order to reach optimal solutions.
- It is well known that the particles in SPSO oscillate in damped sinusoidal waves until they converge to new best positions. With the modification of MeanPSO, the factors $\frac{x_{id}^{localbest}+x_{id}^{globalbest}}{2}$ and $\frac{x_{id}^{localbest}-x_{id}^{globalbest}}{2}$ allows to widen the area under exploration by the particles. So, a substancial increment of the probability to pass over good solutions apears.

Table 4 Best results in relation with the dimension

	SPSO	IWPSO	PSOME	FDRPSO	DPSO	ARPSO	MeanPSO
For dimension equal to 20	8	5	3	9	11	0	44
For dimension equal to 100	7	0	2	10	8	0	47

Table 5 Best results in relation with the swarm size

	SPSO	IWPSO	PSOME	FDRPSO	DPSO	ARPSO	MeanPSO
Any configuracion with 10 particles	7	1	3	10	9	0	45
Any configuracion with 100 particles	8	4	2	9	10	0	46

Table 6 Best results in relation with the number of cycles

	SPSO	IWPSO	PSOME	FDRPSO	DPSO	ARPSO	MeanPSO
For few cycles (100)	2	2	0	7	4	0	32
For a lot of cycles (10000)	8	2	5	8	10	0	27

- The second best algorithm is FDRPSO, obtaining 19 from the total tests, the 14%. However, the results for this algorithm are concentrated in the functions f_7 and f_8, reinforcing the idea of the good selection between the function to optimize and the algorithm elected.
- For the third place, two algorithms obtain 17 best results (13%), they are the SPSO and the DPSO. Specially significant is the fact that the original algorithm obtains 17 best results wining to other more complex modifications; underlining the xx of the original mechanism embedded in algorithm.
- The two following algorithms are the PSOME and the IWPSO with 7 (5.3%) and 5 (3.8%) best results.
- In general, other complex PSO modifications do not deserve improvement in the efficiency of the original PSO algorithm. In this line, the only algorithms which does not obtain any best result is ARPSO.
- As it can be appreciated in Tables: 4, 5 and 6, there are not major differences in the behavior of the algorithms face to different number of particles, cycles or dimension. The number of better results are similar independently of the configuration executed.

As main and final conclusion, it can be said that MeanPSO obtains better results for monomodal functions than for multimodal functions, Table 7. The reason of this result may correspond to the difficulties of MeanPSO to escape fro deep local minimums in multimodal functions. MeanPSO has a great capacity to explore the

Table 7 Best results in relation with the behavior of the function

	SPSO	IWPSO	PSOME	FDRPSO	DPSO	ARPSO	MeanPSO
Monomodal	5	0	1	12	1	0	68
Multimodal	10	7	4	7	18	0	11

local environment of the particles, converging quickly toward good near solutions. This feature arises from the two factors which modify the SPSO: $\frac{x_{id}^{localbest} + x_{id}^{globalbest}}{2}$ and $\frac{x_{id}^{localbest} - x_{id}^{globalbest}}{2}$. The combination of the *global best* and *local best* allows to broaden the area explored in each step or generation. However, this methodology lacks the capacity to escape from local minimum, visiting far away areas where deeper minimums (and, therefore, better solutions) could be found.

Other alternatives, such as: DPSO or PSOME, have the capacity to generate new particles genetically different from the represented ones in the population; or to keep some diversity in the particles presented in the swarm, in the case of FDRPSO. The mechanism adopted to generated new particles or keep a genetic diversity allows to escape from deep minimums in multimodal functions; however, it diminishes the effectiveness of the algorithm to find good solutions. In fact, DPSO obtains more better results that MeanPSO in multimodal functions, see Table 7.

Possibly a different scenario could happen for extremely long number of cycles, where the lesser effectiveness can be mitigated with bigger number of steps. In this case, the major factor to find good solutions should be the capacity to keep some genetic diversity in the swarm, avoiding that all particles became clones of the better one. Moreover, a similar effect could be foreseen for extremely multimodal functions.

The design of a new Particle Swarm Algorithm ought to combine the capacity to explore the closest area to the better particles with a mechanism to generate new individuals or, alternatively, to keep some genetic diversity. In this case, we are close to some mechanisms employed Particle Swarm Algorithm when they are applied to dynamic tasks. These type of approaches propose two kind of particles, some neutral particles which have a behavior similar to SPSO and some charged particles which feel not a attractive force to the other particles, but a repulsion force. This repulsion force is in charge to keep the genetic diversity.

Finally, the following future work is proposed:

- Some improvements over Standard PSO could prove its efficiency under extremely long number of generations. Subsequently, studies with bigger number of generations will allow to better characterize the algorithms.
- Moreover, the study can be broadened including further PSO enhancements. In this way, further studies could cover extra PSO variations.
- Similarly that other Evolutionary Algorithm, the techniques of multipopulations with periodic interchange of individuals could be explored in order to measure the improvements obtained.

Acknowledgement

The research leading to these results has received funding from the European Community's Seventh Framework Programme (FP7/2007-2013) under grant agreement number 211804 (EUFORIA) and grant agreement Number 222667 (EGEE III).

References

[1] Clerc, M., Kennedy, J.: The Particle Swarm: Explosion, Stability and Convergence in a Multi-dimensional Complex Space. IEEE Transaction on Evolutionary Computation 6, 58–73 (2002)

[2] Deep, K., Bansal, J.C.: Mean particle swarm optimisation for function optimisation. Int. J. Computational Intelligence Studies 1(1), 72–92 (2009)

[3] Eberhart, R.C., Kennedy, J.: A new optimizer using particle swarm theory. In: Proceedings of the Sixth International Symposium on Micro Machine and Human Science, pp. 39–43. IEEE Service Center, Nagoya (1995)

[4] Eberhart, R.C., Morgan, Y.S.: Computational Intelligence: Concepts to Implementations. Kaufmann Publishers, San Francisco (2007)

[5] Foster, I., Kesselman, C. (eds.): The Grid: Blueprint for a New Computing Infrastructure, 1st edn. Morgan Kaufmann Publishers, San Francisco (1998)

[6] Kennedy, J., Eberhart, R.C.: Particle swarm optimization. In: Proceedings of the IEEE International Conference on Neural Networks, Perth, Australia, vol. IV, pp. 1942–1948. IEEE Service Center, Piscataway (1995)

[7] Li, B., Baker, M.: The Grid Core Technologies. John Wiley and Sons Ltd., Chichester (2005)

[8] Ozcan, E., Mohan, C.K.: Particle Swarm Optimization: Surfing the waves. In: Congress on Evolutionary Computation, Washington, pp. 1939–1944 (July 1999)

[9] Peram, T., Veeramachaneni, K., Mohan, C.K.: Fitness-Distance-Ratio Based Particle Swarm Optimization. In: Swarm Intelligence Symposium, pp. 174–181 (2003)

[10] Riget, J., Vesterstrom, J.S.: A Diversity-Guided Particle Swarm Optimizer. Tech. R., U. Aarhus (2002)

[11] Xiao-Feng, X., Wen-Jun, Z., Zhi-Lian, Y.: Hybrid Particle Swarm Optimizer with Mass Extinction. In: International Conference on Communication, Circuits and Systems, Chengdu, China (2002)

[12] Xiao-Feng, X., Wen-Jun, Z., Zhi-Lian, Y.: A Dissipative Particle Swarm Optimization. In: Congress on Evolutionary Computation, Honolulu, USA, pp. 1456–1461 (2002)

Using PSO and RST to Predict the Resistant Capacity of Connections in Composite Structures

Yaima Filiberto, Rafael Bello, Yaile Caballero, and Rafael Larrua

Abstract. In this paper, a method is proposed that combines the methaheuristic Particle Swarm Optimization (PSO) with the Rough Set Theory (RST) to carry out the prediction of the resistant capacity of connectors (Q) in the branch of Civil Engineering. The k-NN method is used to calculate this value. A feature selection process is performed in order to develop a more efficient process to recover the similar cases; in this case, the feature selection is done by finding the weights to be associated with the predictive features that appear in the weighted similarity function used for recovering. In this paper we propose a new alternative for calculating the weights of the features based on extended RST to the case of continuous decision features. Experimental results show that the algorithm k-NN, PSO and the method for calculating the weight of the attributes constitute an effective technique for the function approximation problem.

1 Introduction

An interesting problem in Civil Engineering area is to predict the resistant capacity of connectors, the stud type, and the influence of each of the features in the forecast using the gathered information of rehearsals of these connectors. The stud is an essential component of a composite beam, it is responsible of ensuring the connection between the steel section and the armed concrete flagstone. The studs are installed in the upper wing of the steel beam. The connectors ensure that the different materials constituting the composite section have an effect in a combined way.

Yaima Filiberto · Yaile Caballero · Rafael Larrua
Department of Computer Sciences, University of Camagüey, Cuba
e-mail: yaimafiliberto@yahoo.com, yailec@yahoo.com,
       rafael.larrua@reduc.edu.cu

Rafael Bello
Department of Computer Sciences, Central University of Villa Clara, Cuba
e-mail: rbellop@uclv.edu.cu

J.R. González et al. (Eds.): NICSO 2010, SCI 284, pp. 359–370, 2010.
springerlink.com   © Springer-Verlag Berlin Heidelberg 2010

A remarkable number of experimental studies have been developed to deepen in the study of the behavior of the connections. Specifically, the rehearsals of connectors of the push-out type have been an important way for the evaluation of the influence of different parameters in the behavior of the same ones, as well as obtaining the formulations that allow predicting their resistant capacity. In consequence, it is possible to have a volume of valuable derived information of the set of international experimental programs development in the environment of the connections in steel-concrete compound construction, where prevailing studies dedicated to the stud type with head.

At the same time, they have gone evolving the calculation methods. Nevertheless, it has been proven that in some cases the calculus expressions of the resistant capacity of the stud type connectors with head, the main international norms, (AISC-LRFD 2005) & Eurocode 4 (EN-1994-1-1:2004), are underestimate excessively and in other cases they are overestimate; for this reason it is necessary to improve this calculus expressions for this connection type, that is the more internationally diffused. The same happens for other types of connectors, with the added difficulty that the experimental investigation on these connectors is very poor. In spite of the numerous experimental works that have been internationally realized, to reach a better understanding of the behavior of the connections in compound structures of the concrete-steel, exist some aspects that affect significant in their answer and require more study and deepening.

Machine Learning techniques allow to extract underlying knowledge in the information. This learning process can be inductive or lazy learning. In the case of the lazy learning the solution of problems is based on the relationships of similarity among the cases, being the k-NN algorithm the classical method in this family. In this work, we use the Nearest Neighbors algorithm [6], as the prediction algorithm to calculate the resistant capacity of connections in composite structures.

To implement the function approximation based on k-nn algorithm in this work we propose a method for calculating the weights of the features using the Rough Set Theory (RST) [13]. This theory has been widely used in data analysis, particularly in the selection of features and calculation of the weights of the attributes. However, the most results have been achieved in this theory for the case of discrete decision values, which is not the case of function approximation. For that reason, it was necessary to develop a new method for calculating the weight of features in the case of continuous decision systems.

2 Improving the Application of the K-NN Method in the Approximation Function Problem

In order to calculate the weights, a heuristic search is performed. We selected Particle Swarm Optimization's technique (PSO, [8] and [9]) for assigning weights, taking into account the relative ease of implementation, speed in locating the optimal solution, its powerful scanning capabilities and its relative lower computational cost in terms of memory and time. The implementation of an optimization algorithm to calculate different weights for each attribute would free the researcher of the civil engineering area of their definition by using other qualitative or quantitative criteria.

2.1 K-Nearest Neighbours Approximator

The key idea in k-Nearest Neighbours (k-NN) method is that similar input data vectors have similar output values. It is necessary to find a certain quantity of nearest neighbours and their output values, to compute the output approximation. This output can be calculated as an average of outputs of the neighbours in the neighbourhood. Let be a new vector Xh and $N(Xh)$ the neighbourhood of Xh, the output of Xh, denoted by d_h, can be computed from outputs of vector in $N(Xh)$ using expression 1.

$$d_h = \frac{\sum\limits_{Xj \in N(Xh)} d_j}{k} \qquad (1)$$

Where k is the cardinality of $N(Xh)$. One obvious refinement of expression 2 is to weight the contribution of each of the the k neighbours according their distance to the vector Xh such as is analyzed in [12].

In order to built the neighbourhood $N(Xh)$, a similarity measure between two vectors X and Y is defined by expression 2

$$sim(X,Y) = \sum_{i=1}^{N} w_i * sim_i(X_i,Y_i) \qquad (2)$$

The weights w_i, usually normalised so that $\sum w_i = 1$, are used to strengthen or weaken the relevance of each dimension. The function $sim_i(X_i,Y_i)$ assesses the degree of similarity of vector X and Y according to dimension i; frequently the Euclidian distance is used to compute this measure $sim_i()$.

Other alternatives to expressions 1 and 2 are presented in [14]. The similarity is estimated using a weighted distance function given by expression 3

$$d(X,Y) = \left(\sum_{i=1}^{N} w_i * \partial_i(X_i,Y_i)^2 \right)^{1/2} \qquad (3)$$

Where w_i denotes the weight of dimension i and the function $\partial_i()$ defines how values of a given dimension differ. For instance, $\partial_i()$ can be defined as expression 4

$$\partial_i(X_i,Y_i) = \begin{cases} |X_i - Y_i| & if\ i\ is\ continuous \\ 0 & if\ i\ is\ discrete\ and\ X_i = Y_i \\ 1 & if\ i\ is\ discrete\ and\ X_i \neq Y_i \end{cases} \qquad (4)$$

In both expressions 2 and 3 the weight of each dimension is taking into account in order to find the similarity degree between vectors. The adjustment of weights w_i may have a significant influence in the accuracy of estimation, such as showed in [4]. So, to calculate the best set of weights is a key point in this process. The relevance of each dimension may be learned using a heuristic search, as it is proposed in the following section; in which the heuristic value is calculated based in a rough set approach.

2.2 Finding the Weights for k-NN Based on Similarity Relations

The main principle "similar problems have similar solutions", that is, "similar input vectors have similar output values", can be used when employing k-NN method for function approximation, provided that the target function to be approximated can be characterised as locally smooth [16]. According to this principle, the best set of weights must allow to establish a close relation between the similarity according to the input vectors and the similarity between the output real values. This similarity relation can be formulated as following:
    For all vectors X and Y:

$$xR1y \; if \; and \; only \; if \; F1(X,Y) \geq e1 \tag{5}$$

$$xR2y \; if \; and \; only \; if \; F2(X,Y) \geq e2 \tag{6}$$

Where $F1$ and $F2$ are similarity functions to compare vector X and Y, $F1$ includes input data and the weight of each dimension, such as expression 2 or 3, and $F2$ computes the similarity degree between two objects according to the output value, such as expression 4 or 7; $e1$ and $e2$ are thresholds.

$$\partial(x,y) = \begin{cases} 1 \; if \; |x-y| \leq \varepsilon \\ 0 \quad otherwise \end{cases} \tag{7}$$

In order to find the similarity relations $R1$ and $R2$ we could define the sets $N1$ and $N2$ for all vector X by expression 8 and 9, $N1$ and $N2$ of X is the neighbourhood of X according to the relations $R1$ and $R2$ respectively:

$$N1(X) = \{Y \; : \; xR1y\} \tag{8}$$

$$N2(X) = \{Y \; : \; xR2y\} \tag{9}$$

Then, the problem is to find the functions $F1$ and $F2$ such that $N1(x) = N2(x)$, where the equal symbol $(=)$ denotes the greatest similarity between $N1(x)$ and $N2(x)$ given the thresholds $e1$ and $e2$. But, given the comparison functions for each dimension and the output value, the problem is to find the weights w_i.
    In order to solve the problem the measure defined by expression 10 is proposed for each vector X:

$$\varphi(X) = \frac{|N1(X) \cap N2(X)|}{0.5 * |N1(X)| + 0.5 * |N2(X)|} \qquad 0 \leq \varphi(X) \leq 1 \tag{10}$$

Using expression 10 the quality of similarity of the set of M vectors of input-output data in the form $(X, f(X))$ denoted by DS, is defined by expression 11

$$\theta(DS) = \left\{ \frac{\sum\limits_{i=1}^{M} \varphi(X)}{M} \right\} \quad (11)$$

This measure $\theta(DS)$ represents the degree in which the similarity between vectors according to all input dimensions is the same as the similarity according to the output value.

Then, the problem is finding the set of weights $W = w_1, w_2, \ldots, w_n$, where n is the number of dimensions, which maximizes the expression 11.

To find the set W, heuristic methods such as Particle Swarm Optimization (PSO) [8] or Genetics Algorithms (GA) [15] and [5] can be used. In our case, PSO is used to find the best set W, this method has showed good performance to solve optimization problems [10] and [11].

The particles represent the vector W, they have n components (one for each dimension). The quality of particles is calculated by using the expression 11; in this case, we employ the following comparison function 12:

$$\partial(X_i, Y_i) = \begin{cases} \frac{|1-(X_i-Y_i)|}{Max(N_i)-Min(N_i)} & if\ i\ is\ continuous \\ 0 & if\ i\ is\ discrete\ and\ X_i = Y_i \\ 1 & if\ i\ is\ discrete\ and\ X_i \neq Y_i \end{cases} \quad (12)$$

At the end of the PSO search, the best particle is the best weight set W to build the function $F1$; then the similarity relation R1 established by expression 5 can be implemented using $F1$.

2.3 Application of the Heuristic Particle Swarm Optimization (PSO) in the Allocation of Weights to the Attributes

The PSO technique is an optimization technique developed by Eberhart and Kennedy in 1995, based on the behavior of a population such as swarms of fish or birds. Each particle has a measure of quality, as well as a position and a velocity in the space of the search, where the position determines the content of the possible solution. Each particle knows the position of its neighbors, interact with them, "learns" and adjusts its position and velocity in part attracted to its best position so far, and partly attracted to the best position of the swarm (global optimum point).

The general steps of the PSO algorithm are:

The position of a particle i is denote by X_i, where X_i is a vector that stores the values for each dimensions in the search space. Furthermore, we denote by V_i the velocity of particle i, which is also a vector, which contains each of the velocity that

have the particle in each dimension. This velocity is added to the position of the
particle to move the particle from time $t - 1$ to the time t.

The following describes each of the steps involved in this algorithm:

Step 1: Initialize a population of particles with random positions and velocities
in a D-dimensional space.

Step 2: For each particle, evaluate the fitness function in D variables.

Step 3: Compare the current fitness of each particle with the fitness of its best
previous position, pbest. If the current value is better than pbest, pbest receives the
current value, and $P_i = X_i$.

Step 4: Identify the particle in the vicinity (may be the entire set of particles, or
a subset of them) with the best value of the objective function so far, and assign its
index to the variable g.

Step 5: Adjust velocity and position of the particle according to following equations (for each dimension):

$$v_i(t+1) = w * v_i(t) + c1 * rand() * (pbest(t) - x_i(t)) + c2 * rand() * (gbest(t) - x_i(t)) \tag{13}$$

$$x_i(t+1) = x_i(t) + v_i(t+1) \tag{14}$$

Step 6: Check the stop criterion (maximum number of iterations or fitness value
reached), if not go to the Step 2.

It is necesary to prevent the explosion of the swarm using the parameter $Vmax$,
where $Vmax$ is the point of saturation of the velocity, if the velocity of a particle is
greater than $Vmax$ or smaller than $-Vmax$ it is valorized as $Vmax$. If $Vmax$ is too
small there is not enough exploration beyond locally good regions (may fall into
local optimal), if too large can be overcome with good solutions. Other parameters
to consider are:

- The number of particles (swarm size).
- The number of generations or iterations.
- The inertia weight (w).
- The reason for cognitive learning ($c1$).
- The reason of social learning ($c2$).

The recommended values for the parameters are the following:

The number of particles is between 10 and 40. The number of generations is
between 100 and 200. While more higher are the values of these parameters grows
the chance of finding the optimum but grows the computational cost. It is recommended that $c1 = c2 = 1.5$ or $c1 = c2 = 2$, since the low values allow to explore
more regions before going to the objective. The weight of the inertia
(w) controls the impact of the historical velocity; high values facilitate the global
exploration and the small ones the local exploration while an appropriate value produces a balance between global and local search by reducing the amount of generations required. The rule is to give a high initial value and gradually decrease,
$w(k) = Wmax - (Wmax - Wmin)/Ncmax * k$, it is suggested that $Wmax = 1.4$,

$Wmin = 0.4$. Other interesting alternative to prevent the explosion of the swarm is using the restriction coefficient defined by expression 15.

$$const\ coeff = \frac{2}{2 - \varphi - \sqrt{\varphi^2 - 4\varphi}} \tag{15}$$

where $\varphi = c1 + c2 > 4$

3 Experimental Setup

In order to evaluate the quality of the weight set W obtained using the method proposed in this paper the following study was performed.

The database has eight input variables and one output with a total of 66 instances. The input variables are:

1. Area of the connector (area $X10^{-2}(m^2)$)
2. Number connectors (nr).
3. Position of the connector (Pos).
4. Average width of the channel of the Steel Deck (bo).
5. Depth of the channel or the Steel Deck (hp).
6. Height of connector (hsc).
7. Resistance of the concrete to the compression (fc).
8. Tension of fluency of the connector (Fu).

The output for each instance is the value of resistant capacity (Q).

Using the method previously described proposed in this paper the weight for each feature was computed $W=(w_1,\ldots,w_8)$; after that, this W was used to implement the k-NN method to approximate the real function, where the similarity function employed in k-NN was $F1$ defined by expression 2 and using the weights in W. The result obtained using this set W in the k-NN method was compared with other four alternatives to W. The k-fold cross-validation process was employed. K - Fold Cross - Validation divides the original dataset into 10 subsets of equal size where one is used as test set while the others as used as the training set. Then the overall accuracy of the classifier is calculated as the average precision obtained with all test subsets. This technique eliminates the problem of overlapping test sets and makes an effective use of all available data. The recommended value k = 10 was used[7].

Five alternatives of methods to calculate the weights and four values for k were employed for the experimentation. The variants for calculating the weights are: (i) the proposed method in this paper (called PSO+RST), (ii) assigning the same weight to each feature (called Standard), (iii) three alternatives based in the expert criteria. The value denoted by "optimum" it is the value of k that minimizes LOOCE, that it is the error that is made in the prediction, this k is calculated for each instance of the data set. The table 1 shows the five variants with their respective values of weights and k. The first row showed the best set of weighs calculated by our method.

Table 1 Value of weight and k for each variant

# Variant	Values of K	Values of weights for attributes
1 PSO+RST	1, 3, 5, optimum	0.406, 0.034, 0.086, 0.161, 0.117, 0.124, 0.059, 0.013
2 Standard	1, 3, 5, optimum	0.125 one-size-fits-all attributes
3 Variant 1	1, 3, 5, optimum	0.30, 0.075, 0.075, 0.075, 0.075, 0.075, 0.25, 0.075
4 Variant 2	1, 3, 5, optimum	0.25, 0.092, 0.092, 0.092, 0.092, 0.092, 0.20, 0.092
5 Variant 3	1, 3, 5, optimum	0.20, 0.108, 0.108, 0.108, 0.108, 0.108, 0.15, 0.108

The results obtained by these 20 variants were compared with the real value of the resistant capacity according to the experiments that are described next.

Experiment 1: Comparing the results of accuracy of each alternative according to the measures: (i) Mean Absolute Percentage Error (MAPE), (ii) Root Mean Square Error (RMSE), and (iii) the average magnitude of the difference between the desired value and that obtained by the prediction (PMD). These measures are defined by expressions 16-18.

$$MAPE = \frac{\sum_{i=1}^{N} \left| \frac{ai-yi}{ai} \right|}{N} * 100\% \qquad (16)$$

$$RMSE = \sqrt{\frac{\sum_{i=1}^{N} \left| \frac{ai-yi}{ai} \right|^2}{N}} * 100\% \qquad (17)$$

$$PMD = \frac{\sum_{i=1}^{N} |ai - yi|}{N} \qquad (18)$$

Where:

ai is the desired output value belonging to the BD of actual experimentation.
yi is the expected output value for each variant prediction.
N is the number of instances.

The results of MAPE, RMSE and PMD for each variant are summarized in Table 2 shown below.

The results of the errors are expressed in percent (MAPE and RMSE) and the average (PMD) in absolute values. In the Table 2 we can observe that the values of MAPE, RMSE and PMD, for the first variant (PSO+RST with $k = 1$) are smaller than the other ones, you can also appreciate the difference among the values of this variant with the other ones, for what we concludes that PSO+RST with $k = 1$ the is the most effective variant.

Experiment 2: Comparing the results of accuracy of each alternative in order to determine whether there are significant differences in accuracy with respect to the real value by means of the coefficient R^2, the correlation coefficient and standard

Table 2 Summary of results of errors MAPE, RMSE and PMD

Weight	K neighbors	MAPE (%)	RMSE (%)	PMD
PSO+RST	K 1	11,841	14,657	6,432
	K 3	14,807	21,711	7,192
	K 5	17,405	25,966	8,326
	K Optimum	19,304	33,151	9,399
Srandard	K 1	14,7994	19,321	8,348
	K 3	16,6067	22,793	8,155
	K 5	18,680	28,341	8,527
	K Optimum	26,367	44,439	10,572
Variant 1	K 1	15,478	21,236	8,522
	K 3	17,395	26,693	8,626
	K 5	19,142	31,848	9,156
	K Optimum	22,452	37,783	9,962
Variant 2	K 1	13,819	18,658	7,524
	K 3	16,644	23,772	8,305
	K 5	18,341	29,940	8,684
	K Optimum	22,777	39,357	9,625
Variant 3	K 1	14,071	18,218	7,637
	K 3	15,914	22,639	7,884
	K 5	18,117	28,749	8,433
	K Optimum	24,205	40,929	9,961

error; in this case the best combination (PSO+RST with $k = 1$) was used for the comparison.

In Table 3, the test values for the case of $k = 1$ are summarized because they were the best results.

The stadistical analysis yielded the following results: With PSO + RST for $k = 1$, for standard error, the results are significantly lower than the other variants, the R^2 coefficient value is relatively high, above 86 %, and the correlation coefficient is above 0.9.

Table 3 Summary of results statisticians of the experiment 2 for $K = 1$

K1	PSO+RST	Standard	Variant 1	Variant 2	Variant 3
Standard Error	7.7769	10,8661	11,3890	10,0711	9,7370
R2 Coefficient	0,8810	0,7667	0,7437	0,7996	0,8127
Correlation Coefficient	0,9386	0,8756	0,8624	0,8942	0,9015

We used nonparametric tests for two related samples (Wilcoxon test) where RST+PSO with $K = 1$ was compared to the other variants of weight. We applied the Monte Carlo method, confidence intervals 99% and a number of samples equal to 66. We considered:

• significant→significance less than 0.05 and greater than 0.01
• highly significant→a significance less than 0.01

• fairly significant→a result less than 0.1 and greater than 0.05
• not significant→a result greater than 0.1.

In processing the results of the experiments we used the SPSS version 13.0. In Table 4 we show the result of test of Wilcoxon.

Table 4 Test of Wilcoxon

PSO+RST-K=1 vs Others Variants	R+	R−	p-value
Standard K=1	7.73	6.67	.041
Variant 1 K=1	15	9.44	.037
Variant 2 K=1	11.18	8.38	.260
Variant 3 K=1	9.45	6.40	.063

These results show the PSO+RST $k = 1$ obtained better results than the other alternatives to calculate the weights, and in some cases the difference is very noticeable. For the case of the Wilconxon test we can observe that it overcomes significantly to the standard variant of weights and for the remaining ones that it is the case of the approach of experts it overcomes them in some significantly and in others not.

3.1 A Comparative Study between the New Method and Other Classical Methods

As demonstrated above the best results of our method were obtained in the case of PSO+RST with $K = 1$. Next, this method is compared with other classical methods for the prediction of the resistant capacity of connectors (Q). These methods are:

- AISC-LRFD (LRFD) [2]
- Eurocode 4 (EC4) [1]
- NRMC 080: 2007 (CN) [3]

Table 5 shows the results of the comparison parameters between different calculation methods with PSO+RST. We can observe that the arithmetic mean approaches more to the unit for the case of the PSO+RST method, that which means that the

Table 5 Summary of the comparison of the method of PSO+RST with the different calculation methods

Parameters	$Qexp/LRFD$	$Qexp/NC$	$Qexp/EC-4$	$Qexp/PSO+RST-K=1$
Arithmetic Mean	0,88883	1,16046	1,16172	1,00310
Max Value	1,23172	1,83945	1,84486	1,33991
Min value	0,56592	0,72086	0,72088	0,72276
Standard Deviation	0,13291	0,24862	0,24793	0,14302
Correlation Coefficient	0,89425	0,79491	0,79710	0,90715
$0.85 \leq Qexp/Q \leq 1.15$	33	16	16	40
$Qexp/Q < 0.85$	20	10	10	8
$Qexp/Q > 1.15$	3	30	30	8

values obtained by the proposed method approach more to the real value that the remaining methods. When observing the quantity of predictions that are inside the central interval $0.85 \leq Qexp/Q \leq 1.15$ we can say that the PSO+RST method generates more predictions in the interval. On the other hand the maximum and minimum values in the proposed method are among the more approach to the unit and the quantity of elements outside of the central interval for excess and for defect they are the minor. This means that the proposed method is the most stable. For the case of the standard deviation the proposed method is the second of smaller value, and the correlation coefficient is the highest. To compare the experimental results we used nonparametric tests for two related samples (Wilcoxon test) as the previous case. The results of Wilcoxon test are revealed on Table 6.

Table 6 Test of Wilcoxon

PSO+RST-K=1 vs Methods	R+	R−	p-value
EC4	30.57	20.92	.000
NC	30.55	21	.000
LRFD	31.21	22.78	.002

We compared the average modulus of the differences between the actual value and the experimental. It shown highly significant differences between the proposed method and the different methods compared. The difference between the actual and the experimental value is significantly lower with the proposed method.

4 Conclusions

In this paper a new method of function approximation was proposed to solve the problem of the prediction of the resistant capacity of connectors, a classical task in the branch of the Civil Engineering.

The solution to the problem was achieved by means of the implementation of the k-NN method joint to the metaheuristic Particle Swarm Optimization. PSO allows calculating the weights for the input variables included in the similarity function used to recover similar instances in the k-NN method. PSO looks for the set of weights which maximizes the similarity between objects respects to the input variables and the output value; for do this, a new measure based in the rough set approach was introduced.

References

[1] Eurocode 4 (EN 1994-1-1). Desing of Composite Steel and Concrete Structures Part 1.1. European Committee for Standardization, Brussels (2004)
[2] Load and Resistance Factor Design (LRFD) Specification for Structural Steel Building. American Institute of Steel Construction (AISC), Inc., Chicago (2005)

[3] NR 080-2007, Calculation of between floors made up of concrete and steel with soul beams full subjected to load static. Code of good practical: Brunch Norma of the Ministry of the Construction of Cuba (2007)

[4] Stahl, A., Gabel, T.: Using evolution programs to learn local similarity measures. In: Ashley, K.D., Bridge, D.G. (eds.) ICCBR 2003. LNCS, vol. 2689, pp. 537–551. Springer, Heidelberg (2003)

[5] García Martíez, C., et al.: Global and local real-coded genetic algorithms based on parent-centric crossover operators. European Journal of Operational Research 185, 1088–1113 (2008)

[6] Dasarathy, B.V., Sánchez, J.S.: Nearest neighbour editing and condensing. tools - synergy exploitation. Pattern Analysis Applications (2000)

[7] Demsar, J.: Statistical comparisons of classifiers over multiple data sets. Journal of Machine Learning Research 7, 1–30 (2006)

[8] Kennedy, J., Eberhart, R.: Particle swarm optimization. In: Proceedings of the 1995 IEEE International Conference on Neural Networks, pp. 1942–1948. IEEE Service Center, Piscataway (1995)

[9] Kennedy, J., Eberhart, R.: Swarm intelligence. Morgan Kaufmann Publishers, San Francisco (2001)

[10] Parsopoulos, K.E., Vrahatis, M.N.: Recent approaches to global optimization problems through particle swarm optimization. Natural Computing 1, 235–306 (2002)

[11] Reyes-Sierra, M., Coello Coello, C.: Multi-objective particle swarm optimizers: A survey of the state-of-the-art. International Journal of Computational Intelligence Research 2(3), 287–308 (2006)

[12] Mitchell, T.: Machine learning, p. 414. McGraw Hill, New York (1997)

[13] Pawlak, Z.: Rough sets. International Journal of Information Computer Sciences 11, 145–172 (1982)

[14] Lopez, R.L., Armengol, E.: Machine learning from examples: Inductive and lazy methods. Data Knowlege Engineering 25, 99–123 (1998)

[15] Herrera, F., Lozano, M., Sánchez, A.: A taxonomy for the crossover operator for real coded genetic algorithms: An experimental study. International Journal of Intelligent Systems 18, 309–338 (2003)

[16] Gabel, T., Riedmiller, M.: CBR for state value function approximation in reinforcement learning. In: Muñoz-Ávila, H., Ricci, F. (eds.) ICCBR 2005. LNCS (LNAI), vol. 3620, pp. 206–221. Springer, Heidelberg (2005)

Improvement Strategies for Multi-swarm PSO in Dynamic Environments

Pavel Novoa-Hernández, David A. Pelta, and Carlos Cruz Corona

Abstract. Many real world optimization problems are dynamic, meaning that their optimal solutions are time-varying. In recent years, an effective approach to address these problems has been the multi-swarm PSO (mPSO). Despite this, we believe that there is still room for improvement and, in this contribution we propose two simple strategies to increase the effectiveness of mPSO. The first one faces the diversity loss in the swarm after an environment change; while the second one increases the efficiency through stopping swarms showing a *bad behavior*. From the experiments performed on the Moving Peaks Benchmark, we have confirmed the benefits of our strategies.

1 Introduction

Real world is full of problems where uncertainty and dynamism are features that should be taken into account. Many of these phenomena occur in industrial or business environments, where there is a need to find high quality solutions. A particular case are the so called dynamic optimization problem (DOP), whose main characteristic is that their optimal solutions are time-varying. In other words, the objective function changes with time.

Denoting Ω as search space, a DOP can be formally defined as the set of objective functions $f^{(t)} : \Omega \rightarrow \mathrm{IR} \ (t \in \mathrm{IN}_0)$ where the goal is to find the set of global optimums $\mathscr{X}^{(t)}$ in every time t, that:

Pavel Novoa-Hernández
Dept. of Mathematics,
University of Holguin, Cuba
e-mail: pnovoa@facinf.uho.edu.cu

David A. Pelta · Carlos Cruz Corona
Dept. of Computer Science and Artificial Intelligence,
University of Granada, 18071 Granada, Spain
e-mail: {dpelta,carloscruz}@decsai.ugr.es

J.R. González et al. (Eds.): NICSO 2010, SCI 284, pp. 371–383, 2010.
springerlink.com

$$\mathscr{X}^{(t)} = \{x^* \in \Omega | f^{(t)}(x^*) \succeq f^{(t)}(x), \forall x \in \Omega\}.$$

Here, $\succ$ is a comparison relation which means *is better than*, hence $\succeq \in \{\le, \ge\}$. These DOPs can be also understood as tracking problems, where one has to stay as close as possible to a peak that is moving.

It is usually claimed that population based techniques are well suited to deal with DOPs as it is easier to track a moving optimum using a set of solutions than using a trajectory based method. Most existing work uses computational methods (mainly based on evolutionary computing) that have been effective in stationary problems and have had to undergo certain adjustments to ensure proper behavior in dynamic environments [2, 8, 13]. One of these computational paradigms is the Particle Swarm Optimization (PSO)[10]. PSO is based on the social behavior of organisms that live and do most of their activities in groups (eg. bird flocks, fish schools). Because of it is easy to implement and effective in complex problems, PSO has been applied in various real problems, see [3] for a good survey.

In order to adapt a PSO for dynamic environments, two important features must be addressed: *outdated memory* and *diversity loss*. The outdated memory appears when the best solutions obtained so far by the algorithm are no longer true (eg. the global memory), this implies that particles fly will be around false attractors. Diversity loss occurs when the global optimum is shifted away from a converged swarm. In that case, if the swarm is already converged (or has a significant level of convergence), the slow velocities of its particles prevent it to reach the new shifted optimum. This behavior is quite similar to be trapped in local optima in multi-modal optimization problems. While the first of these adaptation issues can be solved relatively easy (eg. updating particle and swarm memories) the diversity loss is more difficult to deal with.

Among the methods to overcome the above issues, effective multi-population approaches had been proposed by Blackwell and Branke in [4]: the multi-swarm charged PSO (mCPSO) and the multi-swarm quantum PSO (mQSO). Both implement an atomic structure for the swarms: neutral (classic PSO) particles are moving close to the current best (nucleus), and charged/quantum particles are moving around the nucleus. In what follow we will call *multi-swarm PSO* (or just mPSO) to the approach (not to a certain algorithm) represented by the algorithms mCPSO and mQSO. Although mPSO had shown good results in certain dynamic environments, we believe that there is still room for improvement.

In this context, the aim of this contribution is to provide two simple mechanisms to enhance the performance of the mPSO approach. This will be achieved through: a new diversity management strategy and swarm control strategy which will detect and stop those swarms showing a *bad behavior*. This paper is organized as follows: Section 2 presents some theoretical foundations of this research. Later, in Section 3 is intended to detail our strategies. Section 4 describes the experiments performed. Finally, conclusions and future work arising from this work are shown in Section 5.

2   Multi-swarm PSO in Dynamic Environments

PSO is a stochastic, population-based technique proposed by Kennedy and Eberhart
[9]. Basically each individual i (called particle) is a candidate solution, whose movement in the search space is governed by four vectors: its position $\mathbf{x}_i$, its speed $\mathbf{v}_i$, the
position of the best solution found individually $\mathbf{p}_i$, and the position of the best solution found in the neighborhood $\mathbf{g}_{best}$. When the neighborhood is represented by the
whole swarm, then the resulting model is called $gbest$, otherwise $lbest$. In this work
we consider the $gbest$ model. Specifically, the formulas that govern the dynamics of
the particles are:

$$\mathbf{v}_i^{(t+1)} = \omega\,\mathbf{v}_i^{(t)} + c_1\eta_1 \circ (\mathbf{p}_i - \mathbf{x}_i^{(t)}) + c_2\eta_2 \circ (\mathbf{g}_{best} - \mathbf{x}_i^{(t)}) \tag{1}$$

$$\mathbf{x}_i^{(t+1)} = \mathbf{x}_i^{(t)} + \mathbf{v}_i^{(t+1)} \tag{2}$$

Where ω is an inertia weight that says how much of the previous velocity will be
preserved in the current one. Besides, c_1 and c_2 are acceleration constants, while η_1
and η_2 are random vectors with components in the interval $[0.0, 1.0]$. Note that the
operator $\circ$ means an entrywise product.

The basic PSO scheme has been used as a basis for developing more sophisticated, problem-specific algorithms. In the context of optimization in dynamic environments, there are several works in literature based on PSO. Among them, we
should highlight the multi-swarm approach proposed in [4] (mPSO for short). The
simultaneous use of several swarms not only allows for an effective exploration of
the search space, but also to follow the optima over time if the DOP is multimodal.
The main steps of the mPSO approach are shown in Algorithm 7:

Algorithm 7. The mPSO approach

---
1   Randomly initialize the particles in the search space;
2   **while** *stopping condition is not met* **do**
3   Apply exclusion test;
4   Apply anti-convergence test;
5   Detecting changes in the environment;
6   **foreach** *swarm s* **do**
7   Move its particles according to their type (neutral, charged or quantum);
8   Evaluate each particle position;
9   Update $\mathbf{p}_i$ and $\mathbf{g}_{best}$;
10   **endfch**
11 **endw**

The exclusion principle avoid multiples swarms exploring the same optimum.
If two swarms are close enough then the worst of them is reset randomly over the
search space. Moreover, the anti-convergence test is intended to explore new areas
of space, through the reset of the worst among all the swarms. For more details
about the mPSO functionalities, please refer to [4].

The use of multiple swarms in PSO in dynamic environments has been suggested not only by the mPSO approach. Other authors have proposed other interesting alternatives. See for example the speciation-based PSO (SPSO) [11] or the Different Topology Multi-swarm PSO (DTMPSO) [14].

3   Improvement Strategies for Multi-swarm PSO

In this section, we describe the two strategies for improving the mPSO approach. Firstly, we propose a diversity management strategy that will be applied after a problem's change. Secondly, we describe a swarm control mechanism that will detect and stop those swarms showing a *bad behaviour*.

3.1   Diversity Management Strategy

The diversity mechanism in mPSO is based on the use of charged and quantum particles. However, recent studies confirmed that the usefulness of these particles strategies is not always satisfactory throughout the run [1]. For example, quantum particles are useful as a response to problem change.

Here, we will just use *standard* particles and apply the strategy shown in Fig. 1. When a change is detected, each swarm is divided into two groups depending on the quality of the particles. A β part will remain fixed, while the rest $(1 - \beta)$ will be diversified. Similarly, the later population is also divided in two groups of sizes α and $(1 - \alpha)$. The particles in the former group will be randomly reset over the whole search space, while those in the later, will be resampled around the swarm best particle (eg. $\mathbf{g}_{best}$).

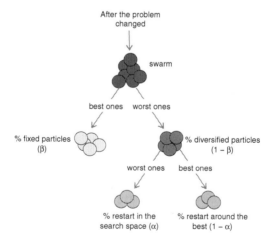

Fig. 1 Diversity management strategy after a problem's change

This resampling around $\mathbf{g}_{best}$ is performed by a Gaussian perturbation of a randomly selected component of the position vector. For example being x_i^k the k-th component of particle i, then: $x_i^k = g_{best}^k + N(0, 1)$, where g_{best}^k is the k-th component of the vector $\mathbf{g}_{best}$, and $N(0, 1)$ stands for a random number generator using a a normal distribution with 0 mean and 1 variance. In what follows we refer to mPSO scheme with diversity strategies described above as mPSOD.

3.2 Swarm Control Mechanism

The idea underlying this mechanism is quite simple: if a swarm is showing a bad behaviour then, we can stop it. As a consequence it will not waste cost function evaluations that could be profitable to other swarms. This idea was also developed in the context of multi-threads cooperative strategies for optimization [12].

In order to model the concept of bad behaviour, and being s a swarm, we resort to the following simple fuzzy decision rule: *IF (the quality of s is low AND s has converged)) THEN Stop s.*

To measure the quality of a swarm s we will calculate its degree of membership to the fuzzy set of *low quality* swarms using the membership function shown in 3. This function takes as argument the fitness of a particular swarm.

$$\mu_{low}(x) = \begin{cases} 0.0 & \text{si } x > b; \\ \frac{(b-x)}{(b-a)} & \text{si } a \leqslant x \leqslant b; \\ 1.0 & \text{si } x < a. \end{cases} \quad (3)$$

Here, a and b are two time-varying parameters. Formally, in each generation both are updated through the following expressions:

$$a = \frac{1}{m} \sum_{i=1}^{m} f^{(t)}(\mathbf{g}_i) \quad (4)$$

$$b = \max_{i=1...m} f^{(t)}(\mathbf{g}_i) \quad (5)$$

where m is the number of swarms in the algorithm. In short, the quality of a swarm depends on the average quality of the set of swarms and on the quality of the best of the swarms. The first part of the rule's antecedent will be *True* when $\mu_{low}(fit_s) \geq \gamma$. The value γ-cut is the threshold to define the *low quality* feature.

It would be unfair to measure the quality of the swarms regardless of their level of convergence. Especially since a fitness value does not has the same meaning or significance at different stages of convergence. That's why we have considered the atomic state as a measure of convergence of the swarm. Denoting the cloud of diversified particles as B_{cloud} and the set of fixed particles as N, then we can formally establish the atomic state by the following function:

$$isInAtomState(s) = \begin{cases} true & \text{if } \|B_N\| > \frac{\|N\|}{2}; \\ false & \text{otherwise.} \end{cases} \quad (6)$$

where $B_N = \{i \in s | \mathbf{x}_i \in B_{cloud}\}$, is the set of neutral particles that are within the cloud. The cloud B_{cloud} depends on a radii r_{cloud}, which will be equal to 1.0 as suggested in [4].

Now, the rule is implemented as:

IF (no change AND isInAtomState(s) AND $\mu_{low}(fit_s) \geq \gamma$) THEN Stop s.

Note that we also added the antecedent *no change*, which means that there is no change in the environment (at least recently). The mPSO approach with the two strategies described above is shown in Algorithm 8.

Algorithm 8. The mPSO approach with two new strategies

---

1 Randomly initialize the particles in the search space;
2 **while** *stopping condition is not met* **do**
3 | Apply exclusion test;
4 | Apply anti-convergence test;
5 | Detect changes in the environment;
6 | **foreach** *swarm s* **do**
7 | | **if** *change* **then**
8 | | | *generateInSearchSpace(s)*;
9 | | | *generateAroundGBest(s)*;
10 | | **endif**
11 | | **if no** *change* **and** *isInAtomState(s)* **and** $\mu_{low}(fit_s) \geq \gamma$ **then**
12 | | | Stop *s*;
13 | | **endif**
14 | | **else**
15 | | | Move its particles according to their type (neutral, charged or quantum);
16 | | | Evaluate each particle position;
17 | | | Update $\mathbf{p}_i$ and $\mathbf{g}$;
18 | | **endif**
19 | **endfch**
20 | Update a and b parameters;
21 **endw**

Note that the diversification strategy after the change is represented by the first *if* block, which contains the *generateInSearchSpace(s)* and *generateAroundGBest(s)* functions. Furthermore, the strategy for increasing efficiency by stopping low quality swarms is contained in the second *if* block.

4 Experiments

The main objective of the experiments is to study the impact of the parameters related to the strategies discussed above. For comparative purposes we selected as test problem the well known Scenario 2 of the Moving Peaks Benchmark (MPB), proposed in [5]. The parameter settings of the Scenario 2 are shown in Table 1.

Table 1 Standard settings for the Scenario 2 of the Moving Peaks Benchmark

Parameter	Setting
Number of peaks (p)	10
Number of dimensions (d)	5
Peaks heights (H_i)	$\in [30, 70]$
Peaks widths (W_i)	$\in [1, 12]$
Change frequency (Δe)	5000
Change severity (s)	1.0
Correlation coefficient (λ)	0.0

Table 2 Algorithm settings indicating the presence ($\checkmark$) or absence (–) of features. The values within parenthesis stands for the number of corresponding particles. Diversity and Swarm Control refer to our proposals

Algorithm	Neutral particles	Quantum particles	Diversity	Swarm Control
mQSO	$\checkmark$(5)	$\checkmark$(5)	–	–
mQSOE	$\checkmark$(5)	$\checkmark$(5)	–	$\checkmark$
mPSOD	$\checkmark$(10)	–	$\checkmark$	–
mPSODE	$\checkmark$(10)	–	$\checkmark$	$\checkmark$

Several algorithms will be used for the computational experiments. These are shown in Table 2, where the main features for each method are described. The reader should note that for mQSO and mQSOE, the diversity is obtained using neutral and quantum particles simultaneously and it would have no sense to also include our diversity management strategy. Nevertheless, the swarm control mechanism can be added to the mQSO method. The basic method mPSO is not included as it is not well suited to deal with dynamic environments.

To evaluate the algorithms performance we used the *offline error* measure, which is an error average through the run [6]. The offline error is always greater or equal to zero, so if it becomes zero means a perfect algorithm performance. This measure has the following expression:

$$error_{offline}^{(t)} = \frac{1}{t} \sum_{t=1}^{n} (f^{(t)}(x_{global}) - f^{(t)}(x_{best})) \tag{7}$$

Unless otherwise stated, we performed 30 runs with different random seeds for the problem and algorithms. Every algorithm has 10 swarms with 10 particles each. Besides, with respect to the PSO parameters we have selected the following values: $\omega = 0.7298$ and $c_1 = c_2 = 1.4960$, as suggested in [7].

4.1 Analysis of the Diversity Management Strategy

Here, we will analyze how the diversity setup (parameters α and β) affects the performance of the algorithm when different severities of the changes occurs. It is important to note that the severity (s) is one the most influential parameter in the algorithm's performance. It represents the magnitude of change (eg. the distance that problem's optimum will be displaced as a result of an environment change).

To better explore the impact of the parameters α and β, we have tested the mP-SOD algorithm on the Scenario 2 but with different severity values $s = \{0, 2, 4, \ldots,$ $10\}$. For each severity value, we tested every combination of α and β in the range $\{0.00, 0.25, 0.50, 0.75, 1.00\}$ leading to 25 different diversity configuration variants of mPSOD.

The results are shown in Figure 2, which contains contour plots for different severities. Each contour plot shows how α and β affect the performance of the algorithm. The color bars on the right side indicates the correspondence of each color with the offline error value. As the severity increases, we can observe that the black area looks like a triangle with the base concentrating around $\beta = 0.50$ and the top vertex goes *down* (the corresponding α values is decreasing). Also the error increases as the severity is greater. This is quite obvious since problem optima (represented by peaks) move to ever larger distances.

From the plots, it can be said that the best results are clustered around $\alpha = 0.00$ and $\beta = 0.50$ (eg. 0% and 50% resp.). This means that after the change, the best strategy is to diversify half the population around *gbest*. These results are in correspondence with those obtained in [4], where the best configuration for each swarm is $(5 + 5q)$ (5 neutral and quantum particles).

4.2 Analysis of the Swarm Control Mechanism

Now we will analyze the impact of detecting and stopping the swarms that show *bad behaviour*. The swarm control mechanism is included in the best mPSOD alternative (eg. $\alpha = 0.00$, $\beta = 0.50$) and also in the mQSO [4] algorithm. The new algorithms will be called mPSODE and mQSOE respectively and both are tested with the parameter $\gamma \in \{0.00, 0.25, 0.50, 0.75, 1.00\}$.

The results over the Scenario 2 of MPB are shown in the Table 3. We have also included the basic algorithms (eg. mQSO and mPSOD) for comparison purposes. In order to ascertain whether our strategy can be used without a predefined value for the γ-cut, we have considered randomly generating this value at runtime ($\gamma \in$ $random(0.0, 1.0)$). In this case, our strategy would not add an extra parameter to the mPSOD approach, which is kept as simple as possible. Table 3 also includes for each algorithm the minimum and maximum offline error, and the improvement rate compared with mQSO.

The superiority of the algorithms including the control mechanism is clear, except for the variants with $\gamma = .00$. This is because, for this γ value, all the swarms are stopped until a new environment change is detected. Interestingly, when $\gamma = rand$

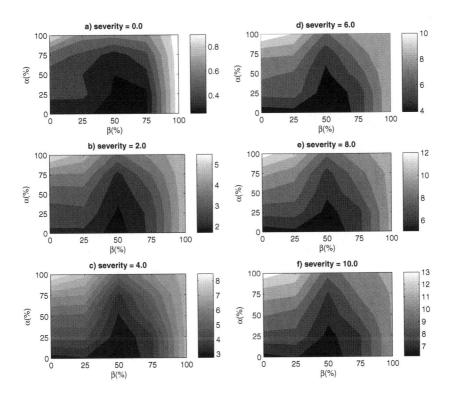

Fig. 2 Contour plot for different instances of the MPB. The bar on the right side associated the color with the offline error

the mQSOE and mPSODE performances are as good as the best alternative (eg. when $\gamma = 0.25$).

To statistically analyze the benefit of our mechanism, we have selected the most representative algorithms (eg. mQSO, $mQSOE_{\gamma=.25}$, $mQSOE_{\gamma=rand}$, mPSOD, $mPSODE_{\gamma=.25}$ and $mPSODE_{\gamma=rand}$). First, we applied a Friedman test to detect differences at group level. The p-value for this test was 0.000, thus suggesting that significant differences exist in at least one pair of algorithms. Then, we applied a Wilcoxon test to detect which pairs of algorithms have these differences. The p-value for each pair of algorithms is shown in Table 4.

Note that almost all comparisons showed significant differences between the algorithms (according a significance level of 0.05). The situation is different for pairs $mQSOE_{\gamma=.25}$-$mPSODE_{\gamma=.25}$ and $mQSOE_{\gamma=rand}$-$mPSODE_{\gamma=rand}$, which have similar performance ($p > 0.05$). In that sense, perhaps the most important conclusions emerging from these comparisons are: first our variants outperform the state-of-art mQSO (even mPSOD that only includes the new diversity strategy), and second the strategy for increasing efficiency can be used without having to set a predefined value for the γ parameter.

Table 3 Offline error for different values of γ in the Scenario 2 of MPB. Value *rand* stands for a random number in $[0.0, 1.0]$

Algorithm	Mean(Std. Dev.)	Maximum	Minimum	Improvement rate
mQSO	1.57(0.32)	1.08	2.29	-
$mQSOE_{\gamma=.00}$	3.15(0.56)	2.16	4.20	*-100.99%*
$mQSOE_{\gamma=.25}$	**0.83(0.29)**	**0.48**	**1.60**	**47.40%**
$mQSOE_{\gamma=.50}$	0.90(0.27)	0.53	1.60	42.63%
$mQSOE_{\gamma=.75}$	1.02(0.32)	0.62	1.80	35.11%
$mQSOE_{\gamma=1.0}$	1.17(0.35)	0.73	2.03	25.68%
$mQSOE_{\gamma=rand}$	0.92(0.29)	0.54	1.71	41.09%
mPSOD	1.17(0.27)	0.68	**1.80**	25.33%
$mPSODE_{\gamma=.00}$	2.57(0.39)	1.74	3.42	*-63.52%*
$mPSODE_{\gamma=.25}$	**0.88(0.33)**	0.48	1.97	**44.17%**
$mPSODE_{\gamma=.50}$	0.94(0.38)	0.48	2.24	40.18%
$mPSODE_{\gamma=.75}$	1.03(0.36)	0.50	1.84	34.32%
$mPSODE_{\gamma=1.0}$	1.20(0.39)	0.56	2.04	23.82%
$mPSODE_{\gamma=rand}$	0.94(0.34)	**0.45**	1.82	39.90%

Table 4 p-values for the Wilcoxon test

	mQSO	$mQSOE_{\gamma=.25}$	$mQSOE_{\gamma=rand}$	mPSOD	$mPSODE_{\gamma=.25}$	$mPSODE_{\gamma=rand}$
mQSO	–	–	–	–	–	–
$mQSOE_{\gamma=.25}$	.000	–	–	–	–	–
$mQSOE_{\gamma=rand}$	.000	.001	–	–	–	–
mPSOD	.000	.000	.000	–	–	–
$mPSODE_{\gamma=.25}$	.000	**.504***	.014	.000	–	–
$mPSODE_{\gamma=rand}$	.000	.000	**.861***	.000	.033	–

To extend our analysis, we have plotted the evolution of the offline error and the optimum tracking over time. To make the comparison more understandable we have separated mQSO from mPSOD, and we have only considered the first 50 changes the problem. These algorithms will be compared with their respective best variants, $mQSOE_{\gamma=.25}$ and $mPSODE_{\gamma=.25}$. Figures 3 and 4 show the plots. As can be seen, there are clear differences between algorithms, which in mQSO-$mQSOE_{\gamma=.25}$ pair are more remarkable than the other one.

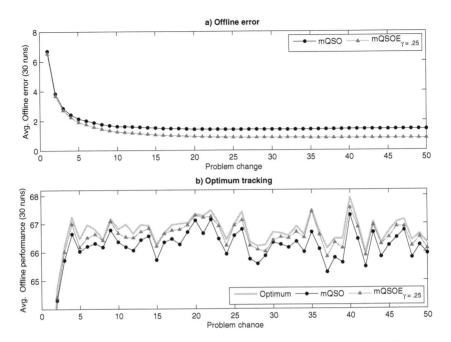

Fig. 3 Evolution of the offline error and optimum tracking. Pair mQSO-$mQSOE_{\gamma=.25}$

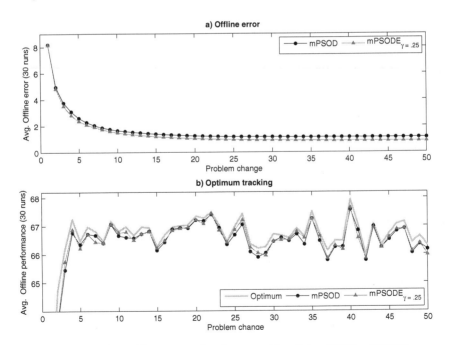

Fig. 4 Evolution of the offline error and optimum tracking. Pair mPSOD-$mPSODE_{\gamma=.25}$

5   Conclusion

In this work we have proposed two simple control mechanisms to improve the performance of a well-known approach for optimization in dynamic environments: the multi-swarm PSO (mPSO). The first of these strategies is intended to diversify the swarm after the problem undergoes a change. Through the experiments performed, it has been observed that the best strategy is to divide half the population of each swarm, a party must remain unchanged while the other must be diversified around the best solution in the swarm. Obviously, this conclusion is valid so far only for instances of problems discussed. Remains to be seen whether this strategy are valid for different dynamic problems as well.

The second strategy is aimed at improving mPSO through stopping those swarms with *bad behavior* and a certain level of convergence. The experimental results confirmed a remarkable improvement in those algorithms using this idea because of a simple reason: the resources (objective function evaluations) are not wasted in unprofitable areas of the search space.

As future work, we consider that self-adaptation is a clear research area to explore. The parameters that govern these control mechanisms can be *carried* by every swarm and then centralized/descentralized sharing information mechanism can be used to promote those parameters configurations that are well suited at every stage of the search.

Acknowledgments

Pavel Novoa-Hernández has the support of the *Coimbra Group Scholarships Programme for Young Professors and Researchers from Latin American Universities*. In addition, This work is supported by Projects TIN2008/01948 from the Spanish Ministry of Science and Innovation and P07-TIC02970 from the Consejería de Innovación Ciencia y Empresa, Junta de Andalucía.

References

[1] Garcia del Amo, I., Pelta, D., Gonzalez, J., Novoa, P.: An analysis of particle properties on a multi-swarm pso for dynamic optimization problems. In: CAEPIA-TTIA (2009)

[2] Angeline, P.: Tracking extrema in dynamic environments. In: Angeline, P.J., McDonnell, J.R., Reynolds, R.G., Eberhart, R. (eds.) EP 1997. LNCS, vol. 1213, pp. 335–345. Springer, Heidelberg (1997)

[3] Banks, A., Vincent, J., Anyakoha, C.: A review of particle swarm optimization. part ii: hybridisation, combinatorial, multicriteria and constrained optimization, and indicative applications. Natural Computing: an international journal 7, 109–124 (2008)

[4] Blackwell, T., Branke, J.: Multiswarms, exclusion, and anti-convergence in dynamic environments. IEEE Transactions on Evolutionary Computation 10(4), 459–472 (2006)

[5] Branke, J.: Memory enhanced evolutionary algorithms for changing optimization problems. In: Proceedings of the Congress on Evolutionary Computation, vol. 3, pp. 1875–1882. IEEE Press, Los Alamitos (1999)

[6] Branke, J., Schmeck, H.: Designing evolutionary algorithms for dynamic optimization problems. In: Tsutsui, S., Ghosh, A. (eds.) Theory and Application of Evolutionary Computation: Recent Trends, pp. 239–262. Springer, Heidelberg (2002)

[7] Clerc, M., Kennedy, J.: The particle swarm - explosion, stability, and convergence in a multidimensional complex space. IEEE Transactions on Evolutionary Computation 6(1), 58–73 (2002)

[8] Dasgupta, D., Mcgregor, D.: Nonstationary function optimization using the structured genetic algorithm. In: Parallel Problem Solving From Nature, pp. 145–154. Elsevier, Amsterdam (1992)

[9] Eberhart, R., Kennedy, J.: A new optimizer using particle swarm theory. In: Proceedings of the Sixth International Symposium on Micro Machine and Human Science MHS 1995, pp. 39–43. IEEE Press, Los Alamitos (1995)

[10] Kennedy, J., Eberhart, R.: Particle swarm optimization. In: IEEE International Conference on Neural Networks, vol. 4, pp. 1942–1948 (1995),
doi:10.1109/ICNN.1995.488968,
http://dx.doi.org/10.1109/ICNN.1995.488968,

[11] Parrott, D., Li, X.: A particle swarm model for tracking multiple peaks in a dynamic environment using speciation. In: IEEE Congress on Evolutionary Computation, pp. 98–103 (2004)

[12] Pelta, D., Sancho-Royo, A., Cruz, C., Verdegay, J.L.: Using memory and fuzzy rules in a co-operative multi-thread strategy for optimization. Information Sciences 176(13), 1849–1868 (2006)

[13] Pelta, D., Cruz, C., Verdegay, J.: Simple control rules in a cooperative system for dynamic optimization problems. International Journal of General Systems 38(7), 701–717 (2009)

[14] Xiangwei, Z., Hong, L.: A different topology multi-swarm pso in dynamic environment. In: IEEE International Symposium on IT in Medicine & Education, vol. 1, pp. 790–795 (2009), doi:10.1109/ITIME.2009.5236313

Particle Swarm Optimization Based Tuning of Genetic Programming Evolved Classifier Expressions

Hajira Jabeen and Abdul Rauf Baig

Abstract. Genetic Programming (GP) has recently emerged as an effective technique for classifier evolution. One specific type of GP classifiers is arithmetic classifier expression trees. In this paper we propose a novel method of tuning these arithmetic classifiers using Particle Swarm Optimization (PSO) technique. A set of weights are introduced into the bottom layer of evolved GP classifier expression tree, associated with each terminal node. These weights are initialized with random values and optimized using PSO. The proposed tuning method is found efficient in increasing performance of GP classifiers with lesser computational cost as compared to GP evolution for longer number of generations. We have conducted a series of experiments over datasets taken from UCI ML repository. Our proposed technique has been found successful in increasing the accuracy of classifiers in much lesser number of function evaluations.

1 Introduction

Data classification has received increasing interest as a consequence of tremendous increase in data generating abilities due to automation. The task of classification can be viewed as labeling unseen data based upon some knowledge extracted from data with known labels. Automated classification algorithms are required to handle the problem of knowledge discovery from large amounts of data. Evolutionary algorithms have been found efficient in solving classification problems due to their stochastic global search mechanism.

"Genetic programming is an evolutionary computation technique that automatically solves problems without requiring the user to know or specify the form or structure of the solution in advance" [1].These inductively learned solutions are efficient in learning hidden relationships among data and discriminate them in a concise mathematical manner. Since introduction of GP, various methods have been introduced to for data classification using GP. These solutions range from derivation of decision trees [2], evolution of classification rules [3] and generation of SQL queries [4]. A relatively new GP based classification method is numeric expression trees [5]. These mathematical expressions trees are evolved using GP as discriminating expression for a certain class using some arithmetic functions and variables defined the in the primitive set. The variables are usually the attributes present in training data and some constants.

Hajira Jabeen · Abdul Rauf Baig
National University of Computer and Emerging Sciences,
Department of Computer Science, Islamabad, Pakistan
e-mail: hajira.jabeen@nu.edu.pk, rauf.baig@nu.edu.pk

J.R. González et al. (Eds.): NICSO 2010, SCI 284, pp. 385–397, 2010.
springerlink.com   © Springer-Verlag Berlin Heidelberg 2010

The expression trees evolved using GP can form arithmetic or logical expressions based upon the primitive set used. In either case these expressions output a single numeric value and this value must be translated into the class labels. In case of binary classification one can simply assign one class to positive values and other class to negatives output values. The challenge arises in the case of Multi-Class classification problems where a single output has to be mapped to more than two classes. For this case methods like Static Range Selection [5] and Dynamic Range Selection [6] have been proposed.

GP presents numerous advantages for classification purpose. GP offers flexible and complex search space to search during classifier evolution. The classifier representations differ in each run, so we can eventually get several different classifiers with same or slightly different accuracy [7]. Easy and fast interpretation of result is possible as only one expression is evaluated to obtain the result [13]. The dependencies inherent in the data can be inducted into the classifier without expert intervention [7]. The classifiers are data distribution free, and able to operate upon the data in its original form [7].

Apart from above mentioned benefits, GP also suffers from the following issues:- GP based classification requires long training time. The classifiers increase in their complexity if necessary measures for avoiding code growth are not taken into account. GP yields different results after each run, both in structure of solution and accuracy.

In this paper we present a novel method for tuning of evolved classifiers making them more efficient and accurate. Several datasets with varying classes and attributes have been used to support the effectiveness of proposed tuning algorithm. Next section gives an overview of classification methods that use GP and an introduction to Particle Swarm Optimization PSO used for tuning. Method section explains the GP Algorithm used for classification and PSO based tuning algorithm proposed in this paper. Results section presents the experimental results followed by conclusions in the end.

2 Literature Review

2.1 Classification Using Genetic Programming

GP's outstanding abilities for the task of classification have been recognized since its inception [14]. Lots of work has been done to solve the problem of classification using GP. The main reason of interest in GP for classification is its ability to represent and learn solutions of varying complexity. In this section we will discuss some of the major techniques used for classification using GP.

Alex Frietas [4] has introduced a GP based classification framework where SQL queries are encoded into the GP Grammar. These are named Tuple Set Descriptor (TSD). The fitness of an individual is the number of rows satisfying the TSD. The framework incorporates lazy learning, i.e. rule consequences are evaluated first and one with the higher fitness is assigned to the rule. The advantages of using SQL based encoding is faster and parallel execution of queries, scalability and privacy. Another method using GP Classifier Expressions (GPCE) for

Multi-Class classification is used by Kishore et al [7]. A 'c' class problem is decomposed into 'c' two class problems. And a GPCE is evolved to discriminate each class from other classes. They define Strength of Association and Heuristic rules to tackle the conflicts arising between classifiers of different classes. They have used incremental learning and interleaved data format for speedup in the learning process. The work done by Bozarczuk et al [8] discovers classification rules for chest pain diagnosis. There are 12 classes present in the dataset and 165 attributes and 138 examples only. They have also evolved GP system for each class separately i.e. 12 times and chose best member as representative rule for each class. The fitness function takes into account classification accuracy, sensitivity, specificity and rule simplicity. At the end predictive accuracy of rule set as a whole and that of individual rules is evaluated. Loveard et al [6] have evaluated five different methods for Multi-Class Classification using strongly typed GP. One is binary decomposition in which the given problem is decomposed into binary problem, where one class is named as desired class and all other classes present in the data are assigned to reject class, the process is repeated for each class. Other method is static range selection, where the real output of a GP tree is divided into class boundaries and classes are assigned to input data based upon the GP tree output. Third method is dynamic range Selection where a subset of data is used to dynamically determine the class boundaries, and rest of the training data is used to evolve the classifiers. Fourth method is class enumeration, where a new data type is introduced into the terminal set of the GP trees. And all trees return a class type which is enumeration of the classes present in the data. The last method used is evidence accumulation, in which each GP tree contains a vector data storage corresponding to each class and these values are updated using new terminal which adds values ranging from -1 to 1 to the vector position particular class. The highest value is declared as the class outcome of tree. The results show that the dynamic range selection method is better for both binary and Multi-Class classification. Loveard et al have proposed two methods for classification of data containing nominal attributes [9]. One method considers splitting of GP execution based upon the value of a nominal attribute (execution branching) and other is conversion of a nominal attribute to binary or continuous attribute. Both methods are found efficient for classification of data containing nominal attributes. An interesting approach for Multi-Class classification using GP has been proposed by Muni [10]. In which collaborative view of classifiers for all the classes is considered. A Multi-Tree representation for classifier is presented. A chromosome has as many trees as there are classes in the data and each tree represents acceptor for samples belonging to its own class and rejecter to samples belonging to other classes. For evolution of this type of classifiers modified crossover operator is proposed. A new notion for unfitness of trees for genetic operations is proposed. A method Oring is proposed to combine results of classifier to achieve better performance. Heuristic rules and weight based scheme are also used to cater for ambiguous conflicting situations. The classifiers are can also output a 'do not know' when confronted with unfamiliar exemplars. A method for addition of weights has been proposed in [11] for Multi-Class object classification. The method for classification is range selection and the gradient descent method is used for searching during the evolution of GP Classifiers. This methodology makes the system more complex but it offers increase in performance of Genetic Programs. Many others methods have also been proposed for data classification using GP.

GP is found a very efficient innovative technique to handle to problem of data classification. GP suffers from an inherent drawback of inefficient code growth (bloat) during evolution. This increases the program complexity during the evolution process without effective increase in fitness. This increase in complexity has to be tackled explicitly by placing a bound on the upper limit of tree depth or nodes of the tree.

Another issue with GP based classification is long training time, which increases many folds with the increase in tree sizes during evolution. In this paper we have proposed a method that eliminates the need of evolving GP for longer number of generations and optimizes the GP evolved intelligent structures using PSO. Next section discusses some basics of PSO algorithm used for optimization in our proposed technique.

2.2 Particle Swarm Optimization

Particle Swarm Optimization Algorithm is originally formulated by Kennedy and Eberhart in 1995 [12]. Although it is also classified as an evolutionary algorithm but it models the sociological principle of bird flocking behavior during flying. The algorithm usually operates upon set of real multidimensional points scattered in the search space. These points move with certain velocity in the search space, mimicking bird's flight in search of optimal solution. The velocity of a particle in a given iteration is a function of the velocity of the previous step, its best previous position in the search space and the global best position. This exhibits a behavior of flying towards better position keeping in view its own best position and exploiting the knowledge of global best particle. The algorithm has been compared with various evolutionary algorithms and found equally efficient. Following are the update equations for particles in standard PSO.

$$X_i = X_i + V_i \tag{1}$$

$$V_i = \omega V_i + C_0 rand(0,1) X_{lbest} - X_i + C_1 rand(0,1) X_{gbest} - X_i \tag{2}$$

Where X_{gbest} is the global best or local best particle and X_{lbest} is the personal best of each particle. The values C_0 and C_1 are problem specific constants.

The Equation (1) is used to update position of a particle and Equation (2) is used to update the velocity of a particle during the PSO evolution process.

3 Methodology

In this section we will explain the algorithm used for classification and our proposed PSO based tuning method. The algorithm used for classification has been proposed by Muni [10]. One of the specialties of this algorithm is its Multi-Tree representation that makes it possible to evolve classifiers for multiclass classification in a single GP run. The evolved trees are in the form of arithmetic expressions elaborating relationships among different attributes of data. Each tree outputs a real value for each data instance. The output of a chromosome is a vector of real values. The tree corresponding to class label is trained to output positive value. Other trees must output negative value for a valid result.

Next section discusses the proposed optimization method that can increase the efficiency of GP evolved classifier expressions.

3.1 Tuning of Classifier Expressions

As mentioned in the previous section the classifiers contain attributes as terminals of the tree and a few constants. We have associated weights to all the terminals present in a tree. Consider a simple tree $((A_1+A_2)/A_3$ where A_1, A_2, and A_3 are attribute 1, 2 and 3 respectively. This tree will become $[(A_1*W_1) + (A_2*W_2)]$ / (A_3*W_3), after weight addition, where W_1, W_2 and W_3 are weights associated to each terminal. As shown in Figure1. The weight chromosome for this tree will be $[W_1, W_2, W_3]$. If the number of terminal nodes present in a tree is $'n'$. The number of nodes added for the sake of optimization is equal to $'2n'$. . Let c be the classes in the data , and t be the terminals in each tree: Then the total number of nodes added to chromosome will be :-

$$\sum_{i=1}^{c} 2 * t_i \qquad (3)$$

In case of multi tree representation, we add weights to each tree in the chromosome. Let c be the classes in the data, and 't' be the terminals in each tree. The weight vector will be :-

$$[W_{ij}] \ where \ i=1:c \ and \ j=1:t \qquad (4)$$

Here each weight is associated to the node j of the tree i. and we are interested in finding optimal value of this weight for each attribute in order to increase the efficiency of classifiers. An important point to note here is that the classifier remains intact if the values of all the added weights are set to '1'. Let CH_0 be the original chromosome and CH_w be the weight added chromosome, then

$$CH_0 = CH_w \ if \ V \ [Wij]=1 \qquad (5)$$

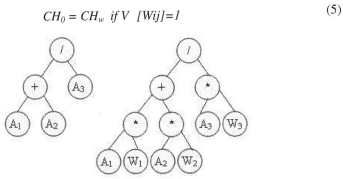

Fig. 1 Addition of weights for optimization

For the sake of optimization, a population of random particles having weights as their dimension is initialized. These weights are assigned random values between -1 and 1. This creates a multidimensional point in hyper space that has as many dimensions as there are weights in a GP chromosome corresponding to each terminal. PSO is used to evolve these weights for optimal values. The fitness of each particle is calculated by putting the values of weights in their corresponding positions and evaluating the accuracy of classifier for training data. We have used cognitive-social model that keeps track of its previous best as well as the global best particle. These weight particles are evolved for optimal position for a few generations until termination criteria is fulfilled.

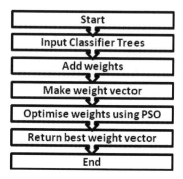

Fig. 2 Proposed tuning algorithm

4 Experimental Details

The data sets used for the classification purpose are taken from UCI repository. These data sets are Iris, Bupa, Wine, Glass and Wisconsin breast cancer. All these datasets are real valued data sets with varying number of classes and attributes. This is to prove the effectiveness of the proposed algorithm.

Each tree in the GP evolved classifier chromosome is appended by weights at its terminals, and the weights are evolved using PSO. The number of generations for evolution in GP is not kept fixed. The system is allowed to evolve until the fitness keeps on increasing. The evolution process is stopped only when the fitness increase is not observed for certain number of generations.

Table 1 GP parameters for Classification

S.No	Name	Value
1	Population size	600
2	Incremental Generations	20
3	Total generations	50
4	Maximum Depth	5

Table 1 lists the GP parameters used for the experimentation all the other para-meters were kept same as mentioned in (10). Table 2 lists the parameters used for PSO. The results reported after tuning are averaged for 10 executions of PSO

Table 2 PSO parameters

S.No	Name	Value
1	No of particles	20
2	Initial value range	[+1 , -1]
3	Number of iterations	30

4.1 Iris Dataset

Iris data set is one of the simple and small data set used for classification. It has 4 attributes and three classes with 150 instances. Figure 3 shows increase in accura-cy of GP classifiers before, and after tuning. It can be observed that PSO based

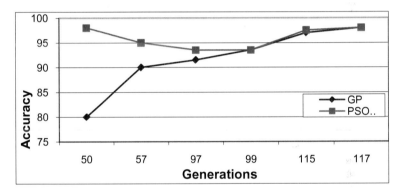

Fig. 3 Increase in Accuracy by PSO for Iris data

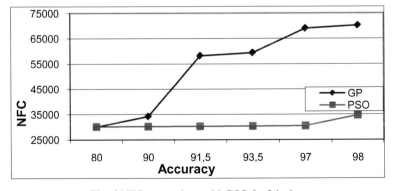

Fig. 4 NFC comparison with PSO for Iris data

tuning offers a considerable increase in most of the cases. Figure 4 compares the
Number of Function calls used by GP and tuning method. GP achieves good results in much larger number of fitness evaluations as compared to PSO based tuning which achieves same accuracy in much lesser number of Function calls. PSO
tuning method is efficient in finding better solutions in lesser number of Function
evaluations. Results of 10 executions of PSO tuning on one classifier are reported
in Table3 where PSO based tuning increases the accuracy in all the cases. Table 4
shows that on average 14% increase in accuracy is achieved.

4.2   Wisconsin Breast Cancer (WBC) Dataset

This data has two classes, 13 attributes and 699 instances. Figure 5 shows the increase in accuracy achieved for different GP classifiers. PSO tuning has increased
the accuracy of simple GP classifiers. Figure 6 shows that PSO based tuning offers
better accuracy with lesser number of functions calls(NFC) for achieving same accuracy as GP evolution process. Average increase in accuracy achieved is 7 %
shown in Table 4. A result of 10 PSO runs is reported in Table3.

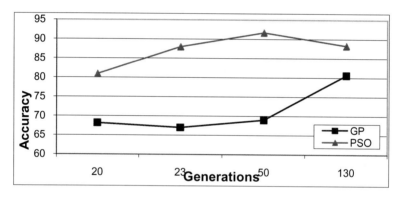

Fig. 5 Increase in Accuracy by PSO for Wbc data

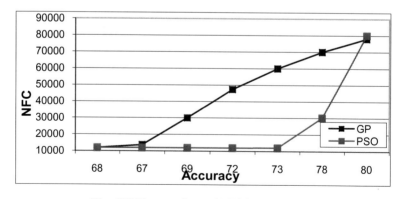

Fig. 6 NFC comparison with PSO for Wbc data

4.3   Glass Dataset

The Glass data has 6 classes and ten attributes having 214 instances. Figure 7 shows the increase in accuracy after tuning of GP classifiers. Figure 8 shows the difference in function calls to achieve same accuracy. PSO offered increase in the accuracy in much lesser number of function calls. Table 3 shows the result of 10 PSO runs and Table 4 summarizes the increase in accuracy achieved that was 1.7%.

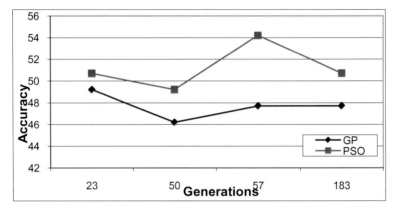

Fig. 7 Increase in Accuracy by PSO for Glass data

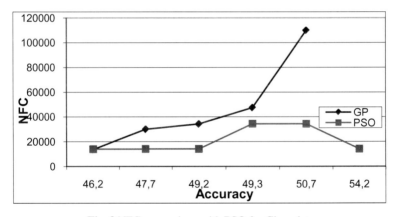

Fig. 8 NFC comparison with PSO for Glass data

4.4   Bupa Dataset

This dataset has 345 instances with 6 attributes and two classes. Figure 9 and Figure 10 show that tuning of weights has offered a prominent increase in the accuracy of the classifier with lesser number of function calls as compared to evolving GP for more number of generations. Average increase in accuracy achieved is

8.4% .Table3 presents results of 10 executions of PSO on one classifier evolved using GP. It is evident that in most of the cases weight tuning method offered an efficient increase in accuracy of the original classifiers. Table 4 gives an overview of increase in accuracy achieved.

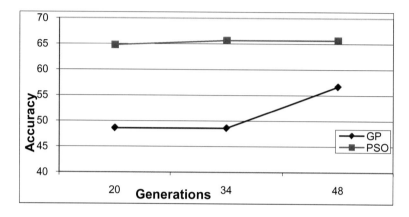

Fig. 9 Increase in Accuracy by PSO for Bupa data

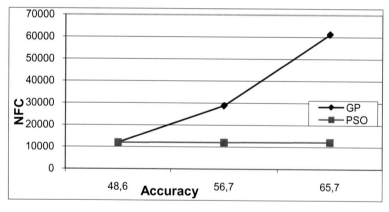

Fig. 10 NFC comparison with PSO for Bupa data

4.5 *Wine Dataset*

The Wine data set has 178 instances in 13 dimensions having three classes. Table 4 presents the result of 10 runs of PSO for the classifier evolved for Wine data. Figure 11 and Figure 12 show the increase in accuracy using the tuning method and difference in number of function calls in achieving the same accuracy. As observed in the previous cases, the PSO tuning method has achieved better accuracy with lesser function evaluations. The average increase achieved is 6.5% shown in Table 3.

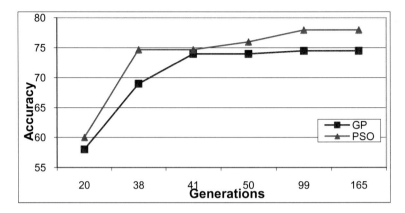

Fig. 11 Increase in Accuracy by PSO for Wine data

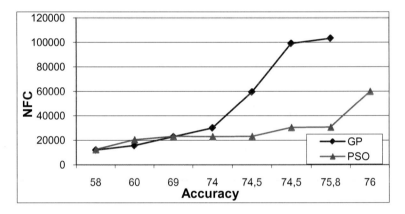

Fig. 12 NFC comparison with PSO for Wine data

Table 3 Increase in accuracy after 10 PSO runs

Datasets	IRIS	GLASS	BUPA	WINE	WBC
GP Accuracy	80.5%	46.0%	56.0%	70.0%	69.8%
PSO$_1$	98.0%	47.8%	65.7%	79.3%	78.4%
PSO$_2$	93.5%	49.2%	64.8%	79.3%	78.4%
PSO$_3$	95.5%	47.7%	65.7%	70.6%	74.7%
PSO$_4$	93.5%	47.7%	65.7%	70.6%	74.7%
PSO$_5$	91.0%	47.7%	65.7%	80.0%	78.0%
PSO$_6$	93.5%	47.8%	65.7%	80.0%	77.9%
PSO$_7$	95.5%	49.2%	64.8%	78.0%	79.0%
PSO$_8$	92.0%	47.8%	65.7%	79.3%	75.8%
PSO$_9$	96.0%	47.8%	64.8%	70.6%	74.1%
PSO$_{10}$	96.0%	49.2%	65.7%	78.0%	77.9%

Table 4 Average Increase in Accuracy

Datasets	Average Increase in Accuracy	Maximum Accuracy Achieved	Minimum Accuracy Achieved	Standard Deviation
IRIS	14 %	98.0%	91.0%	0.02
WBC	7.1%	78.4%	74.1%	0.01
GLASS	1.9%	49.2%	47.7%	0.006
WINE	6.5%	92.5%	87.5%	0.04
BUPA	8.4%	71.6%	65.0%	0.02

5 Conclusions

In this paper we have proposed a new method for the tuning of classifier expressions evolved by GP, it has been shown that this method tends to increase the training as well as testing accuracy of the classifiers. This method can eliminate the need for evolving GP classifiers for longer number of generations in search of better accuracy. It also helps in reducing the number of function evaluations desired for GP evolution. The more number of generations in GP also means increase in GP tree sizes over generation making the task more complex. On the other hand, in case of PSO based tuning we can get better results in much lesser number of function evaluations with increase in depth of trees by only one level. This increase in tree complexity gives an attractive outcome of increase in corresponding accuracy. Future work includes determination of optimal parameters for PSO for tuning and use of different variants of PSO for tuning.

References

1. Poli, R., Langdon, W.B., McPhee, N.F.: A Field Guide to Genetic Programming (2008)
2. Eggermont, J.: Data Mining using Genetic Programming: Classification and Symbolic Regression. Leiden University, PhD Thesis (2005)
3. Bojarczuk, C.C., Lopes, H.S., Freitas, A.A.: Discovering Comprehensible Classification Rules using Genetic Programming: A Case Study in a Medical Domain. In: Proceedings of the Genetic and Evolutionary Computation Conference, pp. 953–958. Morgan Kaufmann, San Francisco (1999)
4. Freitas, A.A.: A Genetic Programming Framework For Two Data Mining Tasks: Classification And Generalized Rule Induction. In: Genetic Programming, pp. 96–101. Morgan Kaufmann, USA (1997)
5. Zhang, M., Ciesielski, V.: Genetic Programming For Multiple Class object Detection. In: Proceedings of the 12th Australian Joint Conference on Artificial Intelligence, Australia, pp. 180–192 (1999)
6. Loveard, T., Ciesielski, V.: Representing Classification Problems in Genetic Programming. In: IEEE Congress on Evolutionary Computation, pp. 1070–1077 (2001)

7. Kishore, J.K., et al.: Application of Genetic Programming for Multicategory Pattern Classification. IEEE Transactions on Eolutionary Computation (2000)

8. Bojarczuk, C.C., Lopes, H.S., Freitas, A.A.: Genetic programming for knowledge discovery in chest-pain diagnosis. IEEE Engineering in Medicine and Biology Magazine, 38–44 (2000)

9. Loveard, T., Ciesielski, V.: Employing nominal attributes in classification using genetic programming. In: 4th Aisa pacific conference on simulated evolution and learning, Singapore, pp. 487–491 (2002)

10. Muni, D.P., Pal, N.R., Das, J.: A Novel Approach To Design Classifiers Using GP. IEEE Transactions of Evolutionary Computation (2004)

11. Zhang, M., Smart, W.: Genetic Programming with Gradient Descent Search for Multiclass Object Classification. In: Keijzer, M., O'Reilly, U.-M., Lucas, S., Costa, E., Soule, T. (eds.) EuroGP 2004. LNCS, vol. 3003, pp. 399–408. Springer, Heidelberg (2004)

12. Kennedy, J., Eberhart, R.C.: Particle Swarm Optimization. In: IEEE International Conference on Neural Networks, pp. 1942–1948 (1995)

13. Engelbrecht, A.P., Schoeman, L., Rouwhorst, S.: A Building Block Approach to Genetic Programming for Rule Discovery, in Data Mining: A Heuristic Approach. [book auth.]. In: Abbass, H.A., Sarkar, R., Newton, C. (eds.) Data Mining, pp. 175–189. Idea Group Publishing, USA (2001)

14. Koza, J.R.: Genetic Programming: On the Programming of computers by Means of Natural Selection. MIT Press, Cambridge (1992)

Author Index